AF765714

Printed in the United States of America

ISBN paperback print book 978-1-939297-31-0

ISBN VitalSource E-Book 978-1-939297-32-7

ISBN Kindle E-Book 978-1-939297-33-4

Cover and Design by Laurie Entringer

Editing by Jennifer Murtoff

HERCHER Publishing Inc.

Plainfield, Illinois

For ordering information see www.hercherpublishing.com

About the Operations Science Institute

Visit www.opscience.org

The Operations Science Institute (OSI) helps organizations eliminate confusion and accelerate performance improvement through the practical application of operations science—the science of working better. The power of operations science is its simplicity in concept and universal application whether in manufacturing, services, or projects.

We equip leaders and teams with practical, science-based frameworks to optimize capacity, inventory, cost, and time while managing variability thereby enabling sustained efficiency and customer satisfaction. OSI bridges the gap between overly complex theory and simplistic "best practices," delivering practical training and coaching that empowers your workforce to achieve lasting results with existing resources. No armies of consultants or new technology or software "solutions" needed.

With training, books, online certification (learn.opscience.org), The Operations Science Podcast, and many articles and videos available from the OSI at opscience.org, there are a myriad of powerful, practical resources available. These enable you to quickly and effectively put the concepts into practice and turn your operations into well-tuned processes delivering products, services, and projects on-time and on cost over and over again.

COVER PHOTO DESCRIPTION AND ATTRIBUTION

While we wish that Dr. Einstein were here to exchange ideas with, we cannot claim him as a direct developer of Operations Science. However, he is, in a sense, a patron saint for the field. His pursuit and description of the laws governing fundamental behavior of the universe is a foundation for our efforts to scientifically understand universal operations b ehavior.

Operations science has been developed and applied for decades to bring it to its current state as an innovative, practical, and powerful field of science. The people pictured on the cover have been part of that effort. The authors are deeply grateful for their contributions and continuing support.

Top Row (L-R)						
Albert Einstein	Mark Spearman	Wally Hopp	Jeff Bell	Ed Pound	Joan Tafoya	Bill Fierle
Middle Row (L-R)						
Todd Zabelle	James Choo	Jenny Snow-Boscolo	Roberto Arbulu	Eric Outley	Uwe Göhring	Mike McBride
Bottom Row (L-R)						
Lisa Elles	Ben Amaba	Jeff Daniels	Mark Graban	Fred Strathmann	Sanjay Huprikar	Mike Gores

Operations Science Applied

Mark Spearman

Jeff Bell

Ed Pound

About the Authors

Mark L. Spearman, with a combination of industry experience and academic credentials over forty years, continues to consult with Fortune 500 companies and to perform research into the fundamentals of manufacturing and supply chains. He is the founder of Factory Physics, Inc., providing the scientific framework to control and optimize the fundamental operations of any manufacturing operation and its supply chain. He is the co-author with Wallace Hopp of the University of Michigan of *Factory Physics*. In 1998, the book won the IIE Joint Publishers Book-of-the-Year Award and has been cited over 5,000 times, used as a textbook in over 200 universities in North America, and has been translated into several languages. This text was followed up with a second book aimed at working professionals, *Factory Physics for Managers*, the first joint effort with two working professionals, and coauthors of this book, Ed Pound and Jeff Bell. Mark holds a PhD in industrial engineering from Texas A&M University and has been a tenured professor at Northwestern University, Georgia Tech, and Texas A&M with over 10,000 citations on Google Scholar.

Ed Pound has more than thirty years of experience in operations, having worked in general management, operations, purchasing, and product development. He is the Managing Director of the Operations Science Institute. His experience areas include aerospace, electronics, pharmaceuticals, cloud infrastructure development, consumer retail, medical devices, healthcare, construction, insurance, health care, and engineer-to-order capital equipment. His career includes two years in Mito, Japan working at JVC Inc.'s Production Technology and Research Center where he was its only non-Japanese employee at the time. He is a 2022 University of Alabama Mechanical Engineering Distinguished Fellow and Past President of the Institute of Industrial and Systems Engineering's Work Systems division. Ed earned a bachelor of science and a master of science in mechanical engineering from the University of Alabama, a master of business administration from the Kellogg Graduate School of Management at Northwestern University, and a master of engineering management from the Robert R. McCormick School of Engineering and Applied Science at Northwestern University. He is a co-author of *Factory Physics for Managers*.

Jeff Bell is the managing partner of Arc Precision, a contract manufacturer for the medical device industry. He is a clinical professor at Dominican University. His career has included roles as vice president at the Rosemount Flow Division of Emerson Electric, where he was general manager for the Magmeter and Vortex business units. He started his career in operations at Procter & Gamble Pharmaceuticals. Jeff has a bachelor of science in operations research from Cornell University, a master of business administration from the Kellogg Graduate School of Management at Northwestern University and a master of engineering management from the McCormick School of Engineering and Applied Science at Northwestern University. He is a co-author of *Factory Physics for Managers*.

Dedications

To my wife, Julia, and my parents, Hays and Karen, for their love and support. To my children – Daniel, Sean, and Maeve – for their inspiration. I am grateful to them for showing me the value of variability in enabling growth. May you always seek a better understanding of how things work and apply that learning to make the world a better place.

Jeffrey H. Bell

To my wife, Blair, who has loved me for 42 years in good times and in hard times. And to my children, Jacob, William, and Rebekah who have blessed me and taught me more than I have taught them. And to all of my grandchildren, Alana, Jake, Remington, Liliana, and Theresa (who was born 6 days before this was written). May they all be blessed and grow in the Lord.

"Call unto me, and I will answer you, and show you great and mighty things that you know not."
Jer 33:3

Mark Spearman

To my family, you have filled my life with joy.

"If God is for us, who can be against us? …Who will separate us from the love of Christ? …in all these things we conquer overwhelmingly through him who loved us."

ROM 8:31-37

Ed Pound

Acknowledgements

The authors would like to thank Wallace J. Hopp for his inspiration and partnership on our journey to see the world with "operations eyes". In 1990 he won the Scaife Award (with Mark Spearman) for the paper with the "greatest potential for assisting an advance of manufacturing practice". It has been an honor to work with him and receive his guidance as we learned the original manufacturing framework he and Mark developed was actually a special case of something bigger and more fundamental. Wally's work in healthcare and other segments helped us understand that operations science applied across a wide range of diverse applications. He showed us how the fundamental operations science framework enabled practitioners to make sound decisions in the face of many complex, real-world situations. This book is a small sample of examples where leaders used the fundamental framework to get

better results in a variety of situations. We hope it inspires others as Wally has inspired us to expand the application of operations science.

As Mark's co-authors, Jeff and Ed want to call special attention to the incredible work that Mark has done in his career in business. His relentless development of operations science in practice has been a bit in the shadows, so to speak, as companies are not prone to share the competitive advantage gained through his groundbreaking applied research. Mark's work in private has provided and continues to provide numerous PhD thesis-level topics for new contributions to science. Since Mark was not in academia for much of his career, he missed out on the opportunity for many of the awards available for pioneering research, but he is no less deserving. It has been a phenomenal experience working along side him.

We thank the contributors to this book for both their expertise in applying operations science and sharing their stories. Thanks to Mike McBride, Margaret Amalfitano, and the team at Delta Tech Ops for their leadership and example of a comprehensive approach to train and mentor an organization to use operations science to manage recovery from a major disruption like the Covid pandemic while simultaneously driving expansion and growth of the business. Thanks to Uwe Göhring for showing us how operations science catalyzes the effective use of information technology in a world that is embracing the use of artificial intelligence. Thanks to Sue Thomas, Mike Severson, and Dale Nugent for reminiscing about the relationships and results forged with operations science during their time in the Rosemount Flow Division. Much appreciation to Kate McKone-Sweet for allowing us to leverage her case study in that chapter. As investors in Arc Precision, we admire the way Jim Lind, James Lind, Dan Baxter, and Mike Severson live the strategies, tactics, and controls of buffer strategy on a day-to-day basis. We thank supply chain professional Luis Dominguez for describing how he uses operations science to lead collaboratively across the supply chain with colleagues Adrian Castro and Tom Norman. We thank Dr. Frederick Strathmann for describing how operations science helped navigate high variability for Clinical Diagnostic Labs during Covid. We enjoyed the opportunity to share insights in applying operations science to project management based on the pioneering work of Todd Zabell, Roberto Arbulu, James Choo and the team at the Project Production Institute. Thanks to the leaders who allowed us to share their stories – Bill Van Sant, Emile Tohme, Joan Tafoya, Mike Severson, Larry Doerr, Bill Fierle, and Lisa Elles. Special thanks to Lisa and Iain Wood for contributing the chapter on shipbuilding.

We appreciate the members of the Operations Science Institute Advisory Board who gave us guidance on the book content. Ben Amaba, Jeff Daniels, Lisa Elles, Bill Fierle, Uwe Göhring, Mark Graban, Sanjay Huprikar, Mike McBride, Jenny Snow-Boscolo, and Joan Tafoya comprise the advisory board. We thank the professionals who helped us with the production of the book - Maureen Gorman, Jennifer Murtoff, and Laurie Entringer. Finally, thanks to Dick Hercher who has been a great advisor from the early days of the Factory Physics textbook to the launch of Operations Science Applied.

OPERATIONS SCIENCE APPLIED

Brief Contents

Part 1 Introduction 18

Chapter 0 About this Book 19

Chapter 1 From Grounded to Great: Operations Science at Delta TechOps 22

Part 2 Operations Science 32

Chapter 2 Why Operations Science 33

Chapter 3 What is Operations Science? 48

Chapter 4 Operations Science of Flow 66

Chapter 5 Operations Science of Stock Inventory 86

Chapter 6 Pulling it All Together 117

Part 3 Operations Science Application 146

Chapter 7 People 147

Chapter 8 A Management Process and Measures 150

Chapter 9 Strategy 169

Chapter 10 Demand 177

Chapter 11 Transformation 187

Chapter 12 Managing Variability 208

Chapter 13 Optimizing Inventory Without Compromising Service 215

Part 4 Stories of Operations Science in Action 228

Chapter 14 Operations Science and Information Technology 229

Chapter 15 Configure to Order: The Vortex Flowmeter Team 243

Chapter 16 Contract Manufacturing: Arc Precision Medical Device 253

Chapter 17 Operations Science in Clinical Laboratories: Lessons from the Pandemic 261

Chapter 18 Operations Science and Project Management 270

Chapter 19 The Maritime Industry: Operations Science Fuels a Sea Change in a Small Ship Repair Business 294

Chapter 20 Operations Science and HealthCare Service Delivery 308

Part 5 Leadership Profiles 323

Chapter 21 Bill VanSant, Emile Tohme, Joan Tafoya, Mike Severson, Larry Doerr, Lisa Elles, Bill Fierle 324

Appendix 00

Additional Resources 00

Full Contents

Part 1 Introduction 18

Chapter 0 About this Book 19

Chapter 1 From Grounded to Great: Operations Science at Delta TechOps 22

Part 2 Operations Science 32

Chapter 2 Why Operations Science 33

Best Practices

Why a Science of Operations

Methodologies as a Proxy for Science

What Is Science? How Can It Be Leveraged for Breakthrough Results?

From Here to Success

Chapter 3 What is Operations Science? 48

Operations Science Basics

The Fundamental Principle of Operations Science

Foundational Relationships

The First Foundational Relationship: Buffers and Variability

The Second Foundational Relationship: Queueing

The Third Foundational Relationship: Little's Law

The Fourth Foundational Relationship: Flow Performance Curves

The Fifth Foundational Relationship: Batching and Cycle Time

The Sixth Foundational Relationship: Efficient Frontier of Inventory Versus Time

Definitions Eliminate Confusion: Descriptive Terms, Quantitative Terms
Appendix: Variability Measures

Chapter 4 The Operations Science of Flow 66

Overview

Flow Fundamentals

Little's Law

Adding Variability to the Flow Performance Curves

A More Detailed Example

Analysis of Production Systems

Utilization of Systems with Multiple Products

Analysis of Service Systems

Conclusion

Chapter 5 The Operations Science of Stock Inventory 86

Stock Inventory Management Case Studies: The Supply Corporation; Bubba's Lumber

Basic Inventory Terms and Data

Inventory Control Data

How Much to Order

Known Order Frequency

Known OOP Cost

Optimal Quantity with OOP Costs and an Order-Frequency Constraint

The Incredible Robustness of the EOQ Formula

Determination of Optimal Inventory Investment

When to Order

Reorder Points and Safety Stock

Safety Stock and Safety Lead Time

Basic Stock Systems

Q, r Models

Summary of Formulas

Execution

Execution with No Visible Customer Demand

Execution with Visible Customer Demands

Conclusion: Known Fill Rate and Order Quantity; Known Fill Rate, OOP Cost, and Holding Cost; Known Order Cost, Holding Cost, and Back-order Cost

Chapter 6 Pulling it All Together 117

Introduction

First Principles

Creating Great Operations Plans

What if Toyota Had Started from First Principles

Starting from First Principles

The Virtual Queue

CONWIP, not Kanban
Few-Piece Flow
Less Capacity Buffer
Operations Science Planning and Execution
Execution Without Detailed Scheduling
Constant Work In Process (CONWIP)
Due Date Quoting
Optimal Batch Sizing
The Economic Production Quantity
Derivation of the EPQ
Adding a Capacity Constraint
The Optimal Utilization
Adding WIP Carrying Cost
Transfer Batches
The Yankee Company
The Good, the Bad, and the Ugly—An Operations Science Parable
The Ugly
The Bad
The Good
Conclusions
Appendix: CONWIP Algorithm

Part 3 Operations Science Application 146

Chapter 7 People 147

Chapter 8 A Management Process and Measures 150

Business Management Systems
What to Manage
How to Manage
Managing the Operations Plan
Management Process Outcomes
Policy Violations
Operations Science Investigations
When Business Conditions Change
Selecting and Using the Right Measures

Two Types of Measures: Leading Versus Lagging

Measures Alignment

Desired Outcomes Measures

When to React to Measures

Summary Guidelines for Operations Measures

Chapter 9 Strategy 169

Business Strategy

Operations Strategy

Business Strategy Drives Operations Strategy

Chapter 10 Demand 177

Quantifying Demand

Cumulative Distribution of Demand

Forecasting

Improving the Utility of Forecasting

Quantifying Forecast Error

When Forecast Error Makes Things Worse

Conclusion

Chapter 11 Transformation 187

Capacity and Utilization

Cycle Time

Cycle Time Elements

Queue Time

Batch Time

WIP Control

Flow Performance Curves

Estimated Actual Performance Curve

End Endless Discussions on Push versus Pull and the Definition of Pull

Operations Science in the Financial Statements

Chapter 12 Managing Variability 208

Variability Buffering Options

The VU Analysis

General Options for Managing Variability

Chapter 13 Optimizing Inventory Without Compromising Service 215

Introduction

Bubba Revisited

SpatialPrint Case Study

Context

Key Measures

Policy Analysis

Change Management

Part 4 Stories of Operations Science in Action 228

Chapter 14 Operations Science and Information Technology 229

Introduction

A Large Commercial Aircraft Manufacturer needs increased production

Addressing Unpredictable Cycle Times

Tackling Bi-Modal Inventory Swings: Insights from a Global Automotive Parts Manufacturer

Avoiding Costly Investments in Additional Capacity Through Simple Process Adjustments

Conclusion

Chapter 15 Configure to Order: The Vortex Flowmeter Team. 243

The Vortex Team

Business Strategy

Operations Strategy: Demand and Transformation

Buffering Options

Tools and Applications

Capacity Planning with a Focus on the Bottleneck

WIP Caps, Due-Date Quoting, and Virtual Queues

Sales Force and Product Line Collaboration

Inventory/Order Interface for Mass Customization

Cycle Time Components

Controls and Measures

Business Results and Lifelong Friendships

Chapter 16 Contract Manufacturing: Arc Precision Medical Device 253

The Arc Team Starts a Company

Business Strategy

Operations Strategy: Demand and Transformation

Demand

Buffering Options

Supply Chain Collaboration

Tools and Applications

CONWIP, Due Date Quoting, and Virtual Queues

Assemble-to-Order

Business Results and Supply Chain Relationships

Chapter 17 Operations Science in Clinical Laboratories: Lessons from the Pandemic 261

Introduction

Utilization and Capacity Planning

Pandemic Realignment and Measures that Mattered

Misapplied Lean Thinking

The Role of Buffers

Realistic Turnaround Times

One Chart to Rule Them All

Conclusion

Chapter 18 Operations Science and Project Management 270

Projects, Products, and Production Processes

Operations Science in Engineering Design

Operations Science and Stage Gate Control

Stage Gate in Major Capital Projects

Operations Science in Major Capital Projects

Rebar Cages for a Major Capital Project

Summary

Chapter 19 The Maritime Industry: Operations Science Fuels a Sea Change in a Small Ship Repair Business 294

Pacific Shipyards International

The Maritime Industry

Impending Change

Self-Assessment

Setting a Vision

Constructing Operations Strategy in Support of Business Strategy

Operations Strategy to Implementation

Design, Implementation and Control: Capacity-Inventory-Time

Senior Management Reorganization

Standing in the Light

Chapter 20 Operations Science and Healthcare Service Delivery 308

A Long and Evolving Journey from Policy to Practice

Strategic Buffers in Healthcare Policy from the Covid experience to Rapidly Evolving Business Models

The Operations Science Behind Success Stories in Practice

St. Mary's Hospital Case Study: Nurse Scheduling for Optimal Labor Cost

Highland Hospital Case Study: Operating Room Capacity Optimization

Part 5 Leadership Profiles 324

Chapter 21 Operations Science Profiles in Leadership 325

Bill VanSant, Emile Tohme, Joan Tafoya, Mike Severson,
Larry Doerr, Lisa Elles, Bill Fierle

Appendix Additional Resources 00

PART ONE

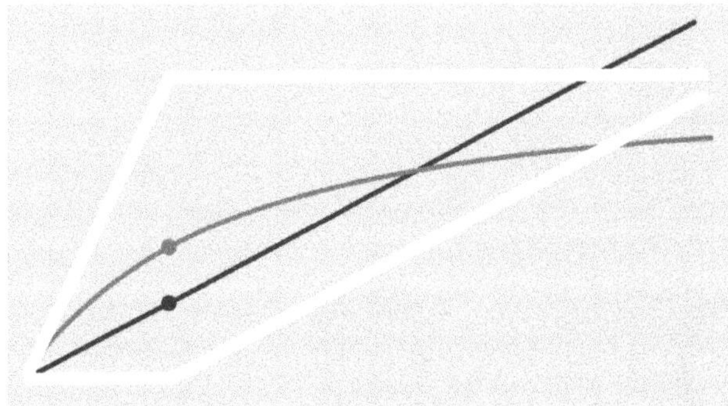

Introduction

To start out, we provide a quick overview of the book and then a powerful success story of operations science in action.

0

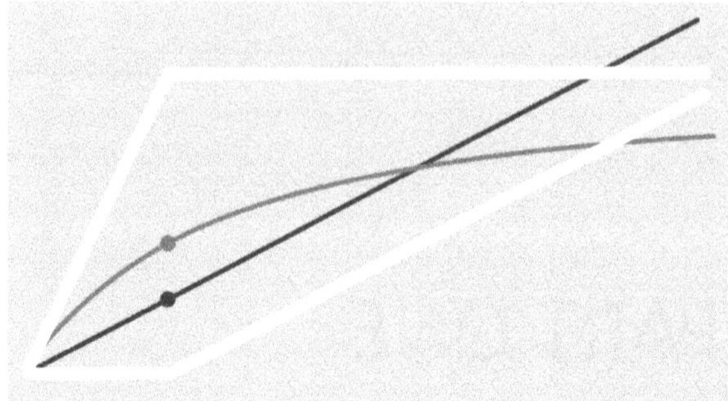

About This Book

What if There Is a Science to Every Business Success (and Failure)?

Have you ever seen something that no one else saw? Our experience has shown over and over that there is great confusion about the drivers of operations performance and, thereby, business results. Success in business might seem like a matter of luck or clever strategy. What if there is a science behind it?

There is. Operations science governs how businesses operate and how they can be understood and controlled to achieve extraordinary results. This book is your guide into the world of operations science, revealing the predictable, measurable principles that drive success in every industry, whether it is semiconductors, airlines, general manufacturing, food, engineering and construction, health care, consumer goods, or fast food.

The book is structured for many types of learners. If you prefer application stories, you will find plenty of those, starting with Chapter 1 and then skipping to Chapters 14 through 20. We recommend at least reading through Chapter 3 before skipping to the application stories in Chapters 14 - 20, so you are appropriately grounded in the concepts to better understand their application in practice.

Here is a quick overview of all the chapters so you can determine your preferred approach.

In Chapter 1, we go straight to the trenches. Hear two operations leaders from Delta Airlines describe operations science's fundamental role in meeting the incredible challenge of bringing one of the world's largest commercial airlines back online after the COVID-19 pandemic while simultaneously rebuilding its workforce and embarking on an ambitious growth plan.

Part Two, The Science of Operations, starts with Chapter 2. In Chapters 2 and 3, we cover the why and what of operations science, without much technical detail, to provide introduction and context for the concepts.

In Chapters 4, 5, and 6, we dive deep into the mechanics of operations science, equipping you with both the technical expertise to build a rock-solid scientific foundation with practical concepts and with tools you can implement right away in your own operations and career, to accelerate success. Chapter 4 focuses on flow, Chapter 5 focuses on stock inventory. Chapter 6 then pulls those together in a comprehensive view of the interactions between flow performance and stock inventory requirements. For those who are in services or project management, we have you covered. Where stock inventory is not possible or plausible, for example, you can't have stock inventory of completed 911 calls, or businesses won't build a stock inventory of completed baseball stadiums waiting to satisfy demand, the technical discussion and the book's examples cover your concerns also. After all, operations science would not be very good science if it only applied in one type of industry. Dr. Spearman, the author of these chapters, has done groundbreaking work in developing the field of operations science. He uses these chapters to describe concepts for total cost optimization of an organization's operations.

Part Three, Operations Science Application, covers operations science from a manager's perspective. It starts with Chapter 7 on People. Chapter 8 provides a management process for understanding and applying operations science to help overcome challenges in your operations and accelerate success in your career. Then Chapters 9 through 13 provide powerful, practical insights and techniques to major elements of operations in the operations science framework: Strategy, Demand, Transformation, Variability, and Inventory.

We know you will have an appetite to hear from people who have "been there and done that" with operations science.

In Part Four, Stories of Operations Science in Action, Chapters 14 through 20, contributors relate their own stories of success in the application of operations science in different industries, including information technology, configure-to-order manufacturing, medical devices, diagnostic lab services, project management in engineering and construction, shipbuilding and ship repair, and healthcare. These chapters comprise a sweeping look at the fundamental operations science concepts applied across industries.

For a final serving of inspiration, Part Five, Profiles in Leadership, Chapter 21 provides profiles of seasoned industry veterans, offering you a glimpse into their personal journeys with operations science and the impact it has had on their careers.

The appendix provides a list of additional resources for your operations science efforts.

At the Operations Science Institute, we are dedicated to promoting, teaching, and coaching the application of operations science. We have found that the concepts and framework in this book eliminate much of the widespread confusion that exists about operations management. More on this in Chapter 2. Learning and using operations science is like getting a pair of high-quality glasses to correct blurred vision. You are still looking at the same things you have always observed but with much greater clarity. How do we know this? Because we hear it repeatedly from executives, managers, and workers, some of whom tell their stories here. For teaching materials and resources in support of the book, join the Operations Science Institute at opscience.org—it is free to join. We have specific resources for academics who want to use the book in their courses.

People in operations enjoy providing services, making things, or building things but do not enjoy the stress and frustration that come with poorly managed operations. At the end of the day, they want to go home to their family and friends energized by their work, not burdened with unnecessary work-related anxiety and worry. The institute works to make that a reality, and we write this book in support of that mission.

The operations science fundamental relationship (introduced in Chapter 3) and other operations science topics continually raise questions for further research. We welcome and encourage efforts to advance the science for the benefit of all.

Once you understand operations science, you'll never look at business the same way again. It is our sincere hope that you will use these concepts to accelerate the success of both your career and the organizations where you work.

1

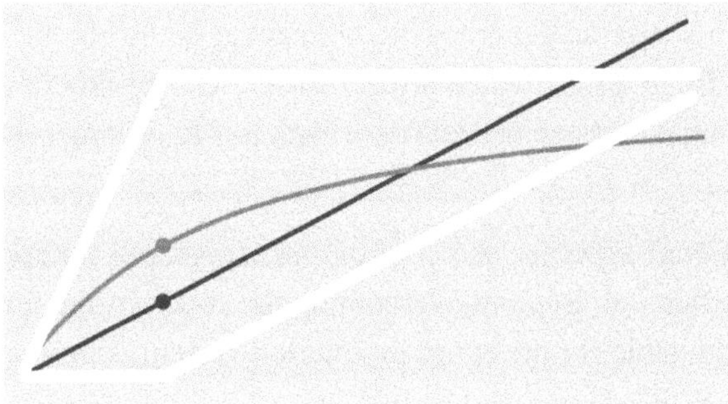

From Grounded to Great—Operations Science at Delta TechOps

Introduction

Mike McBride, Delta Airlines

As the vice president of maintenance operations for Delta TechOps, by the time COVID-19 struck, I was approaching 30 years of experience in the industry. I'd seen more than my share of challenges in this time but restarting the production system at Delta TechOps presented challenges impossible to conceive until they were before us. Parking more than 700 aircraft for months was something we'd never even considered, but bringing the fleet back to life and shortly thereafter embarking on pursuit of aggressive growth has been undoubtedly one of the most challenging experiences of my career. Along with many initiatives and investments, operations science has served as a foundation for successful growth.

Delta TechOps is the third-largest commercial aviation maintenance, repair, and overhaul (MRO) operation in the world and the largest in North America. Delta Airlines is the largest commercial airline in the world and, in addition to supporting Delta Airlines' diverse fleet of 900 aircraft flying more than 3,200 flights per day, the Delta TechOps facilities serve a broad customer base with an array of services including engine overhaul, component repair, engineering services, and other aircraft maintenance activities. Delta TechOps is currently the only MRO facility in the world that has industrialized all three next-generation engine platforms from GE, Pratt & Whitney, and Rolls Royce. The business employs 12,000 technicians and support personnel, with main bases in Atlanta, GA, and Minneapolis, MN, and is considered one of four major revenue streams for Delta.

The COVID-19 pandemic severely disrupted the aviation sector and posed unprecedented challenges for Delta Airlines. This led to major workforce attrition and loss of crucial operations experience. The company offered generous buyout packages during the depths of the pandemic and Delta TechOps estimated it lost nearly 50,000 years of combined experience from departing employees. Bringing a massive aircraft fleet back to life was just step one in recovering from the pandemic.

With Delta's business rebound starting in 2023, Delta TechOps faced the challenge of reestablishing its operational expertise and control mechanisms to support aggressive growth plans and service diversification. Beyond overcoming the chaos of the international supply chain, Delta TechOps had to reset the table for the entire production system, simultaneously addressing the complexity of new products, processes, and people.

MRO Operations Overview

MRO is a complex production environment. Engine repair is similar in some respects to mixed-model production in that we adapt to a number of different engine models, with the added challenge being that each engine's maintenance visit has unforeseen requirements. There is extensive variability. Some engines run fine relative to operating parameters right up until they arrive for a planned shop visit, only for us to discover, upon disassembly, that the engine has significantly more damage than expected. So each engine is its own project, you might say. From a forecasting perspective, what we're trying to do is to ensure we have the appropriate level of inventory and capacity to support the work required, while simultaneously trying to keep inventory levels and labor costs as low as possible.

Allow me to break down the process so it's a little easier to understand. I'll include the application of various strategies we employ to overcome constraints and improve forecasting and execution.

It all starts with a schedule. Every year we build an annual plan for each engine line. At Delta TechOps, our three main engine facilities are organized by original equipment manufacturer (OEM): GE, Rolls Royce, and Pratt & Whitney. Each shop handles multiple engine variants. In the GE shop, we work the CF6 and CFM models as well as the new LEAP engine. In the Rolls shop, we work the four major incarnations of the Trent family of engines. Our Pratt & Whitney shops are divided between legacy products, the PW2000 and PW4000, and the new GTF 1100 and GTF 1500.

Customers want their engines or components back in service as soon as possible. In our business, turnaround time is king. Turnaround time is the time from when the engine or aircraft component enters Delta TechOps until it is shipped back to the customer or, for inventory components, is available in stock for use on an aircraft.

The annual plan for our engine lines is based on production capacity, which is driven by our expected turnaround times. This is a classic operations science problem. Turnaround time is a function of capacity, variability, and work in process (WIP). We have a finite number of engine bays for capacity, so we plan WIP levels using Little's law.[i] We can expand or contract in a given year to ensure that we can meet demand, but we are always focused on achieving the best possible turnaround time.

Every engine has a maximum life expectancy due to life limits for major components. These limits are based on flight hours or takeoff/landing cycles. As an engine approaches these life limits, shop visits are planned to restore the engine to serviceability. The predictability of these intervals makes it relatively easy to build a schedule for these engines. We refer to this schedule as "slot planning." We have finite physical capacity in a given shop, maybe 10 engines at a time, so we plan the slots based on that capacity and the duration (turnaround time) we expect. That said, every engine does not make it to its natural end of life without problems, so we also must plan for unscheduled removals. Exactly when an individual engine is going to ingest a bird, have a major oil leak, or degrade in performance for some other unknown reason is impossible to predict. However, our engineering teams have enough historical data to make accurate assumptions regarding unscheduled removal rates. At the end of the slot planning process, we typically have close to 1,000 engine-shop visit slots planned.

When an engine enters the shop, the first step is inspection at Pre-Induction (P-0). See Figure 1-1. We're looking at both the physical engine itself and the records on the engine to understand its performance since it was last refurbished. Here we first check the materials forecast for any unexpected findings. Since every engine has an expected work scope, it's not difficult to discern if we're going to see scope creep. Minor scope creep related to hardware, tubes, ducts, etc., happens routinely, but major scope creep related to life-limited[ii] parts with issues before their life limit is unusual and causes significant impact to the build schedule. Next, at Induction (I-0), we disassemble the engine to the module level, followed by disassembly to the piece-part level. Again, we have inspections here that ultimately adjudicate parts as either continued time (still fit for service), in need of restoration, or scrap. A repair bill of materials is then created and checked against the materials forecast. We consider all the aforementioned steps gate 1.

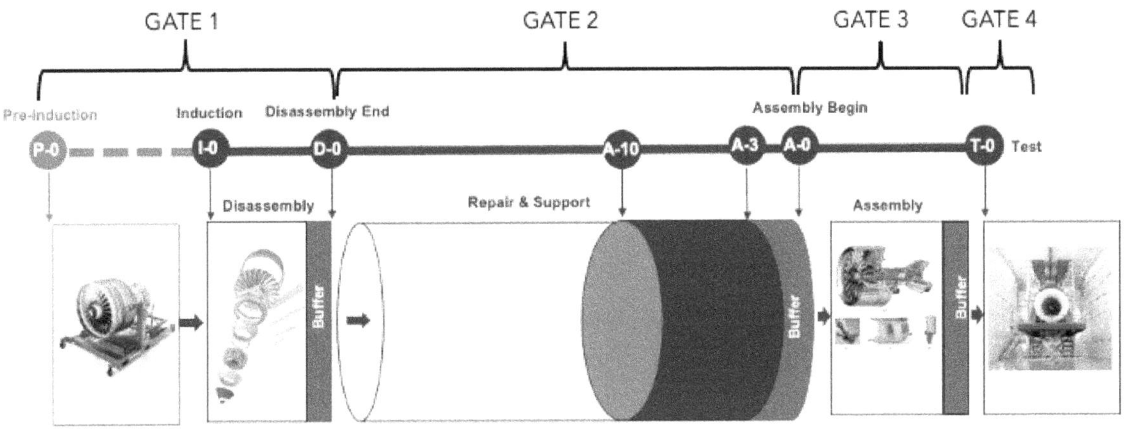

Figure 1-1: Maintenance, Repair and Overhaul (MRO) process

Gate 2 is the repair cycle. Thousands of parts move in many different directions for inspection and repair. Everything from combustion cases to blades and vanes to fuel and electrical components. This is the most complex phase of the process. The whole engine overhaul process is detailed on a Gantt chart, with the gate 2 repair phase generally set for 45 days.

We work the capacity and inventory buffers as described by operations science to try to achieve a 45-day turnaround time. We are not trying to achieve one piece flow or zero waste. We strive to hit our turnaround time and keep our customers happy with minimum cost, high quality, and maximum safety.

We may invest millions of dollars in equipment to increase capacity, or we may work overtime for a more flexible capacity buffer. Since many parts take much longer than 45 days to complete repair, the materials plan must be built to include an inventory buffer to ensure turnaround time targets are achieved. After capacity, inventory, and turnaround time, variability is the fourth component of the operations science portfolio, and we employ Lean, Six Sigma, and all available techniques to reduce process variability. We are not hung up on one initiative or another; *we let the science inform and guide our efforts.*

The goal is to hit the kit date for the engine, meaning all the parts required to start the build process, known as "gate 3", arrive at or before the 45-day mark. We call this kit date A0, which is just a way to tie it to the airline business in general since A0 is the terminology used to define the scheduled arrival time for a flight; if you arrive on time, you "hit" A0.

The build process (gate 3), is typically 8 to 10 days, depending on the engine type. Once the engine is complete, it goes to the test stand, or "gate 4", where it runs for several hours to break in bearings, check performance, and make minor adjustments and/or repairs before it is tagged as serviceable and prepared for installation on an aircraft.

Hopefully this brief explanation of the process makes it clear that the planning and flow of materials through the process is critical to consistently hitting established turnaround times. Over the past few years, Delta TechOps has invested heavily to improve forecasting capabilities. Newer engines produce tons of data every time they fly, and we now capture and monitor the performance of every engine so we can improve the accuracy of our parts forecast. The inventory buffer is critical, but it must be balanced with cost. We can't simply purchase everything on the forecast to hit turnaround times; we must work to improve flow through the shops. We do this using a Lean operations science approach. Lean brings process discipline and stability, while operations science provides a foundational understanding of the relationships between capacity, inventory, turnaround time, and variability. Using operations science, all our employees can speak the same language and quantify results and questions based on an objective, science-based, comprehensive framework to control and improve the performance of our production system. Fast is great, but the trifecta of fast, cost-effective, and safe is the key.

Strategy for Managing Recovery and Growth Successfully

I got involved with operations science early in my career after having worked as a Lean/Six Sigma Black Belt for several years. I had begun to realize that many of the tools I'd been using and teaching to solve process problems had evolved from previous generations across multiple industries and were simply repackaged by clever consulting firms to make them marketable to a new generation. Most of these well-known methodologies contain useful concepts, but after several years of practical application, I had realized that what was needed was simply a scientific approach to understanding production systems and what makes them tick. I knew from experience that the operations science approach would not include the unessential and superfluous information that is often encountered in the more heavily marketed continuous improvement efforts. If one follows the science, the task of prioritizing and building consensus regarding which problem(s) to solve becomes much easier. With this in mind, we had to think through how we intended to map our way from the chaos of COVID-19 to the rigorous application of operations science across the production system.

Our first task was to rebuild the workforce and focus on stabilizing the production system. Without some level of stability, we knew it would be hard to communicate even the most basic objectives to the workforce. By necessity, we had to change many of our processes to accelerate training and bring our new employees up to speed on the technical tasks required of them. Once we got on top of the training curve, we had to reestablish standard production flow, but before we could do that we had to take a step back and reconnect with the basics.

We did this by implementing 5S, a common Lean approach to workplace organization and control. This is something we did originally more than 25 years ago, and which, prior to COVID-19, was embedded into our production system. The chaos and turnover during COVID-19 had all but wiped out this and many other elements of our production culture. Our leadership team knew we wanted to get to a point where we could relaunch operations science and strongly focus our teams on science-driven continuous improvement, but we also knew we had to be patient and create some stability first.

This is in sharp contrast to the common result that many organizations see with Lean efforts, where they do 5S enthusiastically and then efforts fade out. Without operations science, organizations can flounder after the initial initiative enthusiasm of Lean and basic 5S efforts. Without an operations science based approach, improvement selection becomes a scattershot effort with no focus beyond some initially obvious projects. At TechOps, we knew operations science would help unite and lead efforts. That would be easier once we did the straightforward 5S efforts to get our house in order.

In 2021, stability came in fits and starts. The airline was starting to see the resurgence of passenger demand much faster than expected, and we were still dealing with anemic supply chains, new employees, and high demand for services from our customers in the cargo industry. Now suddenly the airline was asking for higher production volumes from us to get more airplanes in the air while we were fighting to maintain a sustainable pace of growth and deliver reasonable economics. And what often goes unmentioned about this crazy time is the fact that this same workforce, charged with restarting the largest airline in the world, was also in the midst of industrializing the production of three new Rolls Royce and two new Pratt & Whitney engine types, representing an investment of over $250 million. Obviously, failure on any of the aforementioned fronts was not an option. We were simultaneously rising from the ashes of COVID-19 and launching the most aggressive MRO growth strategy we'd ever undertaken.

By 2022, though still dealing with myriad challenges from the supply chain, we'd begun to settle into our 5S routines, and the major production milestones were at least measurable again. Our turnaround times in component maintenance had made significant progress toward respectability, but our engine turnaround times were still significantly worse than our pre-COVID-19 performance. Most engine lines were still taking more than 150 days to complete an overhaul, whereas our pre-COVID-19 standard was around 75 days. However, we were as stable as we could be given the circumstances, and the team knew it was time to move forward with our plan to relaunch operations science across the organization to get the knowledge base established and get everyone speaking the same language relative to how we discussed production operations and improving performance.

Operations Science in Execution

Fast forward to 2025, we now have a strong support team teaching and mentoring operations science practitioners across the facility, and we've completed 5S initial and follow-up events in every shop and established daily management routines for each shop. Front-line teams have initiated over 2,000 Kaizen events, and we are using operations science as the foundation to identify, prioritize, and execute on improvement opportunities. Over 500 people have completed internal operations science training courses, and we have developed a suite of reporting tools for shop managers to enable them to use what they've learned to manage their operations more efficiently. To provide the detail on how we got from then to now, Margaret Amalfitano, manager of our Safety, Quality, and Continuous Improvement (SQCI) group, is best suited to tell that story. Margaret is a highly talented manager charged with implementing, among other things, operations science across TechOps.

Margaret Amalfitano

Mike asked me to start the SQCI team in 2021. My team coordinates and directs the engine and component maintenance activities to meet customer and regulatory requirements. SQCI directs process improvement initiatives across all production, engineering, and service delivery areas, with a focus on improving the effectiveness and efficiency of our operations. Then in late 2022, Mike asked me to develop the in-house operations science team to teach those fundamentals to all our production supervisors, support staff, and management.

I have over 30 years of experience in the aviation industry. I started my career as an engineer at Grumman Aerospace (now Northrop Grumman) in the advanced aeronautics division. My work included system design on experimental and classified aircraft. I have a strong background in manufacturing and production systems and am educated in the various continuous improvement and production efficiency programs—Lean, Six Sigma, Theory of Constraints (TOC), etc. After attaining my MBA, I went into the airline industry, where my focus was identifying opportunities to improve operations, impact business growth, and maximize profits through achievements in cost reductions, internal controls, and productivity improvements. I have been with Delta for 10 years now.

Working with Ed Pound, Jeff Bell, and Mike Gores of the Operations Science Institute, we defined the objectives of the training: First, establish a practical science-based understanding of operations behavior among TechOps personnel as a foundation for our standard, sustainable, management framework. Then, apply operations science concepts in TechOps operations planning and control practices in support of our Lean operations strategic objectives and daily performance goals.

With our goals defined, the Operations Science Institute crafted a TechOps-specific training syllabus to train the trainers. The curriculum consisted of six separate modules. To achieve operations science trainer designation status, our Subject Matter Experts (SMEs) had to successfully complete the program, which included lecture sessions, textbook instruction, simulations, one-on-one coaching, practice examples, and written tests. The final stage required that the trainers demonstrate proficiency in delivering the modules to students.

The train-the-trainer content might seem daunting at first, but the training program is broken down into digestible sections. Most of our SMEs have engineering backgrounds, and one trainer is a certified avionic and powerplant mechanic. To be an effective trainer, you need to be comfortable with basic math and algebra. Advanced math, like calculus and statistics, underlies all of operations science but is not a prerequisite to learn and apply operations science concepts. Strong teaching skills are obviously very helpful going in. Some of our trainers weren't naturally comfortable presenting to classes. However, they were willing to take a chance to develop that skill and have turned into strong trainers. The Operations Science Institute's coaches were patient and supportive as the trainers worked through grasping the concepts at different paces.

Interestingly, as we progressed through the program, we found that it wasn't the material content but the teaching role that proved to be an obstacle for some. We started with the full complement of my team as potential trainers and ended with only half that number certified. The team all understood the material and could apply it to the shops, but some were not comfortable in a formal instructor role.

SQCI worked closely with the Operations Science Institute during the training stages. Since the TechOps trainers were going to train both leaders and technicians with varying operational comprehension, it was essential to modify the generic operations science class curriculum. We added more Delta references, shop examples, and expanded explanation of the concepts for ease of understanding. After every operations science session, we provided a survey to the participants. This helped us to gauge the effectiveness of the training. We then utilized this feedback and realized we needed to add more graphical representation of concepts, additional definitions, and even more thorough explanations of the material. It is a live document. We are currently teaching the 11th revision of our initial class workbook.

With our SME's certified, we started rolling out OS training to the leadership in engine and component maintenance. The estimated completion of more than 60 shops in engine and component maintenance is Q1 2025. Our rollout philosophy was to start training the shops that represented different types of functions as a proof of concept and then utilize those lessons learned by spreading to similar shops within the operation.

Overall feedback from SQCI's operations science training has been positive. Technicians, even those that were initially intimidated by the math, appreciate the clear, easy-to-understand presentation of the science. Per the technicians, "Real-world examples from our plasma shop helped to build a clear picture." Leadership is confident that with the framework provided, front-line personnel can apply the principles to design, implement, and control their own specific processes more effectively.

Some students do attend the class with reservations i.e., "Here we go again, another flavor of the month." But this mindset is quickly dispelled as they progress through the class. The training is designed to be highly participatory, with plenty of simple exercises and fun simulations to work through the math. When they work through the basic flow curves and work the dice simulation activities, as the trainer, you can almost see the lightbulbs go on. Class participants realize that this is the science describing the operations behavior they deal with day in and day out. Once they calculate where their operation falls on the curves, they then can apply the appropriate continuous improvement (CI) tools and initiatives (e.g. Lean, Six Sigma, Agile) to optimize their system.

As word of the training spread, we were pleasantly surprised when personnel from other disciplines within TechOps requested to attend. To accommodate, we've added a monthly class, open to any interested employees. One of the disciplines that was highly interested was finance, so much so that they placed a finance team member in the SQCI group to better understand and incorporate operations science into financial planning.

Implementing operations science into an organization as large and varied as Delta is not without its challenges. Many participants in the class were unaware of their shop's true capacity. Some had an understanding due to years of experience but lacked the actual data. When calculating shop capacities, we found outdated, incomplete, or incorrect data regarding setup and process times. Shops were calculating key metrics differently and had varying definitions of throughput and turnaround time. Through our training, we were able to provide standard terminology and a framework to calculate all metrics.

Although we are in the early stages of implementation, we have already seen improvements.

The Plasma Department removes and reapplies thermal spray coatings on parts from aircraft engines, auxiliary power units (APUs), landing gear, components, and airframe. One of the shop's ongoing projects is inventory balancing on the spray booths. Shop employees are working to optimize the throughput by analyzing volume at the booths and identifying where there are constraints and what booths are "starving for parts."

This has led to another study in the plasma shop regarding the first article inspection report (FAIR) process. This production validation process involves verifying that a new or modified process produces conforming parts, which is required on parts for certain customers. The teams are studying the potential of reevaluating parts on multiple booths, instead of one specific booth, to balance the flow more efficiently.

The Case Shop successfully utilized operations science principles in support of our customer. The Case Shop is focused on parts that make up the outer containment and support of the engines for all fleet types managed by TechOps. Rolls Royce requested a utilization-by-part study of our coordinate measuring machine (CMM) on the Trent 1000 engines. The scope of this project was to identify top Rolls Royce repairs that currently require CMM inspection and determine if they are necessary. This list would allow Rolls Royce to better understand where changes are needed when defining MRO requirements around metrology. Rolls Royce has strict quality requirements that lead to CMM bottlenecks in the MRO community.

Operations science application also detected disparities in the Case Shop throughput on the vertical turret lathe (VTL) lines. The Case Shop employees performed capacity studies of VTL lines utilizing the format presented in the training curriculum. Data revealed a decrease in utilization on the weekends from 90% during the week down to 30%.

This analysis brought to light that the weekends were supported by technicians with fewer than a year and a half of experience in machining processes. The product line manager (PLM) implemented a corrective action by training the technicians during the week and backfilling with more experienced technicians.

Mike McBride

In addition to the examples provided by Margaret, there are plenty of anecdotes that reinforce my confidence in our strategy. One such example is from a comment by one of the class participants, a shop technician. He said,

> "We used to think we had 30 problems and 30 solutions because that's about how many people work in the shop, and everyone had an opinion. Now, after operations science training, we have realized we have about four main problems, and we are all focused on solving those problems."

If you are an executive trying to drive change in an organization, that is music to the ears. We are still in the early stages of implementation, but we can see that we are gaining ground daily and generating industry leading performance.

Epilogue: My Personal Journey into Operations Science

I provided a snippet of insight into my personal experience with operations science earlier in the story, but I'll take up some space here to explain my journey in more detail. As I mentioned previously, in the late 1990s and early 2000s, Lean and Six Sigma were all the rage in many companies and served as the basis for many CI programs. I was selected by leadership to participate in the inaugural Six Sigma Black Belt class at Delta. I had a natural knack for problem solving but wasn't very knowledgeable about any of the formal methodologies being employed. That said, I learned a lot through the training I received and went back to the shop floor with a lot of excitement and enthusiasm about using the statistical methods I'd learned to control and improve processes. Shortly thereafter I was given the opportunity to attend "Lean Systems Design" training at the University of Tennessee. Again, I had a great experience with this training and most notably, I learned a lot about the concept of value and how to differentiate between the steps in a process that actually create value and those that don't. Most importantly, I think the training I received and my somewhat successful attempts at practical application motivated me to learn more.

As my skills progressed, I started studying more source material on many of the tools and ideas covered in these methodologies and ultimately began tracing many of these ideas back to their roots. Whether it was Henry Ford, Frederick Taylor, or Taiichi Ohno, I started to realize these were not recent developments, rather, they were ideas that had evolved throughout the Industrial Revolution which had been neatly packaged into digestible and marketable programs to meet the demand demonstrated by the corporate world for CI that impacts the bottom line. I'm sure C-level executives also experienced pressure from their boards to "do Six Sigma." Be that as it may, these methodologies have their place and all have contributed to some amazing results; it just took me digging in a little to understand how they came to be and why.

At about this time I was in a conversation discussing improvement methodologies with someone I don't recall who exactly, who handed me a *Factory Physics*[iii] textbook. It was a heavy book laden with formulas and written in an academic style, so it wasn't for the faint of heart. That said, it was exactly what I'd been looking for. It provided a systematic way of viewing production without all the catchy names and dojo-inspired belt ceremonies. It was a thorough treatment of the science (and math, mostly math) that characterizes any production system, stripped of all the marketing silliness. It was simply physics for the factory. I studied this book from cover to cover, and I count it as one of the most pivotal points in my career. I've read hundreds of books on the subject of CI and production efficiency, but to this day I've never seen the depth of coverage you will find in that book. It enabled me to view and assess most any production system objectively and provided the depth of knowledge I needed at the time to really improve things at the system level. I've used this knowledge to great effect both inside and outside of Delta (I worked in the consulting industry for 10 years in between my two stints at Delta).

I entered the MBA program at the University of Georgia in the fall of 2001. Fortuitously, it was there where I met Dr. Jim Cox, who was teaching the operations classes in the program. Dr. Cox is renowned for his work with Eli Goldratt and the Theory of Constraints International Certification Organization. Jim was an excellent mentor, and we had some healthy debates about TOC, which I found to be a simple but excellent way to think about problem solving in a system, production, or otherwise. But the one thing I would always come back to was this: Once the constraint is identified and you've chosen a course of action, how exactly do you execute? And there it was again, *Factory Physics*! Yes, there are some incredibly thoughtful ways to teach people systems thinking and many are brilliant, but to actually make progress one must understand what to change and, more importantly, how to change it. Thankfully Mark Spearman and Wallace Hopp took the time to compile everything you need to know to execute effectively.

Once I had my hands on *Factory Physics* for several years and was looking for a way to make an impact after I rejoined Delta in 2014, I thought I'd find out who Mark Spearman and Wallace Hopp were and what they would think about taking a look at our production system. After a few inquiries and email exchanges, in 2016 Mark agreed to visit Delta with Ed Pound. I was curious how we could simplify the content of the book such that it could be taught to front-line leaders, giving them the fundamentals of production management and helping them become more scientific when it came to leading their specific shops. We decided to create a two-day training program, which later became mandatory training for everyone on my management team. We worked hard to use consistent vernacular and ensure that our metrics and production meetings were aligned with the factory physics philosophy. I am confident when I say this work was one of the critical success factors for Delta becoming one of the most efficient and reliable airlines in the world prior to the pandemic.

How to get back to that level of performance was the question of the day back in 2021. So having worked with the, now, Operations Science Institute team as I just described, and having maintained contact with several of the principals, it wasn't hard to figure out who I planned to call once we were ready to launch into the next phase of our plan: develop in-house operations science trainers to teach the fundamentals to all of our production supervisors and management. The idea was that we could create a small team of operations science subject matter experts (SMEs), within the group leading the 5S and our Lean Operations Science implementation, which would be self-sustaining relative to the training element and support improvement activities across the organization.

We are well on our way. The journey continues with using operations science as a foundation for developing our people and driving growth and performance improvement throughout TechOps.

---

[i] We will cover Little's law extensively in Chapters 3 and 4. It is the relationship between WIP, Cycle time, (called turnaround time at Delta TechOps) and throughput.

[ii] "Life-limited part means any part for which a mandatory replacement limit is specified in the type design, the Instructions for Continued Airworthiness, or the maintenance manual" (https://www.ecfr.gov/current/title-14/chapter-I/subchapter-C/part-43/section-43.10).

[iii] Wallace J. Hopp and Mark L. Spearman, *Factory Physics: Foundations of Manufacturing Management*, 3rd ed. (Long Grove, IL: Waveland, 2011).

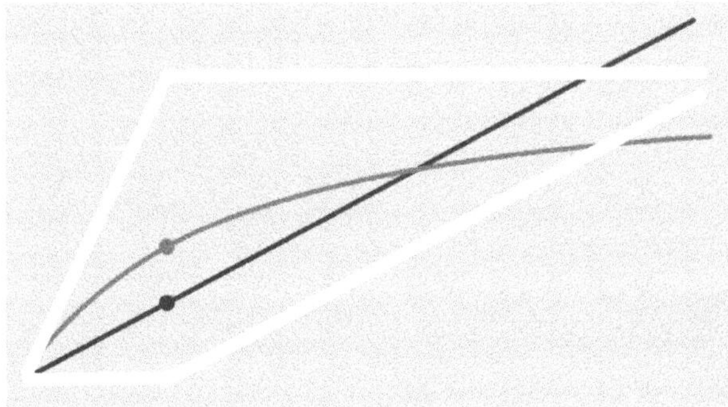

PART TWO

Operations Science

In part two, the core of operations science is detailed. Starting first with answering the question "Why operations science?" 200 years after the start of the industrial revolution, 50 years after the dawn of the computer age, 35 years after the emergence of the Lean and similar continuous improvement efforts, and at the beginning of the age of artificial intelligence. The need for operations science has never been more vital to provide a comprehensive, scientific framework for operations executives, managers, and workers everywhere. Application of the science improves focus, reduces stress and chaos, and allows people to effectively evaluate and apply all the tools and concepts at their disposal to accelerate success.

Chapters 3 thru 6 provide the technical detail to define a science of operations that is universal and can be applied in any organization.

Why Operations Science? (Chapter 2)

What Is Operations Science? (Chapter 3)

Operations Science of Flow (Chapter 4)

Operations Science of Stock Inventory (Chapter 5)

Pulling It All Together (Chapter 6)

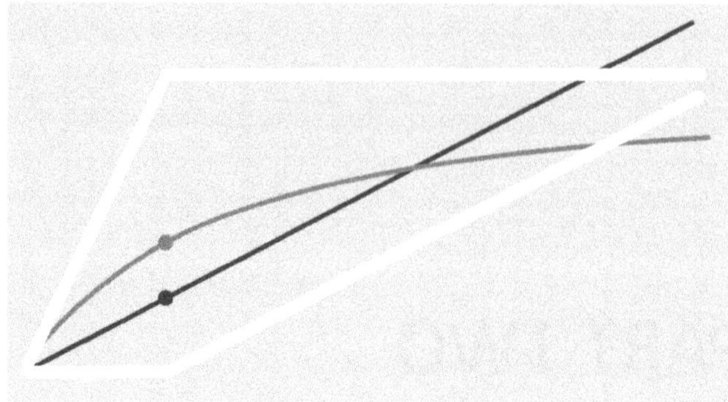

2

Why Operations Science?

Best Practices

Managers and executives spend much of their careers trying to implement best practices to achieve performance goals and advance professionally. There are hundreds of articles trumpeting the best practices for almost anything you can think of—best practices in health care, best practices in higher education, best practices in project management, best practices in software development, best practices for manufacturing, even best practices for analysts who publish best practices, and on and on.

But what is a "best practice?" Why is it "best?" Is the best practice actually going to work for the manager's particular application, and how will the manager know beforehand? It's one thing to read or see how a best practice worked for someone who has already implemented and developed it over time for their situation. Outcomes are uncertain when a manager tries to adopt a best practice and implement it in a different situation. Did the authors really consider all the methods and choose the best one? Seems unlikely. For what situations are best practices the best? How does one know when it is best? Are the best practices for manufacturing cars the same as the best practices for manufacturing airplanes?

Within industries there is contradiction. One article will tout "agile" as a best practice for software development while another recommends "DevOps", and we end up with "Newton's Third Law of Experts": "For every expert there is an equal and opposite expert."

In any case, the one issue common to all of these "best practices" is that they are concerned with optimizing some kind of *operation*.

Indeed, almost anything we do is an operation. Any kind of manufacturing is an operation. Converting a piece bar stock to a pile of screws, using a multi-spindle automatic screw machine is an operation. Drilling an oil well is also an operation. Any service performed is an operation. An engineer designing a product, a bank teller taking a deposit, and a waiter taking an order are all operations. When one drives a car from one place to another, they are doing an operation. In practice, nearly everyone on the planet is performing operation after operation every day of their lives.

Operations are performed any time resources are used to change one or more entities to satisfy some demand. A resource may be a person, a machine, a computer, fuel, information or any other entity required to perform the transformation task. Most of these resources can be made available in advance of the transformation. However, "capacity" resources—also called process centers—such as labor, machines, and other entities that require time to perform the transformation, are not always available. They may be busy doing other transformations for a different demand. The interface between demand and these types of transformations is the heart of operations science.

In order to develop and execute "best practices" that get results for specific situations faced by managers and executives, we need to understand operations. The good news is that there is a science of operations that can help. This is good because the science works across *any* type of operations. Like physics, we don't need a science for manufacturing operations and a different science for hospital operations or for construction operations, just as we don't need one set of physics for the trajectory of baseballs and a different set of physics for the trajectory of rockets. Further, science can be leveraged to achieve breakthrough results. Operations science helps leaders navigate a rapidly changing operations world with artificial intelligence (AI), the Internet of Things (IoT), blockchain, and digital manufacturing. The bad news is many people are not aware of operations science.

Before we get to operations science, we need to precisely define an operation:

> **An operation** is an act that utilizes resources to transform one or more attributes of an entity or set of entities into some good or service that is required to satisfy a given demand.

Thus, an operation must perform some sort of transformation. Transformation is not limited to the conversion of raw materials into some finished product, as in traditional manufacturing. For example, flying on a commercial airliner is an operation because the airline uses resources to change your location, and this satisfies your demand to go to a different place. Likewise for packages with FedEx, the US Postal Service, and UPS. iPhone and Android phone calls are also operations but may be a bit more subtle. The entity they change is the caller or the one being called, and the resources are the entire smartphone infrastructure and its embedded information.

In a perfect world, operation transformations are exactly synchronized to demand. As soon as the finished product or service is needed, it is provided by the required operations. Products or services are delivered 100% on time at minimum cost with resources utilized at 100% so all revenue is met with minimum cost. This provides maximum possible profit and cash flow with perfect on-time delivery.

Unfortunately, this perfect world does not exist. The world we live in has *variability* and it is this variability that prevents demand and transformation from being synchronized. Fortunately, the science of operations provides a way to understand the interactions between demand, transformation, resource utilization, and variability for any operation, be it production or services, including healthcare, transportation, communications, product design, and so on.

With that background, now is a good time to define operations science.

> **Operations science** is the study of the transformation of resources to create and distribute goods and services. It focuses on the interaction between demand and transformation and the variability associated with either or both.

Now you may be wondering why you have never heard of operations science. Quite simply, the reason is that it is seldom taught as a subject in and of itself. Instead, what is taught is a set of methods for doing certain operations—best practices (hopefully). But how can one determine a good method to manage any kind of operation if they do not understand the underlying behavior of the operation? To see this, let's review how such courses are taught and a brief history of how we got here.

Why a Science of Operations?

College engineering and management courses are essentially prescriptive, in that they teach *what to do* for certain given situations. Such courses are essentially different from science courses that describe *how* natural and artificial systems behave under various conditions. Engineering and management courses *prescribe*, science courses *describe.*

But most engineering and management courses assume some level of descriptive knowledge learned by taking particular science courses that are relevant to the applied field of study. In engineering programs these are called, not surprisingly, "engineering science" courses. Chemical engineers study chemistry and thermodynamics before learning how to design distillation columns and fluidized bed reactors. Electrical engineers take courses in electricity and magnetism before they approach integrated circuits or power systems. Business school consumer behavior programs are based on key social science theories that are then applied to marketing practices.

So as professors and students of industrial engineering (IE), operations management (OM), supply chain management, and construction management, it pains us to reveal a little-known fact: Most of the courses in these disciplines are *not based on science* but are almost entirely prescriptive with very little attention paid to *describing* the systems being prescribed. This is not to say that the courses are without rigor. Many involve very sophisticated mathematical models of the systems for which they prescribe. But mathematics is not science.[i]

A brief historical review will help to appreciate why IE and OM are mostly without a scientific basis. Ironically, IE grew out of what was called scientific management, or sometimes called "Taylorism" for its founder Frederick Taylor who wrote around the turn of the 20th century. Although Taylorism was out of fashion by the 1930s, Taylor did use science to develop his system. For instance, he used data to determine the optimal shovel size (too small, and not much gets done; too large, and the worker becomes exhausted).[ii] But by 1913, scientific management was already becoming more management by mathematics than management by science. In that year, Ford W. Harris wrote a seminal paper, *How Many Parts to Make at Once*. In it he writes, "Experience has shown one manager a way to determine the economical size of lots" and then goes on to derive the famous Economic Order Quantity (EOQ) formula.[iii] It was not experience but simple calculus that showed Harris how many parts to make at once. Moreover, the resulting model does not consider capacity or setups, issues that are important in setting lot sizes. Indeed, his model is more useful for *ordering* parts than for *making* parts. A better understanding of the operations involved would have resulted in a more relevant model.

Current events and historical perspective provide insight into the power and perils of raw data without scientific context. Contemporary advances in technology such as AI and the IoT will bring tremendous benefits, but neither of them provides a scientific framework, which will limit the progress possible. Big data is one current example, with Grand View Research predicting annual compound market growth of 14.9% leading to a big data market size of $862 billion by 2030.[iv]

Big data can provide powerful insights, but one should not, and certainly does not, need to rely solely on big data to predict events based on scientific principles. Just as the same law of gravity explains an apple falling from a tree and the orbit of the moon, there is a fundamental operations science relationship that predicts what will happen to response time when work-in-process increases. The potential risk with big data is that, uncoupled from science, it will lead to predictions that do not yield desired results and potentially even violate the laws of science. The same applies to AI. It will be interesting to see to what extent big data and AI will be effectively used in collaboration with science and to what extent results from their use without the context of science will suffer when data bias, inaccuracies, or incompleteness occur.

From a historical perspective, a field related to but distinct from IE called operations research (OR), or operational research in the UK, was developed during World War II. OR was first applied to operational war-time problems during World War II. During the war, these problems involved collecting and analyzing data, using the results to determine optimal policies for a broad range of activities, including anti-submarine warfare, allocation of anti-aircraft guns on merchant ships, and even routine issues such as how many wash-and-rinse tubs were required to minimize wait time in the chow line. The use of data and the performance of experiments made OR a true science of operations (at least, military operations). However, after the war, the idea of optimizing operations using mathematical models began to attract large corporations in America and Europe, and OR became an important part of Industrial Engineering. As this merger occurred, both fields became more mathematical, with a focus on general methods that are divorced from their original application areas. For instance, linear programming is a general technique that has been used in such disparate areas as agriculture, transportation, manufacturing, and energy. By the 1970s, OR had become a field of applied mathematical techniques with almost no scientific inquiry being performed.

Operations Management (OM) has had a similar trajectory. There are certain models and techniques that are contained in most introductory texts such as forecasting, inventory control, linear programming, aggregate production planning, quality assurance, product/process design, capacity planning, and so on. Most research is also conducted in these areas. Even though the OM journals tend to present more empirical data than journals in IE and OR, most of these studies are for unique situations with little emphasis on discovering universal behaviors of business systems. Without the use of operations science, contemporary advances in computing power and developments such as AI risk a continuation of the mistakes that have occurred with the use of technology in operations management.

While it may be an oversimplification, most efforts related to operations management are directed toward general methodologies that can be used for specific problems but that have few if any descriptive models. Queueing theory is a good example. It can be used to describe wait times in service queues of a bank and estimate the total cycle time (CT)[v] through a semiconductor plant. But the specifics of those situations seldom enter the analysis. One assumes a probability distribution for the arrival process and one for the service process and then computes exactly (for a limited set of cases), or approximately for the rest, particular queueing measures such as time in queue, number in queue, utilization, and so on. Moreover, basic relationships such as total cycle time through a series of operations versus the amount of WIP is almost never discussed in the books or taught in the classroom. Indeed, we have seen great confusion regarding such relationships. To make this point, we often ask our audience (usually made up of seasoned operations professionals) the relationship between WIP and total cycle time and between efficiency (utilization) and cycle time.

In this exercise we keep things easy, and forget about variability for a while, and take a simple example of a raw material stock feeding four workstations in series that goes into a finished goods stock. Each workstation takes exactly two hours with no variability, see Figure 2-1. Now we ask a simple question: If I know how much work in process (WIP) there is in the system, can I tell you how much time it takes to go through the system (cycle time)? Do I know the basic relationship between WIP and cycle time? Could I plot it?

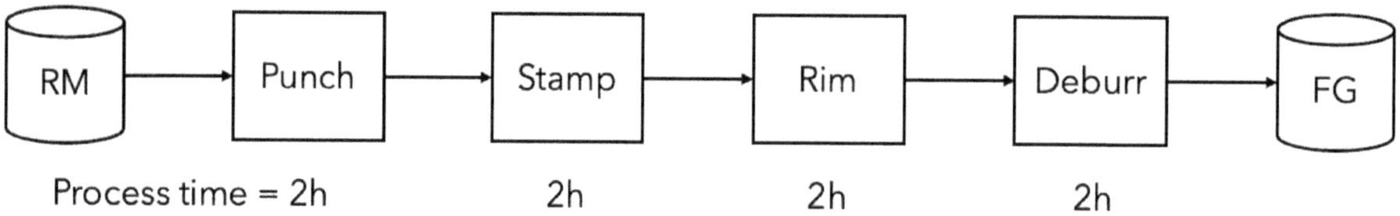

Figure 2-1: Simple production system with no variability

Why is this important? WIP is money tied up in inventory, and it can run into millions of dollars. Cycle time is related to customer lead times and service. If my cycle times are longer than my quoted lead times, I have lousy service, and customers are not happy. If my cycle times are very long and I finish within my long quoted lead times, customers are still not happy with the long lead time. So, the relationship between the two is very important but does not seem to be understood by either professionals or consultants. We have tested this understanding of basic operations science with literally thousands of experienced operations professionals. We describe the process in Figure 2-1 and ask the participants to draw the relationship. Figure 2-2 shows a typical set of responses.

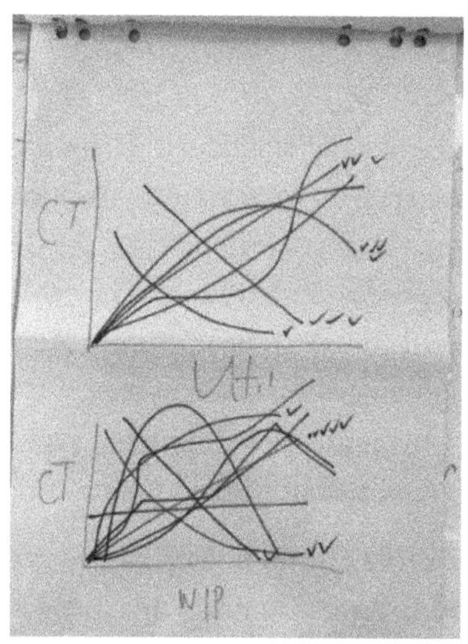

Figure 2-2: Typical responses from operations professionals

The diversity in responses is astounding. Some think cycle time increases as WIP levels increase, some think cycle time decreases as WIP increases, some think it is static, and still others suggest a maximum exists. The same diversity occurs regarding cycle time versus utilization. And almost none of these are correct. For operations behavior, one might expect to receive such a set of responses from amateurs, but to see this level of confusion from highly experienced professional managers, who work with WIP and cycle time every day, is quite remarkable.

Moreover, this confusion has consequences. If I think cycle time increases with WIP, I will want to reduce WIP to reduce cycle time, but if someone else thinks cycle time decreases with WIP, their recommendation will be just the opposite. How would we decide?

Of course, this ignorance is not the participant's fault. These relationships are not widely taught, so it is not surprising that very many people do not know them.

Now try it for yourself. Look at Figure 2-3. These are typical of participant responses when we conduct this exercise. Which is it? A, B, C, D, E, F or G? G is not on the graph but means "I really don't know, it would be just a guess." Write down your answer now. We will come back to this in Chapters 3 and 4.

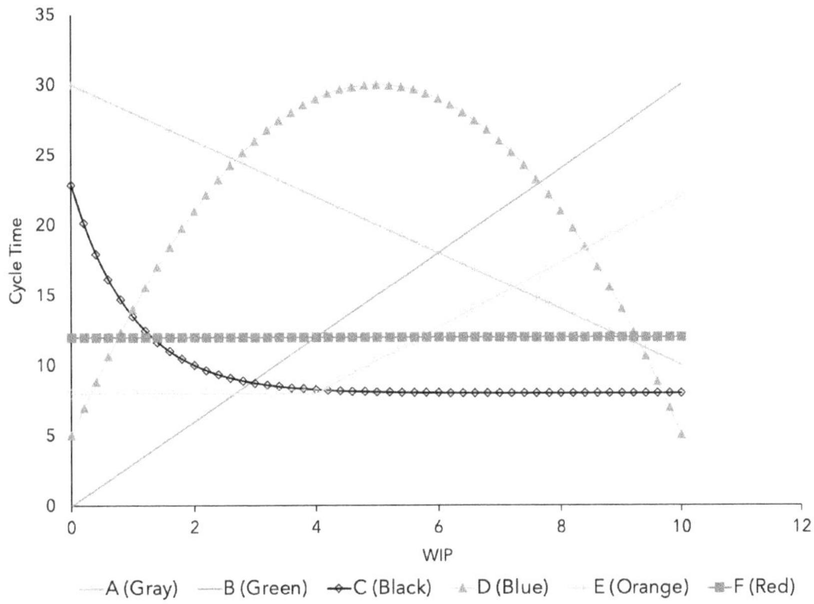

Figure 2-3: Responses for relationship of cycle time versus WIP

As described above, the absence of an operations science approach in industry and academia explains why the results of Figure 2-2 occur very often.

Some interesting questions are Why did operations science not develop sooner? and What will be the challenges in the adaptation of operations science now that it has been developed?

Methodologies as a Proxy for Science

For most engineering disciplines, such as mechanical, electrical, civil, aerospace, or chemical, there is clearly a need to understand the science behind the applications. Imagine if an electrical engineer did not know Ohm's law and designed a power grid that burned up because the current was too high, or if a civil engineer never had a course in the strength of materials and a bridge collapsed. These would be disasters, and there would be an outcry as to why did they not know these things. To avoid such disasters, most professional electrical and civil engineers must be certified in professional engineering, which requires knowledge in the relevant engineering science fields.

However, operations management "disasters" are often taken as "just the way things are." Emergency room wait times are too long, multibillion-dollar construction and engineering projects are late and overbudget, manufacturing jobs don't finish on time, restaurants waste food, and stores have too much inventory. However, unlike a bridge collapsing, there is typically no visible, dramatic disaster (although people often get fired). Consequently, many industrial engineers never get a professional engineer (PE) license, and the certified professional manager (CPM) accreditation involves little science.

Without a true science of operations, the solutions presented in the field of operations management must be based on something other than science—and this something is *methodology*. Consider manufacturing as an example. For more than 40 years there has been a blizzard of buzzwords heralding the ultimate solution to production. In the 1970s came material requirements planning (MRP) and the APICS[vi] MRP Crusade, followed in the 1980s by MRP II and then enterprise resource planning (ERP). Meanwhile, a different, non–IT approach came from Japan, first dubbed *just-in-time* (*JIT*) in the 1980s and then renamed in the 1990s as *Lean Manufacturing*, then just *Lean*, which has had great staying power. While Lean advocates were promoting a very nontechnical approach, others were promoting advanced statistical methods to improve quality. One of the first was total quality management (TQM), which then morphed into Six Sigma which required a relatively advanced level of statistical knowledge to obtain a Black Belt certification. Ultimately Lean and Six Sigma were combined into Lean Six Sigma. All the while various IT solutions were advanced to provide better scheduling methods.

Despite being methodology based, there had to be an underlying model of the production system. Unfortunately, without a solid science to drive these developments, those models were undeniably lacking in fidelity. To see this, consider MRP, MRP II, and ERP. At the kernel of all those generations of IT solutions is a simplifying assumption—the time to get a part finished depends only on its lead time, which is an attribute of the part itself, and not on the state of the factory. No wonder some participants in the seminars guessed that cycle time was completely independent of WIP.

A famous book describing JIT titled *Zero Inventories* indicates that the title "connotes a level of perfection not ever attainable."[vii] Nor should one want to attain it because it would result in zero output. Nonetheless, it has been our experience that some companies have pushed inventories down to the point where they are no longer Lean but anorexic. They have too little inventory and are starving key process centers because of it.

A major underlying model of Lean is that all activities are either valued added or non-value added, from the customer's perspective. Non-value-added activities are the result of improper use of resources of which there are three: *muda, mura,* and *muri.* The first, *muda,* just means waste (anything that consumes resources and does not provide value to the end customer), *mura* connotes unevenness and *muri* is overburdening equipment or labor. These explanations sound somewhat scientific, but, as we will see, they do not pass the test that distinguishes science from pseudoscience.[viii] Later, we will return to examine Lean from an operations science point of view, which makes clear when Lean does and does not work.

The IT approach using sophisticated scheduling algorithms has a somewhat better descriptive model than that of Lean and MRP but is still lacking. It assumes that the production times are known and constant (i.e., there is no variability or randomness). And most of these models assume all jobs are available at the beginning of the period, with no new jobs arriving later. Since most of the important aspects of operations science are related to variability and randomness, we find the scheduling "baby" has been thrown out with the variability "bathwater."

Recent developments in data science such as AI, machine learning (ML), blockchain, the IoT, and so on, only emphasize the point that data science offers very little science and is mostly methodological. The rapidly changing landscape of AI, ML, IoT, digital manufacturing, and big data would benefit greatly from a fundamental understanding of operations science. Current articles on AI and IoT provide the impression that the combination will be able to design, build, and control almost anything.

But if AI/IoT are not informed by operations science, there will be yet another methodology built on the shifting sand of poor descriptive models. For instance, using a neural nets analysis, an early application of AI to soccer, yielded the absurd conclusion that the best chance for a goal is to have all the players in the corner of the pitch. The tracking data that the AI engine used showed players cheering in the corner *after* a goal was made (Figure 2-4). A slight lack of synchronization confused the effect with the cause and yielded the ridiculous recommendation.

Figure 2-4: Soccer goal celebration

The problem with all these methodologies is that there is no standard model to draw upon to determine if something makes sense or to indicate new phenomena. In other fields, standard models have been an important guide for research and to categorize results. Mendeleev's periodic table predicted not only heretofore undiscovered elements but also their characteristics. Likewise, the standard model of particle physics has helped physicists fill in the blanks. For instance, before the standard model was developed, the existing theories had great success in explaining the electromagnetic force and the strong nuclear force. But by the 1960s, all attempts to create a theory that joined these and the weak force (that governs radioactivity) had failed in that they predicted that some particles should be massless when, in fact, they had been discovered to have mass. However, in the late 1950s several physicists discovered a possible way out. Further theoretical work postulated a way to unite the three forces by using a new type of field and a new mechanism, which was later named after Peter Higgs, one of the researchers. This new model did not gain much credence until, during the 1970s, it predicted the existence of several previously unknown particles that were subsequently discovered. After these successes, the existing collection of particles and fields became known as the standard model of particle physics. And, while the Higgs boson was first predicted in 1962, it was not discovered until the technology was able to create energies high enough to detect it in 2012.

If we are going to break the cycle of "new methodology leads to disappointing results," we need a standard model of operations. Without a standard model, we are left to develop improved methods using a trial-and-error approach. Indeed, Taiichi Ohno spent more than 30 years to develop the Toyota production system (TPS). However, as evidenced by the plethora of versions of its offshoot, Lean Manufacturing, and the wide variation of success and failure of Lean efforts, most, including Ohno, have had difficulty in explaining how it works.[ix] Without a standard model that describes operations, companies who want to make big improvements without having 30 years to work on the issue tend to copy others—with mixed results.

When Boeing hired their Japanese consultants to teach them TPS, one of the first questions the consultants asked was, "Where is your moving assembly line?" After some protestation that airplanes were not like cars, Boeing conceded and spent several years and $250 million and shut down the plant during the peak of the business cycle to create a moving assembly line for the 777 aircraft. While we discussed this decision in detail in Chapter 1 of our last book, *Factory Physics for Managers*,[x] we will repeat the bottom line here: After installing the moving assembly line, the cycle time was reduced slightly, from 50 days to 48 days. Larger cycle time reductions were achieved later, though those reductions had nothing to do with a moving assembly line. However, the production rate (planes per day) did not change. In other words, Boeing could not ship more planes to customers. Further, customers did not make buying decisions based on the reduction in lead time. Decisions to purchase new planes occur years before they are delivered. The rate only improved from one plane every three days to one plane every two-and-a-half days after making other changes that were much less expensive than installing a moving assembly line. The takeaway here is that trying to improve production by copying a different industry can be extremely counterproductive. Had Boeing understood operations science, they would never have installed a moving assembly line. A line turning out a Camry every 45 seconds is a much different animal than a line that completes an airplane every 3 days. The fast line has the advantage that whenever it stops because of a problem, everyone knows and takes immediate action. But few notice the line is stopped when it moves only 1.6 inches per minute, *on average* and does not move continuously. Production pacing is important, but there are much easier ways to do it than to put a multi-ton airplane on a moving assembly line. Perhaps most important, business strategy should dictate operations strategy. If customers are willing to wait longer or even feel special because they are waiting for a customized item, then perhaps lead-time reduction is not worth the investment in capacity and/or inventory to achieve it.

In review, we have

- examined operations management practices and the highly common flaw of relying on best management practices without an accurate scientific understanding of the system being managed
- discussed the history of operations management and how a true science of operations did not develop until recently
- explained how when managers apply methodologies without an understanding of the science governing their operations they get unpredictable and often undesirable results

However, before we describe operations science, we will describe the requirements for legitimate science so the reader can understand how operations science improves their intuition about the strengths and weaknesses of any operations management approach they may take.

What Is Science? How Can It Be Leveraged for Breakthrough Results?

The Oxford Reference defines science as "the systematic study of the structure and behavior of the physical and natural world through observation and experiment."[xi]

Based on this definition, simply collecting data is not science. Science involves a "systematic study" or some basic paradigm or model to help understand the data collected. We have discussed examples of such paradigms, including the periodic table and the standard model of particle physics. Another is the biological genome. All these paradigms have been critical not only for organizing the facts of the field but also in guiding research into new areas that have greatly impacted society.

As a preamble to the review in this book of how operations science can be leveraged for breakthrough business results, we find it helpful to reflect on the power of science to deliver amazing changes and the potential peril of society misusing or misunderstanding science. The following timeless quote is from a paper titled "Science and Its Impact on Society" published in 1983 by NASA to celebrate its 25th anniversary, and think of all that has happened since then, with science changing our lives:

> Science and technology have had a major impact on society, and their impact is growing. By drastically changing our means of communication, the way we work, our housing, clothes, and food, our methods of transportation, and, indeed, even the length and quality of life itself, science has generated changes in the moral values and basic philosophies of mankind.
>
> Beginning with the plow, science has changed how we live and what we believe. By making life easier, science has given man the chance to pursue societal concerns such as ethics, aesthetics, education, and justice; to create cultures; and to improve human conditions. But it has also placed us in the unique position of being able to destroy ourselves.[xii]

We also need to realize that the idea of "scientific truth" is a myth and a pervasive myth at that. The myth has been perpetuated by various statements from authority figures that typically run along the lines of "science tells us . . ." or "from science we know . . ." During the COVID-19 pandemic, the statement "we follow the science" is, for most people, equivalent to claiming, "we know what we are talking about."

The myth is an undying misunderstanding that is reinforced by most elementary and even advanced science courses. It comes from the success of scientific theories over the last 400 years. One of the first was Newton's laws of motion, followed by successful theories in chemistry, thermodynamics, atomic physics, biology, medicine, aeronautics, and more. All these theories have made amazingly accurate predictions of the behavior of the natural world that have been verified by experiment. From these successes it appears that they have discovered "scientific truth."

But tomorrow, one could conduct a new experiment that would refute any one of these theories. Of course, some of them have already been refuted. Take Newton's laws, for example. These can make amazingly accurate predictions of the motion of anything from a roller coaster to the planet Mars. But after the planet Uranus was discovered, it was found to not follow the orbit predicted by Newton. But, instead of throwing out all Newtonian mechanics, a search was made for another reason for the wrong orbit, and the result was the discovery of Neptune. So, a single discrepancy is typically not a reason to throw out an otherwise successful theory.

When the orbit of Mercury was found to precess (i.e., a movement of the perihelion, the point in its orbit at which it is closest to the sun) more than expected, another search was made for a hidden planet, tentatively named *Vulcan* after the Greek god of fire and long before Mr. Spock, inside the orbit of Mercury. But no such planet was found. The discrepancy in the precession was small, 43 arc seconds (roughly the angle subtended by one foot at a distance of one mile) but consistent and different from what was predicted by Newton. Many explanations were offered to account for the variation, but none were verified until Einstein applied his new General Theory of Relativity (GR) and came up with the missing 43 arc seconds. This prediction, along with the verification of the prediction of the sun bending star light during the 1919 eclipse, made Einstein an instant sensation. But even these two startling discoveries does not *prove* GR correct. Indeed, numerous issues have been raised from questioning the legitimacy of Einstein's calculations of the Mercury precession to the atmosphere of the sun causing the bending of star light.

The question of whether proof is possible in science and thereby whether there is scientific "truth" is an old one and related to a classic problem in philosophy. The problem of induction was first described in detail by David Hume in 1739 who wrote

> Our foregoing method of reasoning will easily convince us, that there can be no demonstrative arguments to prove, that those instances, of which we have had no experience, resemble those, of which we have had experience.[xiii]

In other words, just because we have seen repeated linkages between a particular action and some seemingly associated effect, there is no logical reason to conclude that the action is the cause that led to that effect.

One idea to get out of this quandary is to repeat the experiment many times and, after many confirmations, conclude that the theory must be true. But that method, as Bertrand Russell's chicken learned, sometimes it does not work out so well. Russell writes:

> The man who has fed the chicken every day throughout its life at last wrings its neck instead, showing that more refined views as to the uniformity of nature would have been useful to the chicken.[xiv]

So, if repetition is not a good measure of the validity of a theory, what can we use? The scientific method.

It is not unusual, from grade school on, to be taught that the scientific method will lead to the true theory. Unfortunately, many descriptions of the scientific method are misleading. A typical inaccurate description is shown in Figure 2-5.

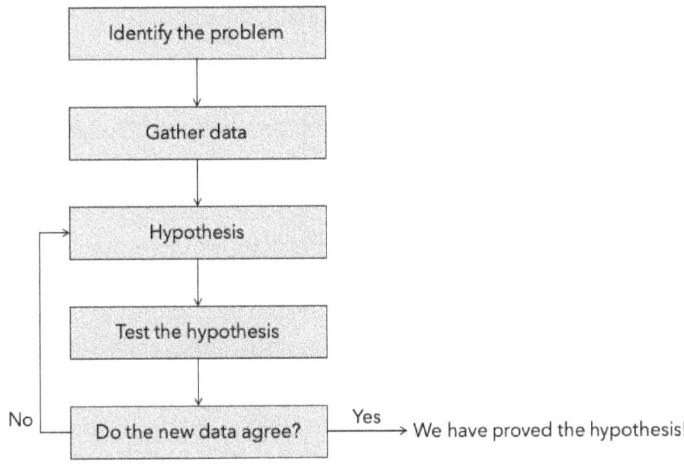

Figure 2-5: Misrepresentation of the scientific method

Figure 2-6 is a more accurate representation of science. We first state the problem we are trying to solve, which may not be a "problem" but a set of observations that we are trying to understand. We gather enough observations to craft a hypothesis based on what we observe. We then create an experiment that will test our hypothesis. If the experiment does not agree with the hypothesis, we must alter the hypothesis and perform more experiments. Once all the experiments agree with the hypothesis, all we can conclude is that the theory is our current best explanation of the phenomenon. However, we cannot conclude that our hypothesis is now the true theory for the phenomenon.

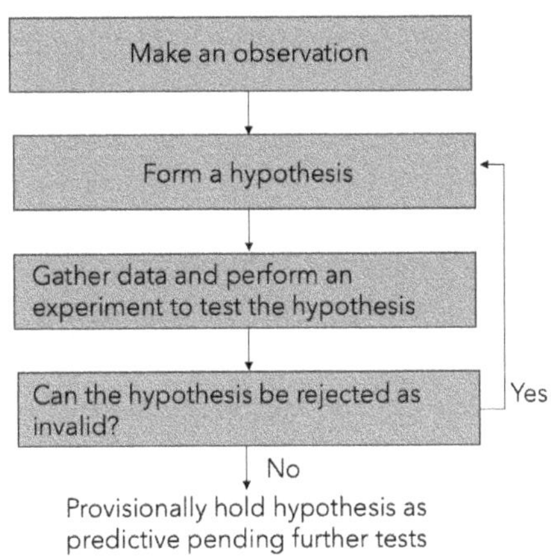

Figure 2-6: The scientific method

And if we were to assert that we had discovered a "scientific truth," we would have committed the logical fallacy of affirming the consequent, a fallacy which, sadly continues to occur too often (and is essential to most advertising!). Logically, it is written as

If P, then Q.
Q is true.
Therefore, P is true.

A simple example illustrates the fallacy.

The facts: (1) One must be over 16 to work at McDonalds. (2) Fred is 17.

Correct logic (classic syllogism)

1. If you work at McDonalds (antecedent), then you must be over 16 (consequent).
2. Fred works at McDonalds (affirming the antecedent).
3. Therefore, Fred must be over 16 (logical conclusion).

Fallacy of affirming the consequent

1. If you work at McDonalds (antecedent), then you must be over 16 (consequent).
2. Fred is over 16 (affirming the consequent).
3. Therefore, Fred works at McDonalds (illogical conclusion).

The same fallacy often occurs when people perform bad science: If Theory T is true, then we should observe O. We perform an experiment and observe O. Therefore, we conclude Theory T is true. This is the same logical mistake as saying that Fred works at McDonalds just because he is over 16.

So how *should* science be done? Instead of committing the fallacy of affirming the consequent, we simply turn things around and use the logically valid method of denying the consequent. Logically, this goes as follows: Theory T implies we should observe O. We do not observe O. Therefore, Theory T is false, and we seek to improve our theory.

In other words, instead of trying to confirm the theory, we seek to disprove the theory and replace it with a better theory that fits the old data as well as the new. This is how good science is conducted. Nobel Prize–winning physicist Richard Feynman stated it this way: "We are trying to prove ourselves wrong as quickly as possible, because only in that way can we find progress."[xv]

But not all science is conducted this way. Indeed, many branches of science are notorious for their confirmation bias, the selection of data that agree with one's hypothesis. Physics has, for the most part, avoided the confirmation bias, and the sociology of the physics community may be one reason. Most physicists consider themselves either a theoretical physicist or an experimental physicist. But the cultures are quite different, with experimentalists seen as less mathematical and more practical and theorists as seen as more mathematical and less practical. Indeed, nothing quite excites an experimental physicist more than conducting an experiment that disproves an existing physical theory. But it is a friendly feud, and both camps recognize the need for the other. Moreover, physicists are aware of the shortcomings of even their most successful theories. While quantum field theory (QFT) has been spectacularly effective in predicting many physical phenomena, its prediction of the vacuum energy is off by a factor of more than 10^{100}, an astoundingly large error! Physicists know that there is something spectacularly wrong with this estimate and with QFT itself, but it continues to be used until a better theory can be developed.

We can use this method of denying the consequent to demark what is and what is not science. Thus, any purported science that does not seek to disprove its theories and instead seeks to confirm its hypotheses is a pseudoscience. And any science that does not generate predictions that can be refuted is not a science at all.

For example, Marx, in his writing, made specific predictions. One of these was that the autocratic rule (i.e., feudalism) must be overthrown by a bourgeois revolution (e.g., the Industrial Revolution) and only after that could the final socialist revolution take place. This sequence of a bourgeois revolution then a socialist revolution was key to the theory of Marxism. The Bolsheviks, under Lenin, insisted on a small revolutionary elite as members of the party who would overthrow the government in a coup d'état for the benefit of the growing proletariat classes. When this set of events occurred in November 1917, the *Socialist Standard* (Socialist newspaper in the UK) condemned the actions of the Bolsheviks suggesting they were acting counter to true Marxism—the Bolsheviks had skipped a step in the predicted sequence. Nonetheless, the Bolsheviks were successful, and the USSR became the first communist state, so Marxism needed to be reexamined. Popper describes this change in Marxism:

> In some of [Marxism's] earlier formulations (for example in Marx's analysis of the character of the 'coming social revolution') their predictions were testable, and in fact falsified. Yet instead of accepting the refutations the followers of Marx re-interpreted both the theory and the evidence in order to make them agree. In this way they rescued the theory from refutation; but they did so at the price of adopting a device which made it irrefutable . . . and by this stratagem they destroyed its much-advertised claim to scientific status.[xvi]

We see similar "scientific" theories of how to better run operations today. For instance, many Lean practitioners make a false dichotomy between one-piece flow and batch-and-queue (e.g., "if you don't use Lean, you must be batch and queue"). But this is not true. What is true is that excessive WIP results in excessive cycle times, which means that the time between when a defect is *created* and the time when a defect is *detected* can be quite long. In such situations, it is very difficult to find root causes of defects and eliminate them. However, a line using several-piece flow can reduce CT almost as much as one-piece flow and without the subsequent loss in output. These subtleties are obvious to one knowledgeable in operations science but can be heretical to one who is trying to emulate the practices of a Lean *sensei* whose experience could be from an entirely different environment.

In spite of these shortcomings, we have seen many instances of operations leaders stating that Lean is the ultimate and only truth for operations improvement. As we will see later, operations science provides a solid foundation for understanding how Lean concepts work and how best to apply them in any particular environment.

From Here to Success

So how does anyone in operations use science to understand and manage operations for best possible performance? This book is dedicated to helping the reader leverage operations science to achieve breakthrough results. We will approach operations from a practical, powerful, and comprehensive scientific angle. The principles in this book have been tested, taught, and implemented in both academia and industry over the last few decades, but there is still plenty of work to do. If the reader finds refutations that help advance operations science, we welcome further discussion—email info@opscience.org. If the reader is in academia and is looking for new and innovative research topics, contact us. If the reader desires, there are plenty of other technical resources including *Factory Physics* and *Factory Physics for Managers*. See also https://opscience.org/resources.

---

[i] Mathematics is a deductive system that works by starting with a set of axioms and then deriving "new" properties that may not have been known beforehand. This does not imply that mathematics is not creative. Indeed, many of the axioms are established to help to understand some phenomenon the "real world" (to see this, consider the history

of probability and the axioms of measure theory). Nonetheless, mathematics is essentially different from science, although most sciences use mathematical models.

[ii] Taylor found that a 21-pound shovel was the most efficient and improved average daily output from 16 to 59 tons. See Frederick Winslow Taylor, *The Principles of Scientific Management* (Harper & Brothers, 1911).

[iii] Ford W. Harris, "How Many Parts to Make at Once," *Factory: The Magazine of Management* 10, no. 2 (1913): 135–36, 152.

[iv] Grand View Research, Big Data Market Size To Reach $862.31 Billion By 2030,

https://www.grandviewresearch.com/press-release/global-big-data-market.

[v] We define *cycle time* as the time to go from the release of a job onto a routing until it is completed. This time includes not only queue time but also batch formation time, down time, changeover time, and so on. Other texts may use cycle time as only the process time (no queueing, down time, etc.) on a resource, typically a machine or labor station.

[vi] *APICS* stands for American Production and Inventory Control Society. It was formed in 1957 at the dawn of the computer age. Over its history, it has merged with a number of other industry organizations and in 2018 changed its name to the Association for Supply Chain Management (ASCM).

[vii] Robert W. Hall, *Zero Inventories* (Dow Jones-Irwin, 1983).

[viii] See Edward S. Pound, "The Value-Added Fantasy—Don't Drive Your Leadership Efforts Astray," Operations Science Institute, accessed April 8, 2025, https://opscience.org/resources/articles/.

[ix] For further discussion, see "There Is a Science Behind Lean," Operations Science Institute, January 18, 2023, https://opscience.org/there-is-a-science-behind-lean/.

[x] Pound, Edward S., Mark L. Spearman, and Jeffrey H. Bell. *Factory Physics for Managers: How Leaders Improve Performance in a Post-Lean Six Sigma World.* (McGraw-Hill Education, 2014).

[xi] Oxford University Press, "Science and Technology," *Oxford Reference*, accessed April 8, 2025, https://www.oxfordreference.com/page/134.

[xii] Donald P. Hearth, foreword, in James Burke, Jules Bergman, and Isaac Asimov, *The Impact of Science on Society* (National Aeronautics and Space Administration, 1985), iii.

[xiii] David Hume, *A Treatise of Human Nature* (John Noon, 1739), 1.3.6.

[xiv] Bertrand Russell, *The Problems of Philosophy* (Williams and Norgate, 1912), 63.

[xv] Richard P. Feynman, *The Pleasure of Finding Things Out: The Best Short Works of Richard P. Feynman*, ed. Jeffrey Robbins (Perseus Books, 1999), 115.

[xvi] Karl Popper, *Conjectures and Refutations: The Growth of Scientific Knowledge* (Routledge, 1963), 37.

3

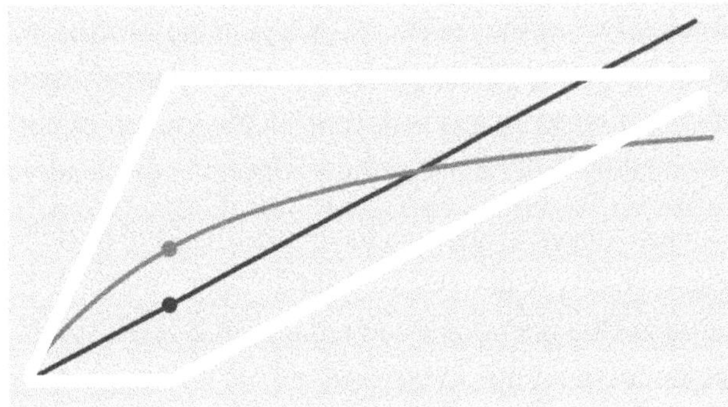

What Is Operations Science?

Operations Science Basics

Now that we know what makes for good science, let's set about to develop a true science of operations. We have given a definition of an operation and will repeat it here.

operation: An act that utilizes resources to transform one or more attributes of an entity or set of entities into some good or service that is required to satisfy a given demand.

Examples include production operations like manufacturing, service operations such as health care, construction processes like foundation framing, design processes such as new product creation, transportation operations, and even broadcasting operations such as radio and television. For an act to be considered an operation, some *transformation* must occur that requires one or more *resources*, and it must be done to satisfy a given *demand.* Since operations are done by and for people, they are intrinsically *artificial* in that they create or make use of artifacts. Consequently, operations science is distinct from natural science. Because they are artificial, operations are a means to an end and so have a purpose. This purpose is important when considering options regarding the operation.

The interesting (and difficult) aspect of operations science occurs at the interface between demand and transformation. If transformation knew exactly what the demand was and had adequate resources to instantly satisfy that demand, the problems in operations science would be trivial—satisfy the demand exactly at the time it occurs. But, because we do not have perfect information or infinite transformation capacity and because there is always some variability and randomness in both demand and transformation, operations become substantially more difficult. If we satisfy the demand before the demand occurs, we create inventory. If we wait until the demand occurs and then satisfy it afterwards, there is some wait time before satisfaction. See Table 1. For service operations, this is the only option because we cannot have an inventory of services (try to imagine a stock of 911 calls that have been addressed *before* they occur). So, depending on who waits for whom, we will have time waiting or inventory—either demand waits on transformation or vice versa. And often in real-world operations, we have *both* inventory and wait time and too much of both. However, we have also observed that having extra capacity will help to reduce both inventory and wait time.

With these observations, we have identified the basic challenge of operations science for any application. There are many examples: Restaurant wait times are decreased with extra capacity in the kitchen. Manufacturing operations can reduce back-order time by having more inventory. Fire trucks and ambulances are quick to respond only if they are not highly utilized (consider what happens when there is a major, widespread disaster, and there are not enough emergency vehicles). Call-center hold times increase when many employees are ill and at home. Inventories fall and back orders rise when demand exceeds capacity (recall what happened during the COVID-19 pandemic).

SEQUENCE	RESULT	EXAMPLES
Demand before transformation	Wait time	Make-to-order, Engineer-to-order, Service operations, Projects
Transformation before demand	Inventory	Make-to-stock, Assemble-to-order

Table 3-1: The two phases of the time-inventory buffer

To understand all these seemingly different interactions, we need a basic model (a primitive element) of an operation. Such an element will include a source of demand, a stock (with inventory and/or back orders), at least one resource, and queue(s) for the resource(s) as per Figure 3-1.

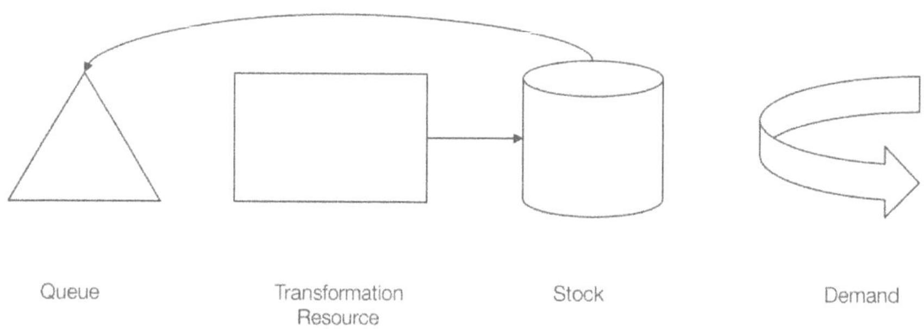

Figure 3-1: A primitive operation

It may seem strange to have a stock for a service operation since we can never inventory a service. But if we consider the stock to be always in a back-order state for service operations, we can generalize our primitive element. And this makes sense because the demand for any service always occurs *before* receiving the service. A stock is a representation of the *time-inventory buffer*, which we will discuss shortly.

Almost any operation can be modeled by a combination of these primitive elements. A simple production flow line is depicted in Figure 3-2 and can be modeled as a series of queues and resources (often called process centers) with a stock at each end. Classically these are raw material stock on the left and finished goods stock on the right, but they could be intermediate stock points. For example, a stock point at the end of a subassembly line feeding a main assembly line or a stock point of surgical supplies as raw materials for an operating room:

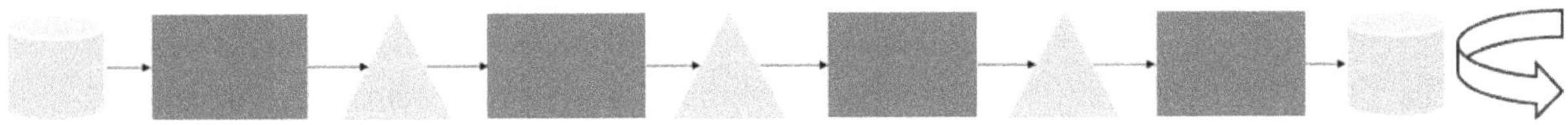

Figure 3-2: Elements of a production flow line

Another example, a moving assembly line, would not have queues between the resources and is depicted in Figure 3-3 (the ∞ appearing in the queue indicates that there will always be items to start).

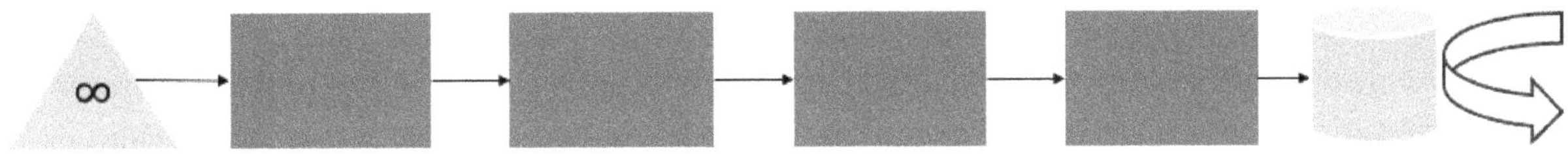

Figure 3-3: Elements of a moving assembly line

These basic elements can be combined to represent extremely complex production systems such as an entire construction project. Figure 3-4 shows the primary supply flows and the installation flow for poured concrete columns.

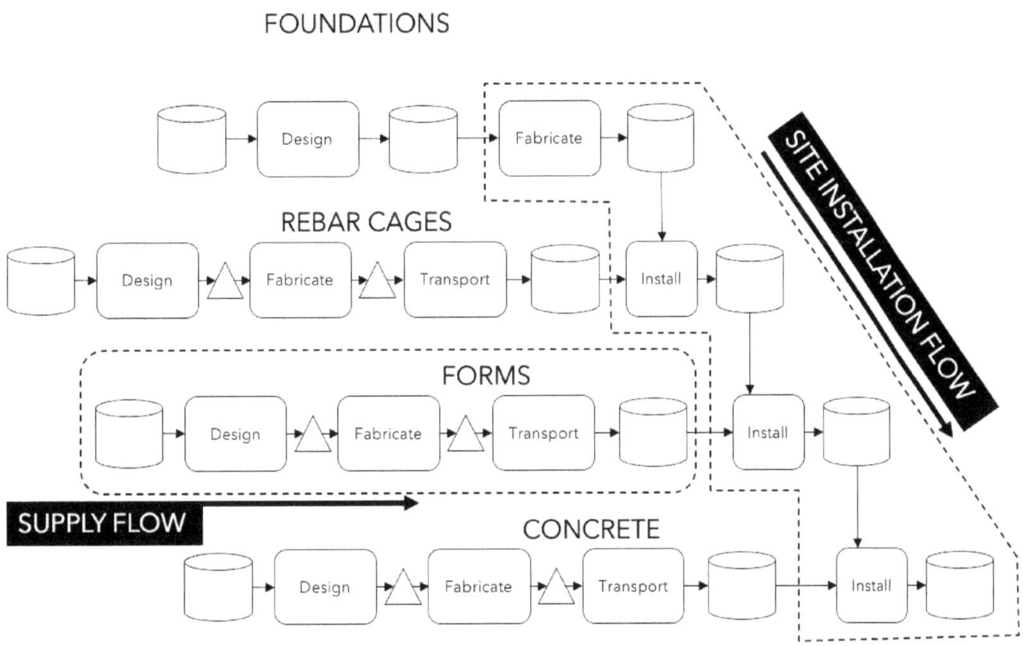

Figure 3-4: Basic production flows in a segment of a construction project

The Foundations supply flow has only Design and Fabricate because the fabrication occurs on the site. Once the foundation is fabricated, it is installed. Then, the Rebar Cages supply flow, consisting of Design, Fabricate, and Transport, feeds the Site Installation flow, and so on. Notice that all of the installation processes have two arrows from two different stocks going into them. This is because both of the supply flow processes never finish at the same time, so one must wait for the other. For instance, we cannot install a rebar cage until the foundation is in place, and we can't install the rebar cage until it is fabricated. One could draw more detailed flows depending on desired level of detail.

The Fundamental Principle of Operations Science

The fundamental principle of operations science is that variability creates *buffers*. Either the demand waits for the transformation (a time buffer), or the transformation waits for a demand (an inventory buffer). These buffers are mirrors of each other. If we increase our inventory target, we increase the inventory buffer but decrease the time buffer. If we decrease our inventory target, the opposite happens. Consequently, we can link the time and inventory buffers as a time-inventory buffer. The size of this buffer depends on the *variability of the net inventory* for a particular type of inventory control system.[i] Net inventory is positive when there is inventory and negative when there are back orders. Now it may seem strange to define the time-inventory buffer as net inventory—where did the time go? The answer is, back orders are demands *waiting to be filled* and so represent time. A good indicator of the time-inventory buffer is the variation in demand during the time to replenish inventory. In other words, how much random demand occurs during the random time to replenish the inventory? This gets a little complicated, so we will postpone the discussion until the next chapter to give a formal derivation. For now, we will give it a name, Δ_{TI}. If Δ_{TI} is large, in order to have few back orders we must set our inventory target high.

Figures 3-5 and 3-6 show plots of the net inventory as a function of time for a production-inventory system with a demand of 10 and a capacity of 10.5 units per day. The variation is fairly high with a squared coefficient of variation (SCV) of 1.0. Figure 3-5 has an inventory target of 20 but has many back orders. Increasing the inventory target to 50, as Figure 3-6 demonstrates, results in no back orders but significantly higher inventory.

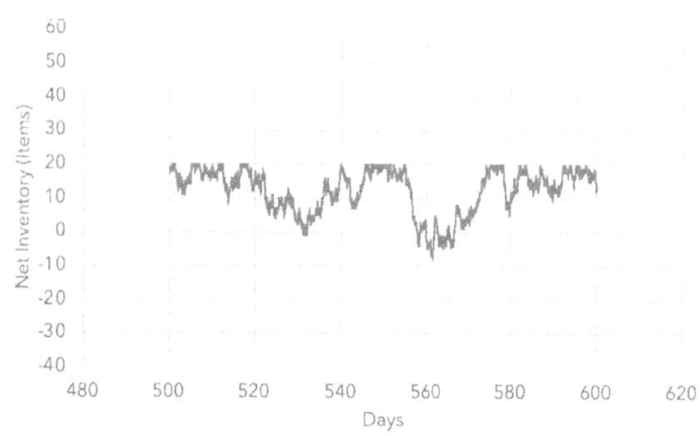

Figure 3-5: Net inventory versus time. Demand = 10, capacity = 10.5, SCV = 1, inventory target = 20

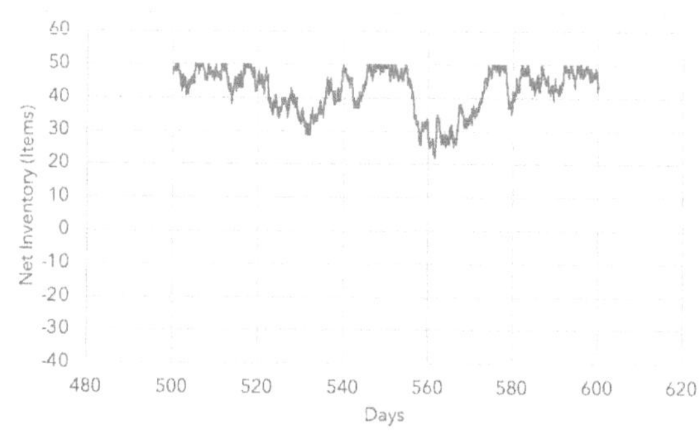

Figure 3-6: Net inventory versus time. Demand = 10, capacity = 10.5, SCV = 1, inventory target = 50

But if the time-inventory buffer, Δ_{TI}, is small, we can have less inventory *and* fewer back orders. The performance of both inventory and customer service depends *only* on the size of the time-inventory buffer and the inventory policy. If we are in a make-to-order environment, we have a zero-inventory target, and every demand is back ordered.

But there is another buffer, extra capacity, which is the amount of capacity that is greater than the demand, Δ_r (r will be used to define capacity in the next chapter). If demand were 10 items per day, and we had a capacity of 12 items per day, the capacity buffer would be $\Delta_r = 2$ units per day.

Note that if we had enough capacity to make anything any demand could want and make it instantly, we would not need a time or an inventory buffer. As soon as demand occurs, we will produce whatever is desired. To do this would require us to have a great deal of capacity available at all times, and this would clearly be much more than the average demand. So our resources would be woefully underutilized most of the time. Obviously, this is not a good policy for production, but it *is* a good policy if you are running an emergency medical service or a fire station! Think of how often a fire engine is used. Most of the time it is sitting but ready—very low utilization.

Figure 3-7 has the same level of variability[ii] as Figure 3-5 and Figure 3-6 but has more capacity, $\Delta_r = 2$. Figure 3-8 has the same capacity, $\Delta_r = 0.5$, as Figure 3-5 and Figure 3-6 but less variability. Over time, any particular inventory target will result in an average level of inventory and back orders depending on available capacity. Thus, the capacity buffer *mitigates* the time-inventory buffer.

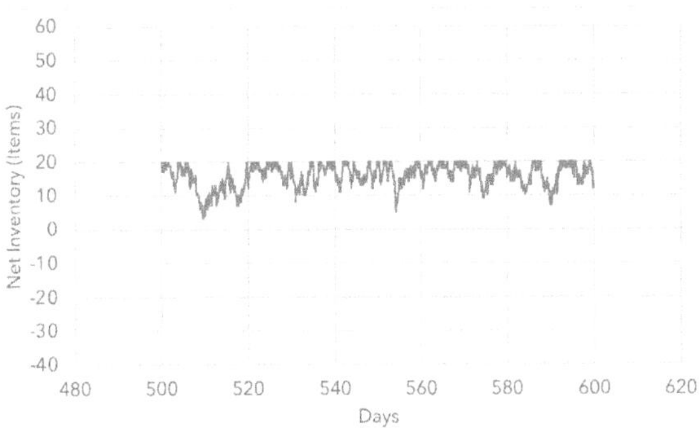

Figure 3-7: Net inventory versus time. Demand = 10, capacity = 12, SCV = 1, inventory target = 20

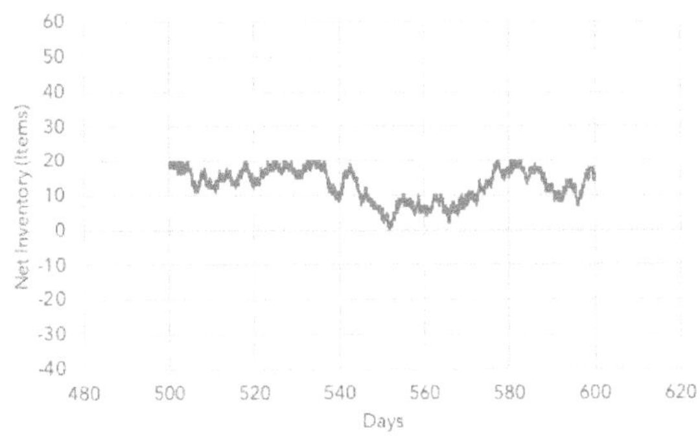

Figure 3-8: Net inventory versus time. Demand = 10, capacity = 10.5, SCV = 0.267, inventory target = 20

In Chapters 4 and 5, we derive the form of the buffers. In the rest of this chapter, we introduce foundational relationships at the conceptual level and provide definitions that will help provide context to the derivations. From an operations science application standpoint, these foundational relationships are the heart of using science to drive business results.

Foundational Relationships

Before providing technical and applications detail on them in Chapters 4 through 13, we introduce the following foundational relationships at a conceptual level.

The First Foundational Relationship: Buffers and Variability

Two important questions are How does the time-inventory buffer interact with the capacity buffer? and How are these buffers related to variability? This interaction is one of the fundamental principles of operations science. We find that for any primitive operation, as seen in Figure 3-1, these two buffers are intimately related to each other and to a measure of the variability of the system, $\bar{V}$, by a simple equation,

$$\Delta_r \times \Delta_{TI} \approx \bar{V}$$

The above equation is sometimes called the operations science fundamental relationship (OSFR), similar to the Heisenberg uncertainty principle of quantum mechanics. This equation is quite informative. If we reduce variability, we can reduce the capacity and decrease inventory and have better service. Or we could reduce capacity and not affect inventory and service. Or any combination of the three. Likewise, if we add capacity, we can reduce inventory and/or increase service. Figure 3-9 below illustrates this interaction.

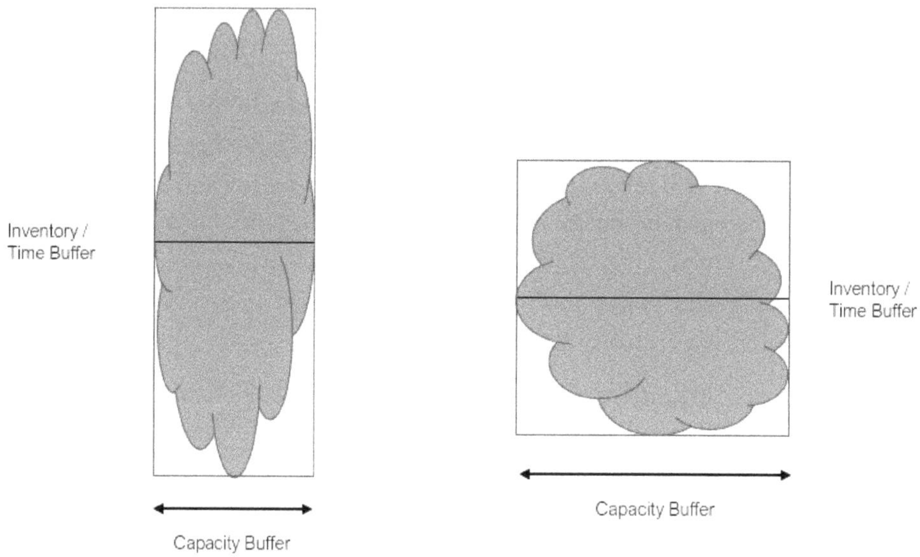

Figure 3-9: Buffer interactions

If we squeeze down on the capacity buffer, the time-inventory buffer blows up and vice versa. Each buffer has a cost—the cost of extra capacity for the capacity buffer and the cost of carrying inventory along with the cost of back orders (i.e., missed promise dates, etc.) for the time-inventory buffer. The goal of operations science is to find the set of buffers that minimizes the total cost of the operation while meeting response time and on-time delivery goals.

Interestingly, when the cost of these buffers is proportional to the buffer size, then the configuration that minimizes the total cost is when the cost of the extra capacity is approximately equal to the cost of the inventory/back orders. The result is exactly equal for a primitive production inventory system composed of a single resource supplying demand for a single item and provides useful guidance for more realistic systems. In other words, one should not seek to have extremely high utilization of resources because it will drive up the cost of inventory, back orders, or both.

Therefore, given the wide variety of subjective definitions of Lean, we recommend the objective, quantitative operations science definition of "meeting customer expectations with minimum buffer costs." Operations science provides a proven approach for achieving this definition of Lean.

The Second Foundational Relationship: Queueing

The second foundational relationship is known as the variability, utilization, and time (VUT) equation. It offers an intuitive formula for the average waiting time in queue, which is the product of a variability factor, a utilization factor, and a time factor. Mathematically, this is written as

$$CT_q = V \times U \times T$$

where CT_q represents the average cycle time in queue, and V, U, and T are the variability, utilization, and time factors. These factors are not the actual variability, utilization, or time but are functions of these and other parameters that we will describe in the next chapter. The important insight is that the U factor grows exponentially as utilization goes to 100%. To see this, consider the fraction of a highway covered at rush hour and at 3:00 a.m. The fraction covered is a measure of utilization. At rush hour, it is bumper to bumper—high utilization—with long travel times. At 3:00 a.m., there are few cars on the road—low utilization and much shorter travel time.

The wait time in a queue is also affected by variability. A conveyor moving at a constant speed can have items "bumper to bumper" and still have short transit times. But the variability of many drivers moving in and out of the roadway, the highway conditions, and the occasional accident greatly increase the variability on a highway. On Saturday morning at 8:00 a.m., the wait at the donut shop far exceeds the wait at that same donut shop on Wednesday at 5:30 a.m., because both utilization and variability have greatly increased as sports parents and their kids line up to pick out donuts. Businesses measuring earned hours incent managers to drive high utilization, often resulting in longer wait/response times.

The last factor, T, is the effective average process time for a process batch and includes not only process time but also the associated down time, setup time, and any other time for a given item. Obviously, if the time to do the process goes up, the queue time goes up as well.

Part of operations strategy is choosing where to be on the cycle time versus utilization curve, Figure 3-10, to meet response-time and profit goals. Ultimately, this requires a specific capacity investment choice. The resulting capacity and time-inventory buffers drive cost and customer service results.

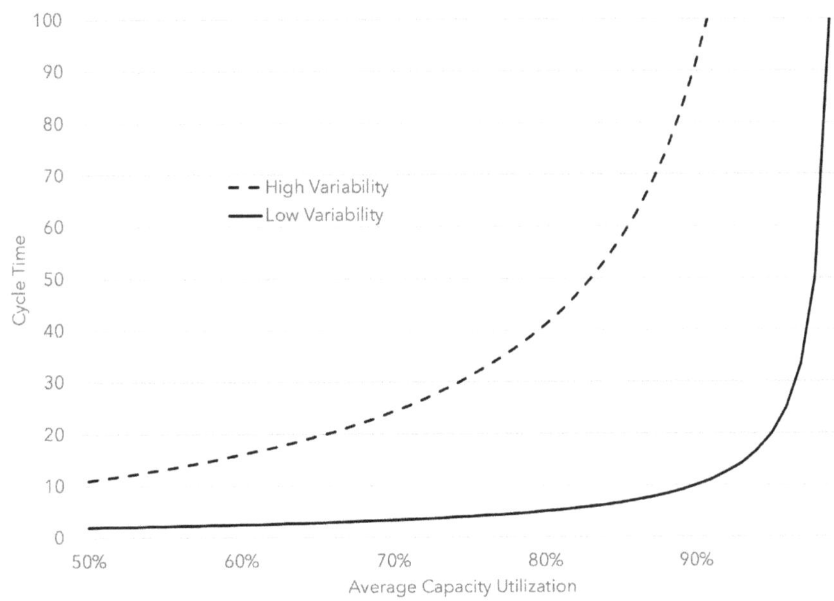

Figure 3-10: Cycle time versus utilization

The Third Foundational Relationship: Little's Law

Little's Law provides the relationship between the averages of demand, response time, and number of entities in the system. John D. C. Little[iii] first wrote it in 1961 as

$$L = \lambda \times W$$

where L is the work in process or entities in the system. λ is throughput, and W is cycle time. We use more direct terms and write it as

$$WIP = TH \times CT$$

where *WIP* is work in process, *TH* is throughput, and *CT* is cycle time. To reiterate, these are all averages. Throughput is the average output of a flow. *Cycle time* is the average time to go through the flow once the item is released, and *WIP* is the average number of items in the flow that have been released but are not yet finished. So *WIP* is inventory, but it does not include *stock inventory*, which is inventory at the end of a flow that is available for meeting demand. More on stock inventory shortly.

Another way to write Little's law is

$$CT = \frac{WIP}{TH}$$

and is quite enlightening so long as throughput, representing the average demand, is less than the capacity of the system. Written this way, Little's law states that cycle time is proportional to WIP and so are essentially the same but with different dimensions. This is extremely useful because we can see WIP but cannot see cycle time. A quick estimate of the average number of days to go through a factory is to count the WIP and divide by the average demand in units per day.

But there is a more important insight as well, since WIP leads cycle time. If WIP increases today, cycle time will increase next week (or month, depending on how long the cycle time is). While it is difficult to control cycle time, it is easy to control WIP, and controlling WIP in a systematic manner is a key to successful operations management. We will provide an informal proof of Little's law in the following chapter and devote an entire section to the control of WIP.

The Fourth Foundational Relationship: Flow Performance Curves

Once we understand Little's law and the VUT equation, we can create a set of curves that involve all three variables, WIP, CT, and TH. These curves provide the guidance needed by those who confuse the relationships, as discussed in Chapter 2. The correct curves are shown in Figure 3- 3-11. The practical worst cases (PWC) curves represent performance with variability. We will explain practical worst case performance, also called marginal case performance, in detail in Chapter 4. Stock inventory has a different set of performance curves, which we will discuss shortly.

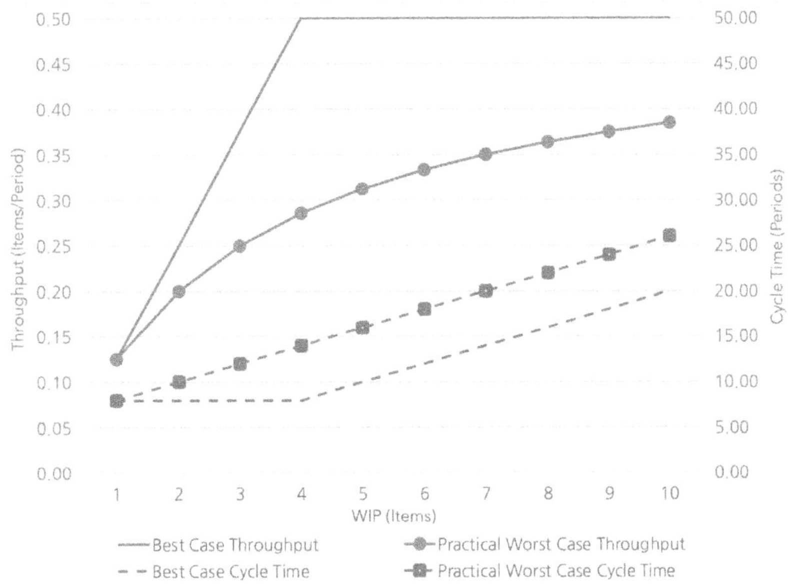

Figure 3-11: Flow performance curves

The curves are drawn for the simple process in Figure 3-2, where each resource (the blue rectangles) has an average process time of two hours. Here we see throughput and CT are plotted as functions of WIP. This necessarily implies that WIP will be the control variable. In Chapter 6 we discuss a method known as CONWIP whereby we keep the WIP fairly constant and let throughput and cycle time vary.

The two sets of curves are for the best case (the solid curves) and the marginal case (the dashed curves), sometimes called the practical worst case. . The best case shows the maximum throughput and the minimum CT for a given system with a bottleneck rate of 0.5 unit per hour and a raw process time of eight hours (see the definitions for these parameters in the next section). The best case has zero variability in process times, so they are always exactly two hours for any level of WIP. In contrast to the best case, the marginal case shows the effect of introducing variability with random process times.[iv] Consequently, the process times always vary, and so the performance varies as well. For that reason, we plot the average CTs and the average throughput. As a result, the marginal case serves as a boundary between good and poor performance. Because the variability of the marginal case is quite high, it serves as a suitable boundary between good and bad performance. The two circles are the current performance showing the throughput (red circle) to be somewhat below this boundary and the CT (blue circle) to be slightly above. This indicates the performance has a significant level of variability and offers an opportunity for improvement.

In order to create these two graphs with the two operating points, we need only four parameters:

1. Bottleneck rate (BNR)
2. Raw process time (RPT)
3. Current WIP
4. Current throughput (demand)

This makes it easy to consider the possibility of improvement for any production flow. We will discuss these terms and concepts in more detail in Chapter 4 and in some of the case studies that follow later in the book.

The Fifth Foundational Relationship: Batching and Cycle Time

In operations science there are two kinds of batches, transfer batches and process batches. The amount of material that is moved from one process to another is a transfer batch. The number of items produced between changeovers (e.g., setups) is a process batch. While many people set the two batch types to be equal, they do not need to be and should not be. If a process batch is formed by a number of smaller transfer batches, CTs will be reduced.

But how large should the process batch be? If it is too small, one performs too many changeovers and runs out of capacity, utilization goes up, and the VUT equation predicts an explosion in both queue time and total CT. This behavior is seen in the left side of the optimal batch size in Figure 3-12. On the other hand, if the process batch is too large, the process time becomes very long, and again CTs increase, as observed on the right side of the optimal batch size in Figure 3-12. So there must be an optimal process batch size that minimizes CT and therefore, via Little's law, minimizes WIP. We will discuss how this is done in Chapter 4.

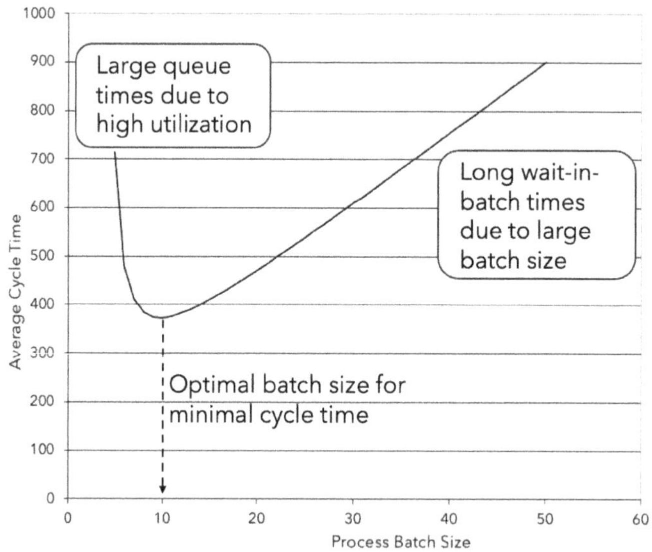

Figure 3-12: Cycle time versus process batch size

The Sixth Foundational Relationship: Efficient Frontier of Inventory Versus Time

We mentioned at the beginning of this chapter that stock inventory performance depends only on the time-inventory buffer, Δ_{TI}, and the inventory policy applied. The sixth foundational relationship states that there is an efficient frontier between the stock inventory buffer and the time buffer. Practically, this means that for any required time buffer and replenishment frequency, there is a policy that minimizes the inventory investment.

The time buffer can be the expected back-order time, expected back-order time when it is back ordered, the probability of a stockout, or others. We will use one of the others that is a kind of inverse time buffer: the probability of having inventory on hand for demand, also known as the fill rate. If the fill rate is 0.98, then there is a 2% chance of a stockout.

Figure 3-13 presents an example of a set of time-inventory efficient frontiers: These are the stock inventory performance curves. Here we see three order frequencies: 0.5, 1, and 2 replenishments per week. Note that the required inventory to achieve a given fill rate decreases as the replenishment frequency increases but with diminishing returns.

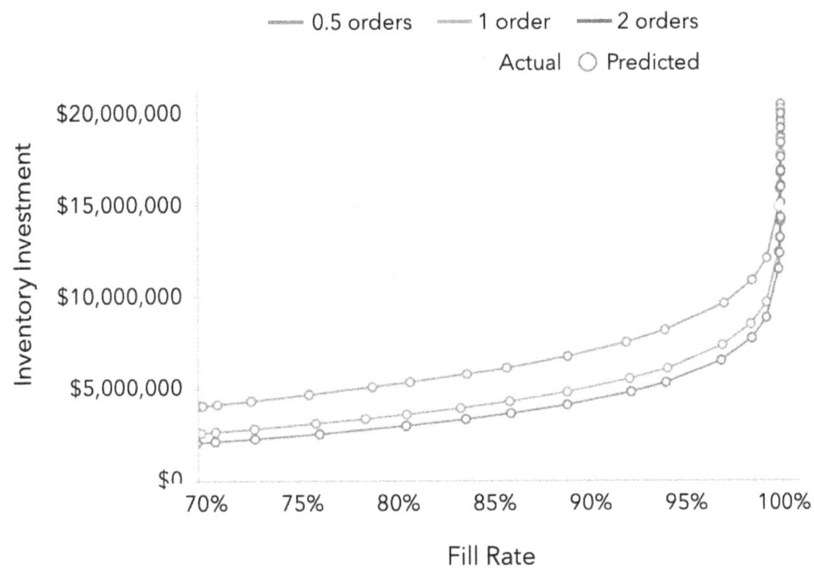

Figure 3-13: Time-inventory efficient frontier (stock inventory performance curves)[v]

Each curve corresponds to an optimal inventory policy for that particular fill rate and order frequency. A key insight is that inventory needs to increase asymptotically as fill rate approaches 100% because the inventory is trying to cover all possible cases, including extreme ones like where demand is spiking and, at the same time, suppliers are delivering late. Best policies have relatively high fill rates such as 95% or 98% but not 100%.

The analysis also provided the expected back-order time *when back ordered.* For local suppliers, this is typically quite low—maybe less than one day. But for distant suppliers (e.g., China), the wait may be much longer. Consequently, one usually performs such inventory optimization on a collection of items with similar replenishment times.

Definitions Eliminate Confusion

Now that we have gone through the foundational relationships, we must do the hard work of defining our terms. Definitions may be boring, but without them we end up confusing ourselves. In this section, we provide the basic definitions needed to explain the fundamentals. These and more definitions are included in the book's glossary.

We defined *operation* and *operations science* in Chapter 2, and in the previous section we defined *WIP*, *TH,* and *CT.* Now we will formalize these definitions and add other important terms. Note that descriptive terms are defined first to establish what an operation is. Afterwards, qualitative terms are defined to understand how operations performance is measured. For ease of reference, all terms are listed alphabetically in the glossary.

Descriptive Terms

operation: An act that utilizes resources to transform one or more attributes of an entity into some good or service that is required to satisfy a given demand.

resource: Anything required to perform a transformation. Examples include raw materials, subassemblies, components, labor, process centers, information, intelligence, energy, fuel, and so on.

process center: A resource that takes time to perform the transformation and may require other resources, e.g., labor, machines, workstations.

routing: A specific sequence of transformations required to create an entity to satisfy a demand. Routings always end at a stock.

flow: A collection of routings. The specific assignment of routings to flows is done for convenience (e.g., parts using a specific set of process centers). Since a flow could be one routing, the words *flow* and *routing* are sometimes used synonymously.

item: The unit that is moving through the flow. An item need not be a physical entity. CAD drawings moving through an approval process are items. Additionally, an item can be a service, such as a patient diagnosis.

stock: A physical or virtual location representing items, such as material, resources, or information, that are waiting on demand.

Quantitative Terms

demand (d): A requirement for a particular good or service to satisfy some need. Demand is usually presented as a quantity required per unit of time, e.g., parts per hour, square yards per day, silicon wafers per month, calls per hour, pipe spools installed per week, patients seen per week, completed designs per week, and so on.

throughput (TH): The average output per unit time at some point in a flow. The dimensions of throughput are the same as demand, quantity per unit time.

availability (A): A measure of a resource's time for use, as affected by unplanned outages or failures.

A = MTBF/(MTBF + MTTR)

where

MTBF = mean time between failures and

MTTR = mean time to repair

process time (*t*): The average time it takes for a resource, also called a process center, to perform its transformation. Dimensions are those of time.

effective process time (t_e): The average time it takes for a process to perform its transformation given preemptive outages.

$$t_e = \frac{t}{A}$$

We describe more complex situations, such as those with setups, in the next chapter.

number of resources (m): The number of parallel resources available for a process (dimensionless).

capacity (r): The average of the maximum number of items one resource can produce per unit time (dimensions same as TH). Given the two previous definitions, the relation between r, t_e, and m is

$$r = \frac{m}{t_e}$$

Throughput will always be less than or equal to capacity.

load (L): The throughput that is "seen" at a point in the system. Offhand, one would think the load would be the same at all points, but whenever there is rework (items requiring multiple visits to the same process) and/or yield loss (items scrapped at a process), the load will be different.

utilization (u): The fraction of capacity being used in transformation. Utilization is the load divided by capacity and so has no dimensions,

$$u = \frac{L}{r} = \frac{L \times t_e}{m}$$

Note also that utilization can be computed by dividing the time required to complete the demand for a given period by the time available in that period. If D is the number of items required in each period of T_{avail}, then the demand is $d = D/T_{avail}$. We also need to consider rework and yield loss. If f is the fraction of demand that visits the given process center, then the load will be $L = f\ d$. For example, if there is a 30% chance of having one rework and a 70% chance of zero rework, then $f = 1.3$. Likewise, if 10% of the items were scrapped (i.e., a 90% yield) before reaching this process, the $f = 0.9$. If there is no rework or yield loss, then f is simply one. The time required to complete $f\ D$ items is $T_{req} = f\ D/r$. Then

$$u = \frac{T_{req}}{T_{avail}} = \frac{f\ D/r}{T_{avail}} = \frac{\left(\frac{f\ D}{T_{avail}}\right)}{r} = \frac{L}{r}$$

demand variability factor (SCVa or c_a^2): We characterize variability with a dimensionless number known as the squared coefficient of variation, which is the variance of the time between arrivals divided by the square of the time between arrivals. Poisson demand[vi] has a SCVa = 1 (the *a* stands for arrivals to a queueing system).

production variability factor (SCVe or c_e^2): Like the demand variability factor, the production variability is also a dimensionless number representing the variance of the *effective* process times divided by squared mean of the *effective* process time. The effective process time includes the actual process time making the part, performing the service, or completing the task, plus any associated setup and downtime. The total variability increases or decreases when either SCV increases or decreases. For a complete discussion of these variability factors, see the appendix at the end of this chapter.

queue time (CT_q): Also known as average cycle time in queue, CT_q is typically measured for one process. For a resource with only one process (e.g., one workstation, one machine), the CT_q is approximately

$$CT_q = \left(\frac{c_a^2 + c_e^2}{2}\right)\left(\frac{u}{1-u}\right) t$$

This is sometimes written as $CT_q = VUT$, where the average of the variability coefficients is V, the utilization factor, $u/(1-u)$ is U and the average process time is T. It is important to note that for CT_q, the T factor is the time for a process batch, not for a single item. This expression is exact for Poisson demands (i.e., $c_a^2 = 1$) and is only approximate for other cases.

cycle time (CT): The average time for an item to traverse a flow from the time it is released until it is completed. The dimensions of CT are the dimensions of time, e.g., seconds, hours, days, weeks, and so on. CT is the sum of all of the process times, queue times, batch formation times, down times, changeover times, etc.

inventory (I), stock: A set of completed or partially completed items after a flow and/or between flows. The value of stock inventory is usually just the number of items and so is dimensionless, although it can be measured in terms of the value of the stock (e.g., USD), its weight, area, volume, and so on.

Here we use *stock* as an adjective of *inventory*, but there is also a noun *stock*, indicating a location where physical inventory is kept. If there is no physical inventory because there are back orders or because the process is a service process and never has physical inventory, the stock location contains the status of orders required to satisfy demand.

inventory, work in process (WIP): The average number of items in a flow between stocks. The value of WIP is usually the number of items and so will be dimensionless. However, it could be measured in terms of its value, weight, and so on.

changeover (s): The act of switching from the production of one type of item to another usually requiring some nonzero time during which no production occurs (setup time, s).

process batch (Q_p): The number of items completed between changeovers.

transfer batch (Q_t): The number of items completed before moving between processes. Transfer batches need not be equal to the process batch and are often less.

bottleneck rate (BNR): The average capacity of the throughput limiting process center. It will be the process center with the highest utilization. People often think the bottleneck is the slowest or most expensive process center. However, scrap and rework effects can turn the fastest process center into a bottleneck. The bottleneck is also known as the constraint.

raw process time (RPT): The average CT of a transfer batch when released one transfer batch at a time. This time includes downtime, setup time, and so on, and is the minimum average CT that can be achieved.

critical WIP (CW): The amount of WIP required for maximum throughput (i.e., TH = BNR) with minimum CT (i.e., CT = RPT) in a production system with zero variability. Critical WIP is always the product of the BNR and the RPT and may or may not be an integer.

In Figure 3-2, if the process time at each resource is 2 hours, the RPT is 8 hours and the BNR is 0.5 items per hour, yielding a CW of 4, the same as the number of stations. This equality is due to the fact that the line is perfectly balanced (i.e., all processes have the same capacity). For unbalanced lines, the CW is always less than the number of WIP positions. A WIP position is equal to one for each sequential process center (i.e., process centers that work on one item at a time) and is equal to the batch size for a simultaneous process center that works on the entire batch at a time (e.g., an oven).

Appendix: Variability Measures

Not surprisingly, variability measures deal with statistics. This appendix will attempt to explain where these measures come from with as little mathematics and statistics as possible. We will define the term in plain English and then define it mathematically. For some, mathematics improves their intuition; for others, not so much. So, if the math helps, read it. If not, ignore it.

mean (various): The expected value of a random variable. In other words, it is the long-term average. Examples from this chapter include d, r, TH, CT, WIP, I, BNR, and RPT, which are the mean values of demand, production rate, throughput, cycle time, work in process, stock inventory, bottleneck rate, and raw process time.

The usual statistic estimating the mean is the arithmetic average, $\bar{X}$, given by

$$\bar{X} = \sum_{i=1}^{N} X_i / N$$

Formally, the mean, μ, for discrete random variables is defined as

$$\mu = \sum_{j} p(j) x_j$$

where we sum over all the possible values of the random variable, and $p(j)$ is the probability that the random variable takes on the value, x_j. For a continuous random variable, we must integrate the product of the random variable and its probability density function, $f(x)$, over all possible values,

$$\mu = \int_{-\infty}^{\infty} x f(x) dx$$

Because the expected value of $\bar{X}$ is equal to the true mean, $\bar{X}$ is said to be unbiased.

variance (σ^2): A measure of variability. Intuitively, the variance is a measure of dispersion of random values about the mean. Its unbiased estimator is S^2:

$$S^2 = \frac{\sum_{i=1}^{N} (X_i - \bar{X})^2}{N - 1}$$

Formally, it is the expected value of the squared variation of a random variable from its mean. For continuous random variables,

$$\sigma^2 = \int_{-\infty}^{\infty} f(x)(x - \mu)^2 dx$$

As such, the variance typically has dimensions that are the square of the dimension of the random quantity. However, this is not the case for the variance of demand or throughput (see the discussion of statistics of demand and throughput below).

standard deviation (σ with variations): The square root of the variance. Its estimator is typically the square root of the variance estimator, but this estimator is biased (i.e., the expected value of the estimator is not the standard deviation). An estimator with less bias (usually) is

$$S = \sqrt{\frac{\sum_{i=1}^{N}(X_i - \bar{X})^2}{N - 1.5}}$$

coefficient of variation (c or CV): The standard deviation divided by the mean, a dimensionless measure of variability. Used to characterize the variability of time-based randomness, such as time to perform a task or time between demands.

squared coefficient of variation (c^2 or SCV): A dimensionless measure of variability that is the variance divided by the mean squared.

variance-to-mean ratio (VMR or ψ^2): The variance divided by the mean. Used to characterize the variability of rates such as demand and production. It, too, is a dimensionless parameter because the variance of a rate has the same dimensions (1/time) as the mean rate (it also is dimensionless, see the discussion below).

Discussion of the Statistics of Demand and Throughput

Demand and throughput have dimension of 1/time, and so one would think that the variance of demand and throughput would have dimension of $1/\text{time}^2$. But instead, the variance of these variables also has dimensions of 1/time. The reason involves a fair amount of probability theory, so if you are not up for that, just trust us.

Otherwise, here is a brief explanation that does use some probability theory but is still not a rigorous explanation. For those wanting to see a more rigorous treatment, we recommend Cox, 1962.[vii]

Let S_i be the time until the ith demand after some starting time, $t = 0$, with $S_0 = 0$. Also let

$$X_i = S_i - S_{i-1}$$

The X_i random variables represent the time between demands, and we assume that the values of the variables are all independent and identically distributed, with a probability density function of $f(x)$ having a mean μ and a variance σ_t^2. A common example of a random variable with a distribution with these characteristics is the exponential distribution with a mean of μ,

$$f(x) = \frac{1}{\mu} e^{-t/\mu}$$

In this case, it is easy to show that the mean is equal to the standard deviation (i.e., $\mu = \sigma_t$). The exponential distribution also possesses the memoryless property, in that the time until the next demand is independent of how long it has been since the last demand and is characteristic of a system having many independent entities creating demand.

Then the squared coefficient of variation for the interarrival time of the demands, c_a^2 (a for time between *arrivals*) is

$$c_a^2 = \frac{\sigma_t^2}{\mu^2}$$

Now, let N_t be the number of demands in $(0, t)$. Notice that

$$N_t < r \text{ if and only if } S_r > t$$

The random variable N_t converges to a normal random variable with mean t/μ and variance $\sigma_t^2 t/\mu^3$ as $t \to \infty$. Thus, we approximate the mean and variance of demand per unit time as

$$d = 1/\mu$$

$$\sigma_D^2 = \frac{\sigma_t^2}{\mu^3}$$

Note that these quantities have dimensions of t^{-1} (e.g., per day, per year, per second, etc.). Also, since $d = 1/\mu$, we can write the variance-to-mean ratio (VMR), ψ^2, as

$$\psi^2 = \frac{\sigma_D^2}{d} = c_a^2 = \frac{\sigma_t^2}{\mu^2}$$

Thus, the VMR of the demand converges to the squared coefficient of variation of the times between demand as the time from the start goes to infinity. This is a handy result because we seldom have data for time between demand, but we often have data for number of demands per period (e.g., demand per week).

A result from probability is that whenever times between individual demands are exponential, the number of demands in a given time is Poisson. In our notation, if X_i is exponential with mean μ, then N_t will be Poisson with mean *and variance* of t/μ. Since the mean and standard deviation of the exponential distribution are equal, i.e., $\sigma_t = \mu$, so that, $c_a^2 = 1$, and we obtain

$$\sigma_D^2(\text{Poisson}) = d$$

which matches what we know about the Poisson distribution.

The analysis for production is essentially the same:

$$\sigma_P^2 = c_e^2\, r$$

So not only are the squared coefficients of variation (e.g., c_a^2) dimensionless, but the variance of demand divided by the mean demand is also dimensionless and is the VMR, sometimes called the index of dispersion.

Consequently, both the mean and the *standard deviation* of time random variables have dimensions of time units (e.g., hours, days, weeks). But it is the mean and *variance* of rate random variables that both have dimensions of reciprocal of time units (e.g., parts per hour, linear feet of pipe laid per week).

---

[i] The inventory control system is known as a base stock system, which keeps the net inventory plus all on-order inventory equal to a given level, called the base stock, hence the name. We will describe this system in detail in Chapter 5 when we focus exclusively on stock inventory.

[ii] The measure of variability is the squared coefficient of variation (SCV). We will cover this term later. See definitions of *demand variability factor* and *production variability factor* at the end of this chapter.

[iii] John D. C. Little was a longtime faculty member at the MIT Sloan School of Management who had a distinguished career spanning five decades. Professor Little passed away in September 2024 at the age of 96. His seminal paper is J. D. C. Little, "A Proof for the Queuing Formula: L = λW," *Operations Research* 9, no. 3 (1961): 383–87.

[iv] In this case, the process times are exponentially distributed where the mean and standard deviation are both two hours.

[v] Screenshot courtesy of Strategic Project Solutions, Inc.

[vi] Poisson demand is the result of exponential times between demands. Such demands are said to be memoryless in that the time until the next demand does not depend on how long it has been since the last demand. Consequently, it is often used to model external demand made up of thousands or millions of independent individuals.

[vii] D. R. Cox, *Renewal Theory*, Methuen's Monographs on Applied Probability and Statistics (Methuen, 1962).

4

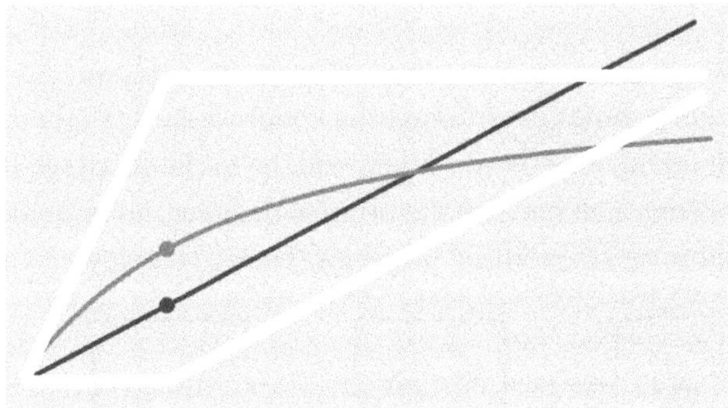

Operations Science of Flow

Overview

In Chapter 2, we described an exercise that demonstrates the widespread confusion among operations professionals and project managers about fundamental operations science relationships. In Chapter 3, we introduced the five relationships and four graphs that form the foundation of operations science.

For this chapter we are concerned only with the flow components of an operation and their interaction with demand, as illustrated in Figure 4-1. These are demand (and throughput), process centers, WIP, and queues.

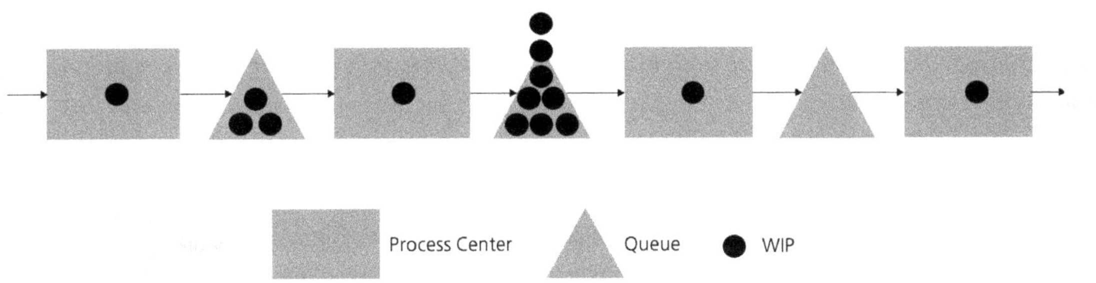

Figure 4-1: Flow components of an operation

We derive one of the five relationships, Little's law, and its associated flow performance curves. We do this first with no variability and then with variability added to the process times. If one is to be successful in any type of operation, knowing this relationship is critical to understanding how the operation or project behaves. We provide a detailed example of application in practice then provide a general description of the calculations needed to describe flow performance for production and for service systems.

Flow Fundamentals

We keep things easy, forgetting about variability (for a while), and consider four workstations in series, each taking exactly two hours with no variability (see Figure 4-2). This little production line is known as the Penny Fab because the original demonstration involved moving physical pennies across a set of squares on an overhead projector (back in the day).

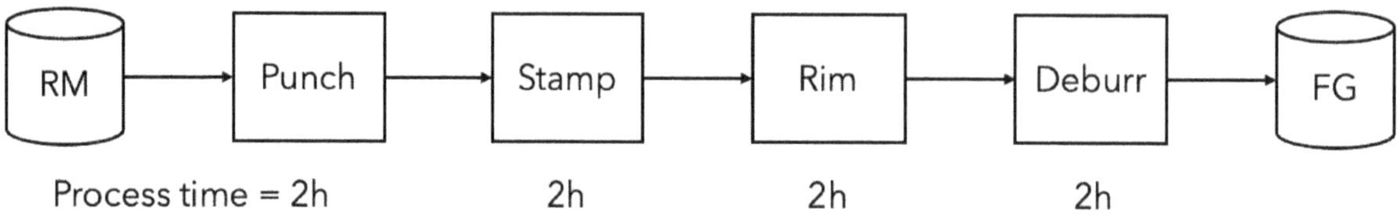

Figure 4-2: The Penny Fab—A simple production system with no variability in process times

Now we get back to our original question from Chapter 1: If I know how much WIP there is in the system, can I tell you how much time it takes to go through the system (cycle time)? Do I know the basic relationship between WIP and cycle time?

Let's review the definitions. First, the phrase *cycle time* has different meanings in different texts. We take ours from *Factory Physics,* first published in 1995, as the average time it takes an item or items to go from release to completion. In Figure 4-2, this is the time from release to Punch through finishing at Deburr. WIP is the number of items that have been started but that are not yet completed in a given flow. The item can be a final assembly, subassembly, or just a component. Because operations science is applicable to any kind of operation (not just classic manufacturing), the item does not have to be a physical entity. It could be something like a CAD drawing in a data set or the number of completed tasks that go into a yet-uncompleted design. The use of tasks as WIP is often a requirement when looking at operations in work areas such as design engineering, or in service operations such as call centers or maintenance. Cycle time includes queue time, down time on machines, changeover times, batch formation times, wait-in-batch times, and so on. Thus, it has dimensions of time, e.g., hours in the case of Figure 4-2. This quantity is sometimes called flow time or throughput time in production literature and sojourn time in queueing theory.

Another common definition of cycle time is the process time required at a single work station (a definition often used in Lean). However, this definition only includes the process time and so is more related to capacity than to time in system.

Next, throughput (TH) is the average output of a workstation or an entire flow, depending on the context. As such, it has dimensions of items/time such as parts per hour (as in Figure 4-2), calls answered per day, pipe spools installed per day, patients seen per week, completed designs per week, and so on.

We now do a virtual simulation of the cycle time and throughput for various WIP levels, starting with WIP at a single item. In other words, a single item starts, and when it finishes, the next one starts. Obviously, for Figure 4-2, the cycle time is 8 hours, the sum of the process times. But what is the throughput? In this case, it is clear that one item comes out every 8 hours. So the throughput is 1/8 or 0.125 items per hour.

With two items it is a bit tricky because there is a transient event. As you see in Figure 4-3, if we start with two items and limit WIP to only two items in the system at any time, we will initially have a queue.

But after 2 hours, this queue goes away and never occurs again. Since we are interested only in the steady-state performance, we ignore the first two hours, and after the second item starts, the cycle time is always 8 hours. And then the configuration at 10 hours will look just like the one at 2 hours, and so on.

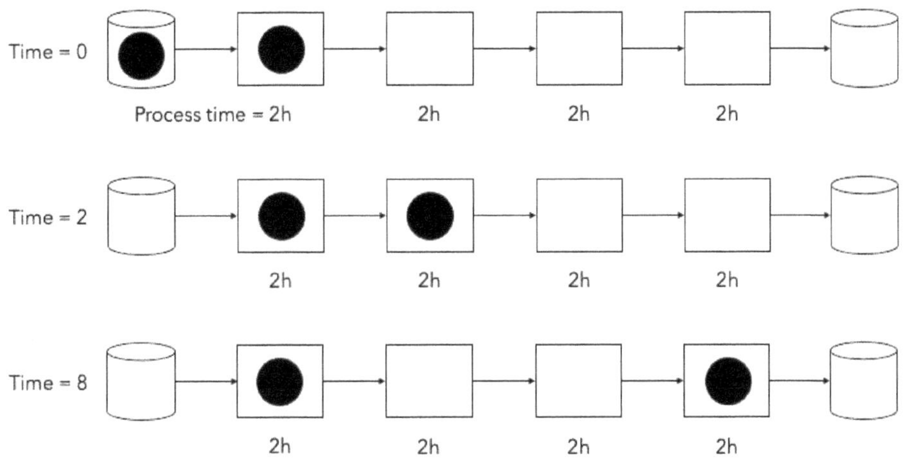

Figure 4-3: Penny Fab with constant WIP level of two

So the cycle time, after the transient, is still 8 hours. Another way to see this is to remember, in this case, cycle time will be process time plus queueing. After the first 2 hours, there is never any queueing, so the cycle time continues to be 8 hours. The throughput will be 2 items every 8 hours or 0.25 items per hour. Likewise, with 3 items, after 4 hours, the cycle time becomes 8 hours and never changes. But the throughput becomes 3 items in 8 hours or 3/8 items per hour.

With WIP equal to 4, the performance becomes quite boring (see Figure 4-4). The transient lasts the first 6 hours, but after that the system always looks the same. But, again, the cycle time will be 8 hours since there is no queueing. Now the throughput is 4 items in 8 hours or 1/2 item per hour.

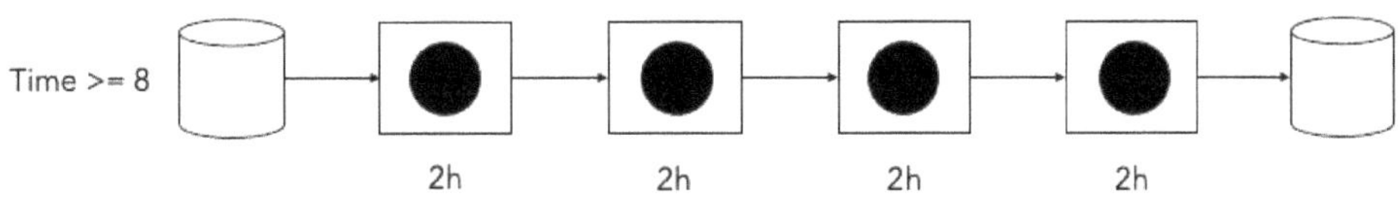

Figure 4-4: Penny Fab with constant WIP level of four

At this point we see that the throughput is equal to the capacity of the system. We define the capacity of a linear flow such as this as equal to the smallest capacity of any of the resources. In this case, the process time of all the resources is the same—2 hours—so the capacity is also the same 1/2 item per hour and their utilization is the same, 100%.

So far we have seen throughput increasing with each additional item but the cycle time staying the same. The cycle time remains constant for the simple reason that, after a transient, there is no queueing, and the sum of the process times is always 8 hours. But if we add one more item, as in Figure 4-5, there must be a queue as there will be 5 items but only 4 workstations.

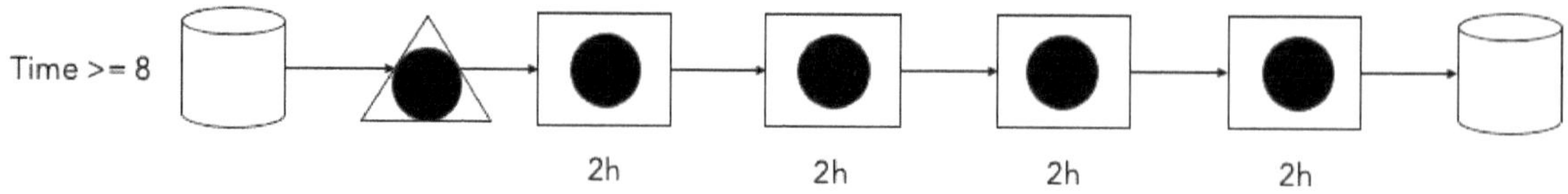

Figure 4-5: Penny fab with constant WIP level of five

Now the cycle time is 10 hours (8 in process and 2 in queue), but the throughput remains at 1/2 item per hour. If we have 6 items, the cycle time becomes 12 hours and, again, the throughput remains the same. We summarize the performance in Table 4-1.

WIP	CT (HOURS)	TH (ITEMS/HOUR)	TH × CT
1	8	0.125	1
2	8	0.250	2
3	8	0.375	3
4	8	0.500	4
5	10	0.500	5
6	12	0.500	6
7	14	0.500	7

Table 4-1: Performance of the simple production system

Using this data, we can determine the correct answer to the quiz from Chapter 2, Figure 2-2, along with a plot of throughput versus WIP, see Figure 4-6. We define BNR as the bottleneck rate (i.e., the rate of the station with the highest utilization, in this case, the station with the longest process time[i]), RPT is the raw process time (i.e., the sum of the average process times in this zero variability case)[ii], and CW is critical WIP which is equal to BNR × RPT. This is the WIP level at which maximum throughput is achieved in the best case (zero variability)[iii].

Thus, we can write

$$CT = \max\{RPT, WIP/BNR\} \qquad \textit{eq 1}$$

and

$$TH = \min\{BNR, WIP/RPT\} \qquad \textit{eq 2}$$

So, the cycle time is 8 hours with 4 or fewer items and then becomes WIP/BNR with 4 or more items. This is not surprising if you realize that cycle time can never be less than the sum of the average process times, i.e., RPT. Also, if there are WIP units in the line, they cannot go out at a rate more than the bottleneck rate, BNR, requiring WIP/BNR time.

Also notice that there is a simple relationship between WIP, TH and CT:

$$WIP = TH \times CT$$

This is Little's law, and it is true for any system that is stable (i.e., where TH is strictly less than the capacity of the line). Here we see it is true for a case with no variability (even though the throughput is equal to the capacity). In the following section, we will derive Little's law for the general case with random process times and random arrivals.

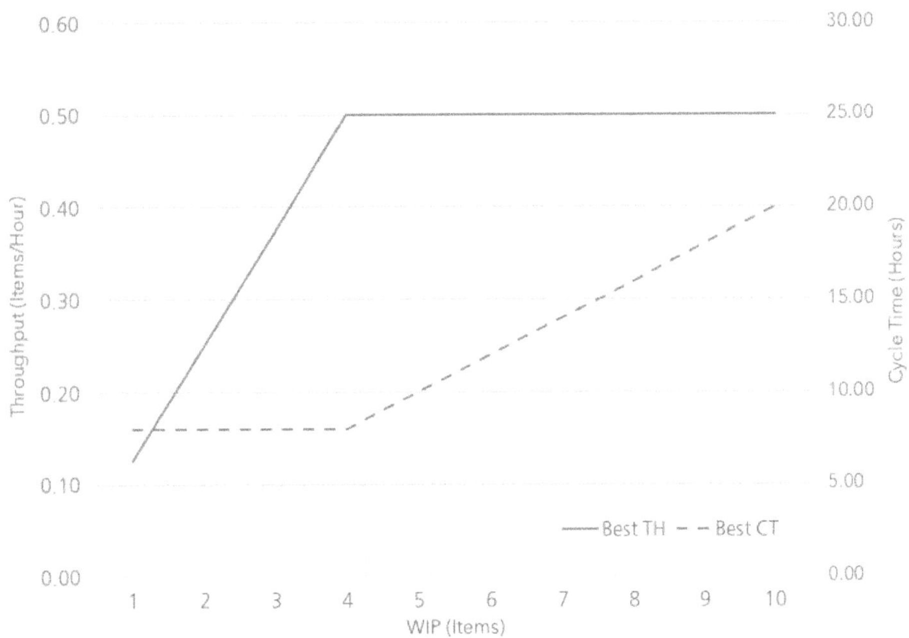

Figure 4-6: Correct plot of CT vs WIP and TH vs WIP

For our very simple model with no variability, there is an optimal WIP level. Above this level, adding more WIP only increases cycle time with no increase in throughput. A common misunderstanding is to think that one should start work as soon as possible ("If I don't start it, how can I finish it?"). But, from this simple example, we see that starting more WIP *when the bottleneck is (nearly) always busy* only increases cycle time with no increase in output. Misunderstanding this simple relationship has been the downfall of many a manager and on occasion, entire companies.

Little's Law

The above exercise demonstrates that Little's law, $WIP = TH \times CT$, is true for systems with no variability. We will now demonstrate that Little's law also applies to a system with variability. In this second system, there are three stations and each station's process time has an average of 8 hours and *also* a standard deviation of 8 hours, so significant variability. The throughput is set by the arrival rate with an average and standard deviation of 9 hours between arrivals. We run a simulation of the performance of the system and then use the data produced from the simulation, as if it were our real process, to demonstrate Little's law calculations.

Figure 4-7 provides a cumulative plot of the number of items released into the system versus time (top curve) along with the number of items completed versus time (bottom curve). At the beginning of the simulation, more items are being released than completed until the system fills. This is expected and represents a transient that always happens at the start of a simulation but will not occur again. Since we desire steady-state results, we will ignore this transient and begin with the 30th item and end with the 50th item (although you may need many more than 20 to get good statistics).

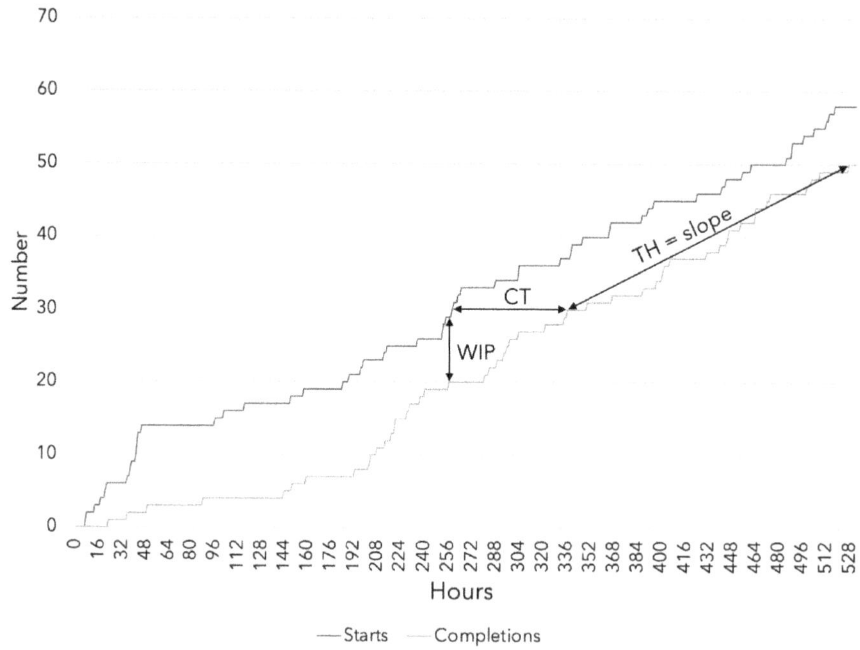

Figure 4-7: Little's law demonstration

The 30th item started at hour 259 while the 50th item started at 461 hours, yielding an expected time between arrivals of (461 – 259)/(50 – 30) = 10.1, which is around the value of 9.0, used in the simulation.

The 30th item completed at hour 337, while the 50th completed at hour 527. The actual throughput from our process was (50 – 30)/(527 – 337) = 0.11.

The cycle time for the 30th item was 337 – 259 = 78 hours.

Figure 4-7 shows a WIP of 13 units at 277 hours, a point in cycle time of the 30th item when WIP was at its highest. Using Little's law, we get

$$TH = \frac{WIP}{CT} \approx \frac{13}{78} = 0.17 \text{ items/h}$$

which is significantly above the average throughput of 0.11. This illustrates the hazards of using a small dataset for Little's law calculations.

If we use all the simulation data and average the WIP level from when item 30 starts to when item 50 finishes, we get an average WIP level of 8.67. If we average all the cycle times of items 30 through 50, we get a cycle time of 82.6. That calculation for throughput yields

$$TH = \frac{WIP}{CT} \approx \frac{8.67}{82.6} = 0.105 \text{ items/h}$$

which is very close to our first throughput calculation of 0.11, as we would expect.

Recall that the number of arrivals or completions could be physical or virtual. For example, we could be looking at the number of arrivals of not only items released to a manufacturing line but also patients to an emergency department or engineering design requests to an engineering contractor.

Over time one might see significant shifts in throughput. The operations science approach is to examine whether the variability in throughput is due to natural variability, i.e., noise, or to a particular cause that indicates the process has shifted" so that a new average throughput should be used to describe the process. If it is noise, no response is needed, and, indeed, responding to noise simply creates more noise, making things worse. However, if there is a statistically significant process shift, it should be addressed. Causes of a process shift could be an increase or decrease in demand or a drop in output due to process problems.

Assuming the system is stable, we can derive Little's law. Define $S(t)$ as the number of starts and $C(t)$ the number of completions at time t after the transient period. Once we reach steady state, the average number of starts will be simply

$$S(t) = TH \times t$$

The number of completions is similar, except the first does not come out until after one average cycle time:

$$C(t) = TH \times (t - CT)$$

At any point, the current WIP is simply

$$\begin{aligned} WIP(t) &= S(t) - C(t) \\ &= TH \times t - [TH \times (t - CT)] \end{aligned}$$

$$WIP = TH \times CT \qquad \textit{eq 3}$$

which is Little's law. And since TH is the demand, the implication is that average cycle time is always a linear constant times the WIP—essentially equivalent with different dimensions.

Since WIP is visible and CT is not, we should try to control WIP, not CT. This is easily done by maintaining a level of WIP in the line. Moreover, if we increase WIP today, we will not see an increase in CT until the first newly added WIP is completed. In other words, WIP leads CT and offers yet another reason to control WIP and not CT.

Adding Variability to the Flow Performance Curves

A couple of decades ago we worked with a prominent semiconductor manufacturer and attended a weekly production meeting. An engineer reported on his production output: The goal was 500 wafers/day. His chart showed the goal line and daily output above and below the line over the last week. The engineer's comment was, "We're above the goal some days and below the goal some days but can't seem to hit 500 every day." He then sat down to no comment. This was surprising to us, as it seemed that the engineer and others in the room had the expectation that the goal of exactly 500 per day was an attainable result. This is a common misunderstanding.

It should go without saying but *variability means there will be some variation!* If the average production rate is 500 per day, it is highly unlikely that one will ever see exactly 500 produced on any given day. Yet companies constantly talk about building to demand or providing service to demand each day. "Set your takt time equal to demand" is something that we hear often. If we try to set takt time equal to demand (or, more accurately, to the reciprocal of demand) then around 50% of the time, the line will not meet demand, and around 50% of the time the line will make more than demand. If the demand is 500 units per day and we work 2 shifts, 16 hours with 2 hours for breaks and lunches, we would have around 50,000 seconds of productive work. The takt time would be 50,000/500 = 100 seconds. But this would not account for variability and, consequently, would fail to meet demand more often than not.

As an example of not meeting demand, the Golden Gate Bridge was built to support 7,700 pounds per linear foot. Before the celebration of its 50th anniversary, the organizers expected some 80,000 people to attend, only to have more than 250,000 people show up on the bridge. Even though the crowd was much larger than anticipated and there was virtually no room for any more, its load reached only about 5,400 pounds per linear foot, or around 70% of its designed capacity. While no one would build a bridge to support exactly the maximum weight it could ever see, people are always trying to run a factory at its maximum capacity. So, for the scenario in the previous paragraph, taking a less conservative 85% of the 100 second demand takt time yields a practical takt time of 85 seconds, which is around 588 per day, providing an 88-unit safety factor.

If there were no variability, it would be possible to run at or near 100% utilization. But variability exists and always reduces efficiency. So, let's now describe the effects of variability on the Penny Fab example in Figure 4-2. The variability we add is to distribute the process times exponentially, much like the Little's law example. In this case, the standard deviation of the process times is equal to their mean, so the SCVs (c_e^2) are equal to one. Recall that the exponential distribution is memoryless. A process with exponential process times does not remember how long it has been since it started. So even though the average process time is 2 hours, the expected time to complete after it has worked a given amount of time is still 2 hours. So if it has just started, the expected time to complete is 2 hours. If it has been working for 2 hours, the expected time to complete is still 2 hours. Even if it has been working 100 hours, the expected time to complete is 2 hours! While strange, it does make it easier to compute cycle times. We call this case the marginal case as it is between the best and worst cases. However, it is also called the practical worst case.[iv]

With the previous deterministic example (zero variability in process times), the Penny Fab with one penny takes, on average, 8 hours because the mean of the sum is the sum of the means. From Little's law we obtain a throughput of 1/8 per hour. So, the throughput with zero variability and with variability is the same with only one item of WIP because there is no interference from other WIP. But if there are more than two pennies, it gets more complicated quickly.

The second penny "sees" the system as if it had one penny in steady state. The steady-state WIP with one penny is one, and it is equally likely to be at any one of the four stations. So there is an expected queue of 1/4 penny at each station. Since it does not matter how long the process has been working, the expected remaining time is still 2 hours. Thus, the cycle time at any station will be $CT = (1 + 1/4) \times 2 = 2.5$ hours, or 10 hours for all four stations. The same analysis can be done now for 3, 4, 5... pennies, yielding 12, 14, 16, and so on. And the throughput is simply the number of pennies divided by these cycle times, 1/8, 1/5, 1/4, 2/7, 5/16 and so on. In Figure 4-8, we see the throughput drop from 0.5 to 0.28 at WIP = 4 (red circle) and the cycle time go up from 8 hours to 14 hours (solid blue square). These changes are due only to variability because we have the same *average* process times and therefore the same bottleneck rate and the same raw process time.

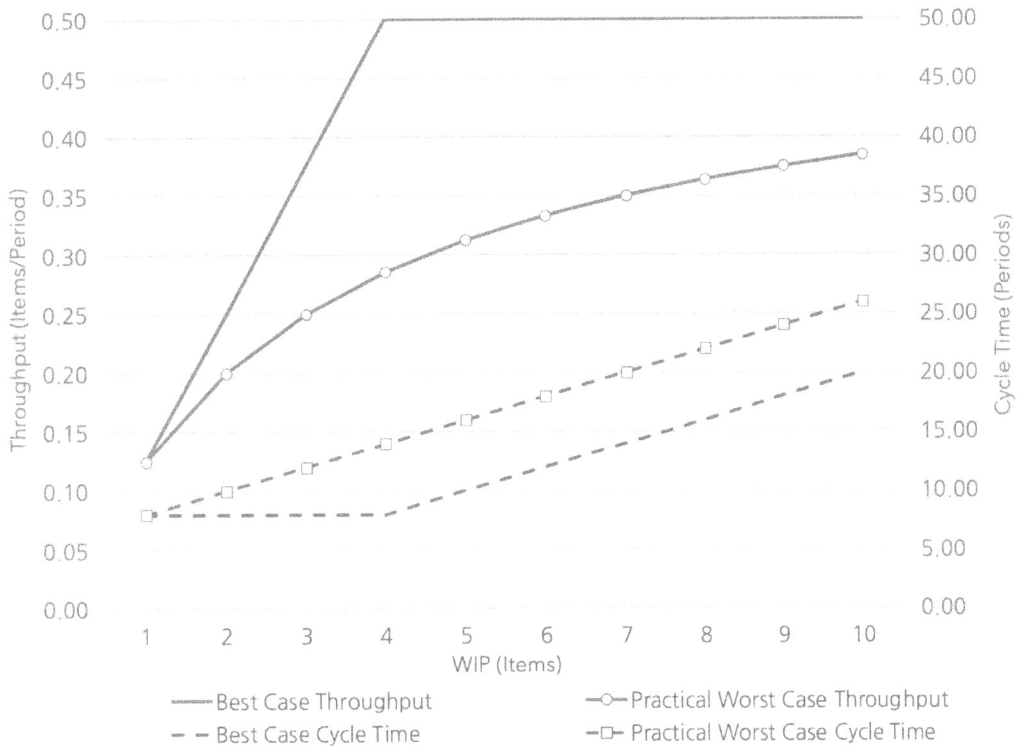

Figure 4-8: Penny Fab Best Case and Marginal Case

Using the above logic, we can derive the equations for the marginal-case performance curves.

$$CT_{MC} = RPT + \frac{w-1}{BNR} \qquad \textit{eq 4}$$

$$TH_{MC} = \frac{w \times BNR}{CW + w - 1} \qquad \textit{eq 5}$$

where w = the fixed WIP level. See section 7.3.3 in *Factory Physics, 3ed Ed* for detailed description of their derivation.

We now have a powerful analysis for designing and evaluating flow performance. A flow cannot have throughput greater than its bottleneck rate or an average cycle time shorter than its raw process time. At the same time, a flow should perform better than marginal case conditions.

A More Detailed Example

Now let's apply the analysis to a more complex example. To complete the poured-concrete part of a large construction project, 2,240 pilings must be fabricated and installed in 250 days. To install these pilings, reinforcement bar, or rebar, cages will be needed. Rebar cages often are highly engineered structures of metal bars usually wired together in complex shapes to support concrete structures, see Figure 4-9.

Figure 4-9: Rebar cages on a construction site

The rebar cage is placed, forms are built around it, and concrete placed in the forms and allowed to dry to create the final shape and strength required. Figure 4-10 presents a schematic of a rebar fabrication flow.

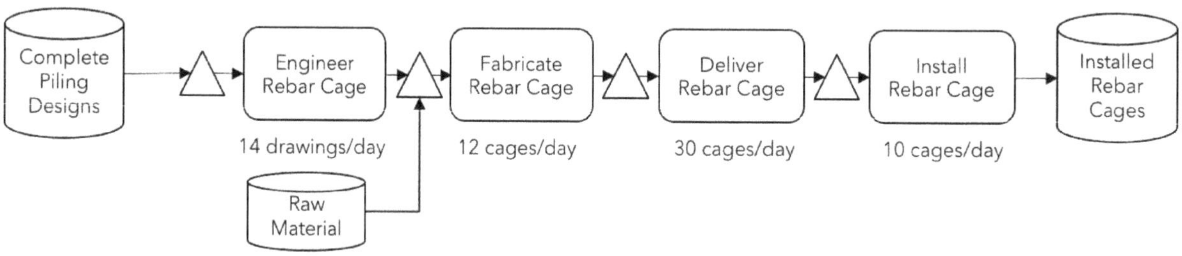

Figure 4-10: Rebar cage supply flow in construction

While this is a construction project and not a production system per se, we will see that it has all the properties of a production system. Indeed, when those details are neglected, the construction project is usually in trouble—finishing late and over budget—as we will see shortly.

Modeling the rebar fabrication and installation system as a production system provided powerful insights. The data in Figure 4-10 were used to estimate the capacity needed for completing the project within the required 250 days. If all went to plan, it seemed like there was no problem: 2,240 installations at 10 per day means the project would be delivered in 224 days, well ahead of the allotted 250 days.

However, looking at a steady fabrication of 12 cages per day and an installation of 10 cages per day means that onsite WIP of rebar cages would peak at 370 cages of WIP onsite before eventually declining to 0 with completion of the project. This presented a major problem because site laydown area (land used for storing material) was extremely scarce.

As the project kicked off, things went from bad to worse as the demonstrated capacity for rebar cage fabrication was only 10.8 per day (instead of 12) and site installation demonstrated capacity was only 7 per day (instead of 10). At this rate, using a simple deterministic analysis shows that WIP would peak at over 700 cages on site, and project delivery would take 321 days instead of 250, see Figure 4-11.

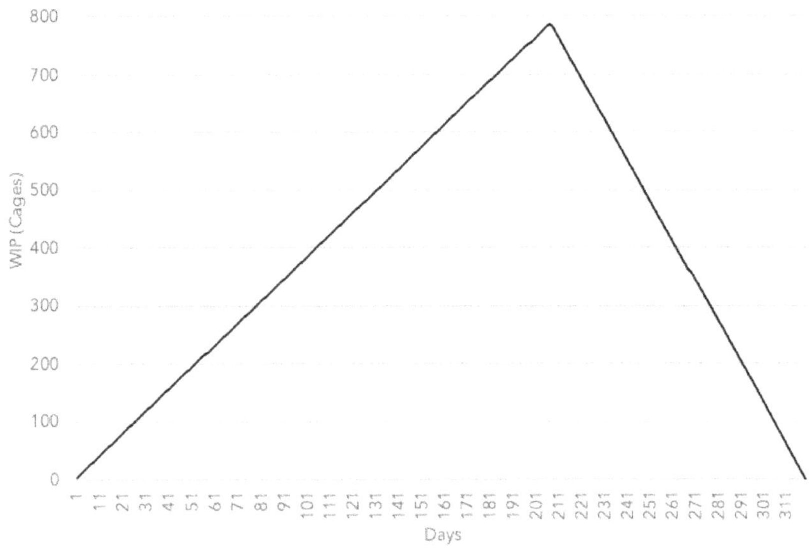

Figure 4-11: Analysis of WIP levels onsite over project duration

There is a saying in construction management: "I cannot install it if I don't see it." Many construction project managers would be fine with having all the WIP on site because it allows them to keep everyone working on something, whether it is needed at the time or not, and show high efficiency of resources; however, WIP overload causes many unintended problems.

First, there is the issue of finding a location to put the material and then rehandling it when needed. With the demonstrated capacity (10.8 cages/day) and uninterrupted production of rebar, the average cycle time will be 56 days, which is extremely long considering that the time to complete all the operations (raw process time) is fewer than 2 days. There is also potential damage caused by the excess handling. With so much material lying around, some may be lost or used out of sequence because of the confusion. Material may be cannibalized from one cage to make modifications to another cage, thereby rendering the first cage in need of rework or useless. Additionally, production of such a large amount of WIP also can create obsolescence waste because of design changes occurring after the cage has been fabricated.

So is there a better way? Operations science shows that once the bottleneck has enough WIP to keep it running, adding more WIP to the system only increases cycle time without significantly increasing throughput and throughput is going to be either the demand or the capacity, whichever is less. Recall that WIP is visible cycle time and that WIP is a leading indicator while cycle time is a lagging indicator. So controlling WIP is a better way to control cycle time than trying to control throughput by setting a takt time for releases.

But how much WIP should we have? Many companies have focused on one-piece flow as a goal, only to find that they have achieved very short cycle times but also have, at the same time, greatly reduced output. Look at the marginal-case throughput curve in Figure 4-8. As WIP is reduced to one piece per station, throughput decreases precipitously. This is a common effect in all production systems. When WIP is too low, the bottleneck is constantly starved, so the throughput suffers. We can use our operations science flow performance curve analysis to help us determine that decision with confidence that we will not decrease WIP so low as to reduce throughput below its required rate. At the same time, we don't want so much WIP as to cause long cycle times for all the reasons stated previously.

So how do we find a good level of WIP without needing to create a huge Monte Carlo discrete-event simulation that takes months to construct? While it is less accurate than a full-blown simulation, the flow performance curve analysis provides a powerful quick pass, and sometimes it is all that is needed to pinpoint performance improvement opportunities. If the flow performance results do not yield satisfactory answers or if the performance in question is of very high value or needs great precision, simulation provides a powerful tool for higher precision and accuracy.

For the question of what the WIP level should be, we will use the marginal-case equation to solve for an upper bound for WIP, eq 5, and the best-case equation for a lower bound on WIP. We have data from the facility that RPT is about 1.6 days, and we use 10 cages per day for the BNR, since that was the original planned capacity. This means CW = 1.6 × 10 = 16. If installation can do only 7 cages per day, we set the throughput to 7 and solve for WIP

$$TH(WIP) = \frac{BNR \times WIP}{CW + WIP - 1}$$

$$7 = \frac{10\,WIP}{16 + WIP - 1}$$

Solving yields, WIP = 35 cages.

With a WIP of 35 cages and a throughput of 7 cages per day, we get a cycle time of 35/7 = 5 days, which is quite an improvement from the original conditions of 56 days. So we think that is the most WIP we will need to maintain in the system at any one time because the marginal case has a significant amount of variability. But if the operation has less variability in process times than the marginal case's SCV of 1 (exponential distribution), our WIP should be less. But how much less? To address this, let us find the theoretical least amount of WIP that would be required to produce 7 cages per day. This would be the best case. Let's do that calculation. The best case for throughput is given in eq 2, $TH = \min\{BNR, WIP/RPT\}$. Since our throughput is less than the BNR, we solve the second expression in the braces,

$$7 = \frac{WIP}{1.6}$$

Thus $WIP = 11.2$. But WIP is always an integer, and 11.2 is the minimum possible value; therefore, the minimum possible integer that is at least 11.2 is 12 cages.

Now we have boundaries for our WIP level of rebar cages. What level should we use? A good approach is to start at 35, monitor throughput to make sure the installation rate of 7 per day is being maintained. Then gradually reduce the number to fine-tune the WIP level if desired. The reader may skip ahead to Chapter 18 to see the actual numbers and how operations science was put into action to great success on the project.

Analysis of Production Systems

While variability is a very important aspect of operations science, we should not forget the first concern is to have sufficient capacity. In the example above, we needed to install 2,240 pilings in 250 days, or 8.96 units per day.

There appears to be enough capacity in Figure 4-10 with the four stages having 14, 12, 30, and 10 units per day. The bottleneck is Install, at 10 units per day. Its planned utilization is $u = 8.96/10 = 89.6\%$. While that seems sufficient, it is actually tight. Anything over 85% has a chance of having problems, especially when the variability is high and the duration is short (recall the relationship of CT versus utilization, Figure 3-10 in Chapter 3). What saves this system is the extra capacity at all the other stations and the long duration that allows the law of averages to dominate.

Using our previous definitions, average demand (d) is the average number of items that will be required to satisfy demand for a given time interval (e.g., hours, days, weeks, etc.). Utilization of the resource is given by

$$u = \frac{d}{r} = dt_e$$

If there are m resources in parallel, each having a capacity of r, then the utilization is

$$u = \frac{d}{mr} = \frac{dt_e}{m}$$

Referring back to our example, we first compute the utilization for engineering, $u = \frac{d}{r} = 8.96/14 = 64\%$. If there were two engineers, the utilization would be, $u = \frac{8.96}{2 \times 14} = 32\%$.

Once we have enough capacity, we must consider variability. In this case, the variability of both demand and production have SCV values of 1, and there is only one resource at each transformation step.

Now that we have the utilization and the variability coefficients, we can obtain the average queue time. A reasonable approximation for the expected queue time with a fairly high utilization is the Kingman equation, or the VUT equation, as introduced in Chapter 3:

$$CT_q \approx \frac{c_a^2 + c_e^2}{2} \times \frac{u}{1-u} \times t_e \qquad eq\ 6$$

which is exact for exponential times between arrivals (i.e., Poisson demand). For situations with more than one resource, the approximation is a bit more complicated:

$$CTq = \frac{c_a^2 + c_e^2}{2} \frac{u^{\left(\sqrt{2(m+1)}-1\right)}}{m(1-u)} t_e \qquad eq\ 7$$

Note the content of the upper parentheses is the exponent of the utilization, u. For low utilizations, these approximations suffer, but then there is not much queueing for low utilization. Recall that the average process time, t, is $1/r$. Then, for engineering,

$$CTq = \frac{c_a^2 + c_e^2}{2} \frac{u}{1-u} t$$
$$= \frac{1+1}{2} \frac{0.64}{1-0.64} \frac{1}{14} = 0.13 \text{ days}$$

So the total cycle time at engineering is

$$CT = CTq + t$$
$$= 0.13 + \frac{1}{14} = 0.2 \text{ days}$$

and the total WIP at engineering is

$$WIP = TH \times CT$$
$$= 8.96 \times 0.2 = 1.79$$

Of this WIP, on average the WIP in queue is

$$WIPq = TH \times CTq$$
$$= 8.96/\text{day} \times 0.13 \text{ days} = 1.16$$

If we repeat this analysis for the rest of the operations, we obtain the results in Table 4-2. We encourage the reader to create a spreadsheet and duplicate these results.

Throughput (cages/day) 8.96

DATA	Engineer	Fabricate	Deliver	Install
SCVa	1	1	1	1
SCVe	1	1	1	1
r	14	12	30	10
t	0.07	0.08	0.03	0.10
CALCULATIONS				
u =	0.64	0.75	0.30	0.90
CTq =	0.13	0.25	0.01	0.86
CT =	0.20	0.33	0.05	0.96

Total CT = 1.54 days
Total WIP = 13.77 cages

Table 4-2: Initial data and analysis for the construction project

As we saw, the planned production was not realized. Let's look at that scenario.

Let's change the throughput to 7 per day, since that has been demonstrated from initial operations. We'll change the capacity of install to 7.2 since it is working at very high utilization. This will yield a 97% utilization. We will change the capacity of fabricate to 10 since that is what it has demonstrated it could produce. The results are shown in Table 4-3. Note the nearly 260% increase in cycle time and a 180% increase in WIP from the higher levels of utilization.

Throughput (cages/day) 7

DATA	Engineer	Fabricate	Deliver	Install
SCVa	1	1	1	1
SCVe	1	1	1	1
r	14	10	30	7.2
t	0.07	0.10	0.03	0.14
CALCULATIONS				
u =	0.50	0.70	0.23	0.97
CTq =	0.07	0.23	0.01	4.86
CT =	0.14	0.33	0.04	5.00

Total CT =	5.52	days
Total WIP =	38.64	cages

Table 4-3: Data and analysis for the construction project (TH = 7, Capacity = 7.2)

You may have other ideas for the numbers and make different assumptions. If so, put these into the spreadsheet created for the initial data. Being able to quickly try various scenarios is a big benefit of operations science. And remember that some of the calculations, like utilization, are exact while others, like CT_q are approximations (except for special cases).

Another benefit of operations science is that it informs intuition. One can gain more intuition by tinkering with one of these models for an hour than by spending a month watching a production line. This sounds crazy, but you are less likely to do something extreme in the factory, whereas you can do anything you like with the model with no negative consequences. And if your "great idea" blows up the model, you just don't mention it to anyone; however, if that same idea were implemented in the actual factory, everyone would know! So when making guesses with operations science, they are truly "educated" guesses.

Many times in working with clients, we will get zingers such as "You didn't think about rework," or "We have more than one rebar producer; this doesn't show that." All fair points. The response usually is "Good point; let's put it in the model."

Indeed, these situations are the rule rather than the exception. It has been our experience that the first model of the system works *much better* than the actual system. We then have to work hard to make the model behavior as bad as the actual system. We also have found that if we keep notes on the things we need to change in the model to make it worse, many of these can be done in reverse to make the actual system better.

We worked with a company that had a 15-year-old production line they had been trying to improve since it was installed. In the first two days onsite, we used the type of operations science analyses described above to identify a 30% increase in throughput. This was implemented, and they realized the increase in about two months. With the concepts and techniques in this book, we believe you will achieve similar results.

Although we learn a lot from creating the model, *the model is not the goal!* It is only a tool, and trying to make it perfect can result in extremely diminishing returns. We want the reader to understand the concepts of operations science and have a reliable way to apply them. The goal is to improve your intuition, that gut instinct commonly referenced in decision-making, to *make better decisions faster.*

Utilization of Systems with Multiple Products

In Chapter 3, we gave a definition of utilization and provided some simple examples. The simple definition is

$$u = \frac{\text{load}}{\text{capacity}} \qquad \textit{eq 8}$$

With no rework or scrap, load is the same as demand, the number of units in a given time available while capacity is the run rate of the process. So the number of units/run rate = time required to produce demand, thus,

$$u = \frac{\text{time required to produce demand}}{\text{time available to produce demand}} \qquad \textit{eq 9}$$

So if we have multiple products, we simply add up the times required to produce the demand for each product and divide by the total time available for all the resources (e.g., the number of parallel processes).

The best way to understand this is to do an example. Table 4-4 shows the data for the example.

Item Data

Item	Wkly Dmd	Batch Size	Rate per hr	Setup Time hrs
900	100.00	400	5.00	1.00
910	20.00	40	6.00	2.00
920	150.00	1200	7.00	1.50
930	76.00	152	5.00	0.50
940	42.00	84	4.00	0.50
950	54.00	108	5.00	0.75
	442.00			

Process Data

No. Mach	Hrs Per Day	Days per Wk	hrs to fail	hrs to repair
2	10	6	50	4

Table 4-4: Multiproduct data for utilization calculation

Our demand period is one week. The time available to do all of this demand is the product of the number of machines, hours per day (hpd) and days per week (dpw):

$$T_{avail} = 2 \times 10 \text{ hpd} \times 6 \text{ dpw} = 120 \text{ hours per week}$$

The outage times are of the machines and are not deducted from the time available.

Next, we will compute the number of batches per week, which is just the weekly demand divided by the batch size. Then we compute the average batch time given by adding the time to make one process batch plus the setup time. The time to do the batch is the batch size divided by the *effective* run rate. The effective run rate is the nominal run rate times the availability. So, the average time to complete a batch is

$$t_{batch} = s + Q_p/(rA)$$

where s is the setup time in hours, Q_p is the process batch size, r is the run rate, and A is the availability that is computed from the mean time to fail (MTTF) and the mean time to repair (MTTR), viz.

$$A = \frac{MTTF}{MTTF + MTTR} = \frac{50}{50 + 4} = 0.93$$

Note that we use the availability to degrade the production rate (as seen in the t_{batch} equation), but it does not increase the setup time, nor does it take away from the time available. We model it this way because the failures are machine failures, not failures in the setup, which is mostly labor. We could have modeled it by reducing the time available if the setups were affected.

The time used by one item will be the batches per week times the average batch time,

$$T_{used} = d/Q_p(s + Q_p)/(rA)$$

$$= \frac{ds}{Q_p} + \frac{d}{rA} \qquad eq\ 10$$

These calculation results are shown in Table 4-5 for the first three items, leaving the last three for you to fill in,

Availability
92.59%

Item Data

Item	Batches /wk	Batch Time	Time used
900	0.25	87.40	21.85
910	0.50	9.20	4.60
920	0.13	186.64	23.33
930			
940			
950			
		Total Time Used	90.08 hrs

Time Avail	120	hrs
Utilization	75.1%	
Capacity	588.84	units/wk

Table 4-5: Results from multiproduct utilization calculations

The utilization is simply the total time used divided by the time available, while the capacity is the sum of the weekly demand divided by the utilization.

Note that the capacity depends strongly on the product mix. To see this, swap the demands for items 920 and 940. If we do this, the utilization goes from 75% to 86%, while the capacity drops from 589 to 515 per week. We encourage you to create the spreadsheet and experiment with different mixes of products to see how the capacity and utilization are changed.

Analysis of Service Systems

In many production systems, a stock of finished goods is awaiting an external demand to purchase or consume them. However, the construction project production system in Figure 4-10 is modeled as if it had a stock of installed rebar cages awaiting demand. The demand comes from the next operation that sets forms around the rebar cages. In other words, there is no actual stock of installed rebar cages, but there could be rebar cages that were finished early. So there may be a *virtual stock* even when there is no real stock.

For service systems, such as a restaurant or an emergency room, there is no real stock. One cannot have a stock of meals eaten or repaired broken arms before the desire for a meal or the healing of an injury was presented. But, like the construction case before, there can be a virtual stock. In fact, in the case of a service system, the virtual stock is always *back ordered.* In other words, the demand for the service always occurs before the service can be completed.

So an upscale restaurant would have an empty stock between transformation and demand. The empty stock would represent the back orders of customer meals (i.e., customers ordered their meal and then are waiting for it to be prepared and served). The stock would never have any inventory since upscale restaurants do not prepare meals beforehand. However, for a fast-food restaurant, there is usually some inventory between transformation and demand. But if one were to model the arrival of restaurant supplies and groceries, it would be represented by another queue and with a stock at the restaurant. The demand to the supplier would be generated by the restaurant in question as well as other restaurants.

So the analysis for a service system such as a restaurant is similar to that of a production system, except that we have only cost for capacity and a cost for time spent. In this case, the cost for time spent is probably not linear with time. For instance, suppose if there is no wait, the cost is zero. But one would expect that customers having to wait for an hour would cost much more than four times the cost of a 15-minute wait.

Consider a high-end restaurant that prides itself on having excellent cuisine *and* no waiting for a table. To do this, they promise that you will have a table immediately or receive a discount on your dinner. They decide to pay $25 if a party has to wait for 1/4 hour and $200 if they wait for an hour. A continuous function for the discount versus time is given below as $k(t)$,

$$k(t) = 133.33\, t(t + 0.5)$$

During the dinner period, about 10 customer parties arrive every hour and spend around one hour for dinner. We want to determine how many tables we need to minimize the cost of the tables plus the cost of a party not having a table immediately upon arrival.

Given an arrival rate, $d = 10$ per hour and an average process time of $t_e = 1$ hour, we find that we need at least $m = 11$ tables so that, $d \times t_e/m < 1$. The SCV of arrival is estimated to be 1. The SCV of the time spent is estimated to be 0.5. If the average wait time is time spent in queue, then the cost per hour for waiting with m tables will be

$$TC(m) = dk\big(CTq(m)\big) + 50\, m$$

where CT_q is given by our queueing equation for more than one resource,

$$CTq = \frac{1+0.5}{2}\frac{u^{\left(\sqrt{2(m+1)}-1\right)}}{m(1-u)}1 \qquad eq\ 11$$

and $u = d/m = 10/m$. Running the numbers shows a minimum of 13 tables having a cost of \$713 per hour, although 14 tables is not much more at \$730 per hour. Having only 11 tables costs around \$1,251 per hour.

This analysis provides an example of a particular situation that can be optimized using operation science principles. Of course, there could be a very different wait-time cost function and even a nonlinear cost of tables, but the analysis would look very similar. The idea here is not to derive the *single correct way to solve the problem* but to show how to set up a problem using basic operations science concepts.

Conclusion

This chapter covers the basics of flow analysis and describes ways to model both production and service systems. The reader now understands the relationship between WIP, throughput and cycle time and can compute the capacity and utilization for single and multiproduct systems. She can evaluate her operation to see if it operates near the best case, near the marginal case, or somewhere below that. Moreover, she now knows what must be done to improve the operation. The reader can now estimate the cycle times and WIP for single and multi-station queues, whether in a production setting or for a service.

Ignoring these results and going with one's instinct can lead to disastrous consequences in terms of higher inventory, lower customer service, and greater costs. Luck is a good thing to have in a casino but cannot be trusted to run an operation.

In the next chapter, we will address stock inventory and the issues of how much is enough, reorder frequency, back orders, and the probability of having stock when needed (also known as the fill rate). We will discuss the very simple base stock system, of which kanban is an example, and the more sophisticated reorder point and reorder quantity systems and provide methods to optimize all of these systems. Moreover, we will relate these to the ubiquitous Material Requirements Planning module that resides in almost every ERP system and how those modules can be optimized.

Once we understand flows from this chapter and inventory in the next, we will be ready to pull it all together in Chapter 6 with a methodology that will minimize all the relevant production and inventory costs while providing desired customer service levels.

---

[i] In this simple case, all stations have the same utilization since they all have the same process time and experience the same throughput and therefore are all bottlenecks. In more complex systems, one might have a reduction in throughput. This will lead to utilization due to scrap or an increase in utilization due to rework, or a change in stations' maximum utilization due to demand mix change. Therefore, the maximum utilization station might not be the slowest station.

[ii] The general definition of RPT is the time it takes one transfer batch to complete the flow. This includes any setup time and downtime that might be encountered. RPT is more practical than value-added time (VAT) because RPT includes those real-world effects whereas classic VAT is only the processing time. We discuss transfer batches later.

[iii] A common mistake is to think of critical WIP as a target WIP level for processes. In the presence of variability, the CW is too low and results in lost throughput. CW is only a characteristic of the process; it is occasionally used in calculations but is not a goal for WIP level, unless there is zero variability.

[iv] The worst case is when the part selected is always last in a queue of all the other parts. This *is* the theoretical worst case, but it is so bad that it is hardly worth considering. For practical purposes, the practical worst case is a better lower bound on performance than the worst case, although we have seen plenty of cases worse than the practical worst case.

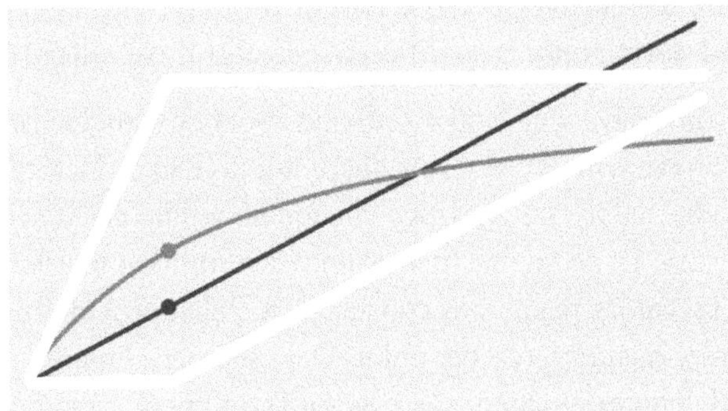

5

Operations Science of Stock Inventory

As we described previously, there are two primary types of inventory, WIP inventory and stock inventory. In a perfect world with no variability, there is no need for stock inventory because transformation would satisfy demand as soon as the demand happened. This chapter focuses on stock inventory, represented by the green cylinder icons in Figure 5-1. Stock inventory is the inventory at the beginning or end of a process—it is not WIP. Note that we use the terms *stock*, *stock inventory*, and *inventory* interchangeably.

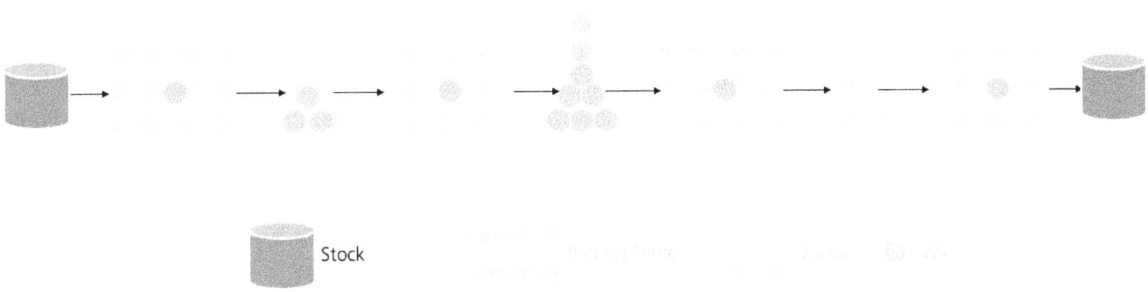

Figure 5-1: Stock inventory in the basic operations construct

This chapter will require some knowledge of probability and statistics. If those fields are not in your wheelhouse, then learn the concepts and let someone else do the math for you. But do not make short cuts that provide simple solutions: there are very many on the Internet, and these often do not work (as we will show with some examples).

We first introduce the primary topics for inventory management of how much to order and when to order. For how much, we derive a formula that satisfies an ordering capacity constraint while minimizing on-hand inventory. We find this is similar to the classic economic order quantity but more practical because we do not need to specify a cost of capacity. This is important because when our ordering capacity is fixed, there really is no difference in cost if we order 100 times per month or 101 times per month.

We then have a general discussion about how to determine when to order using "inventory position," which we will define. From there, we dive into the technical details of determining reorder points and, thereby, safety stock, with a caution about the dangers of some of the safety stock formulas found on the Internet. To do so, we first explore determining reorder points and fill rate for a base stock system where the inventory position is constant. This then provides the groundwork to expand to the more sophisticated reorder quantity, reorder point (Q, r) strategy. Finally, we provide a general algorithm for reorder point (which gives us safety stock as well) for the Q, r model. The result will be a sound method for the reader to evaluate the trade-offs between the buffering choices time (back-order time) and inventory (average on hand). However, instead of specifying the desired average back-order cost (which is very hard to specify), we specify a desired customer fill rate. Fill rate is the probability of having stock on hand when there is demand. The (Q, r) model works well for steady demand but not so well for lumpy demand, so we provide translation of the concepts to time-phased reordering approaches (e.g., MRP) using safety stock and other MRP methods. This covers most of the options for modern operations science–based inventory planning systems.

We conclude by summarizing the options and formulas so the reader can determine the best application for their purposes.

Stock Inventory Management Case Studies

The trick to stock inventory management is to have what you need when you need it but not have too much when you don't need it. That is more easily said than done. It is all about ordering enough and getting it before running out. So there are two decisions: when to order and how much to order. Simple, right? Consider an example from the real world.

The Supply Corporation

The Supply Corporation (not its real name) provides building supplies to retail suppliers such as Lowes and Home Depot. They have warehouses all over the United States. The warehouse in Nashville has more than 500 items. Currently it holds around $2.4 million in inventory and the service level (fill rate) is 91%, with a replenishment frequency of 333 orders per week. In Chapter 3 we described an efficient frontier graph depicting the optimal value of inventory versus the resulting fill rate for a given order frequency. We used this tool to model the Supply Corporation's inventory policies. Table 5-1 summarizes the options displayed on Supply Corporation's inventory efficient frontier, Figure 5-2.

INVENTORY DATA	DESCRIPTION	ORDERS/WEEK	FILL RATE	INVENTORY
ACTUAL	Actual results from inventory management practices	333	91%	$2,362,000
PREDICTED	Expected results if policies in planning system are used	333	83%	$1,402,000
SELECTED OPTIMAL	Option available to improve performance	167	96%	$1,926,000

Table 5-1: Operations strategy options for inventory

Although there appears to be only one graph plotted, the graph for the current order frequency of 333 per week falls directly under the more efficient graph with half as many orders per week. The graph also shows that the company could accomplish a higher fill rate, going from 91% to 96% with less inventory, going down from $2.36 million to $1.93 million and fewer (167) orders per week. We also note that the performance predicted by the current policies (83% fill rate, $1.40 million inventory investment and 333 orders per week) is not matched by what is actually happening. While it appears that the predicted values are actually better than the optimal, the predicted point is below the curve, we find that the fill rates under the current policy are highly variable. The optimal policy constrains the fill rates to be 80% or above, while 35% of the items using the predicted policies have fill rates below 80%, and 20% are actually below 50%. This huge variation in fill rate indicates how poor the current policy is.

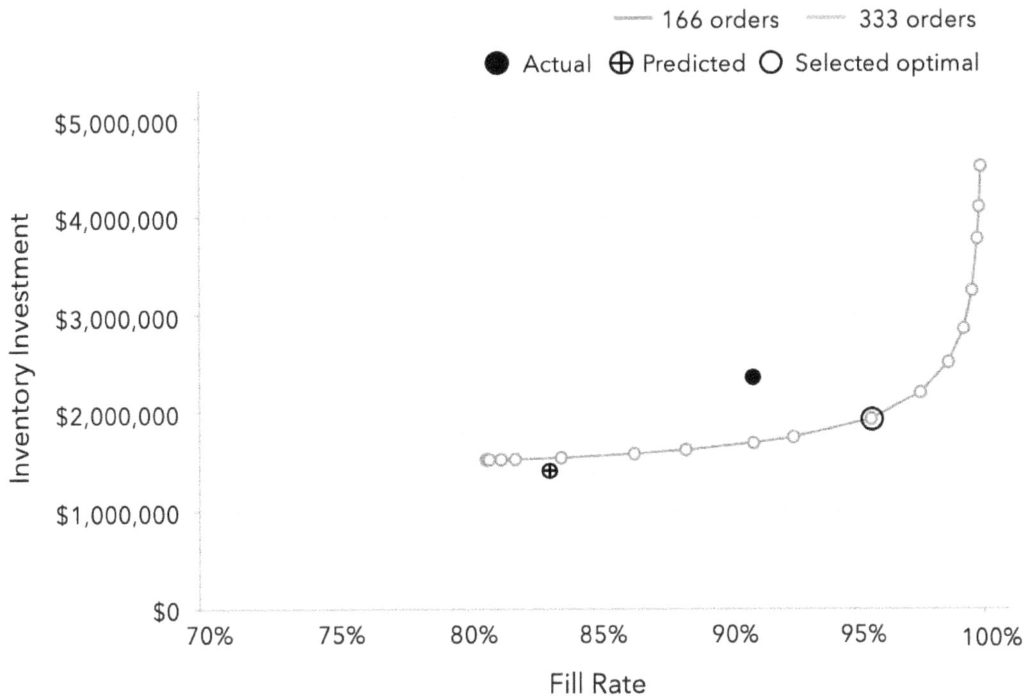

Figure 5-2: Efficient inventory frontier for Supply Corporation

Fortunately, the planners are not following these policies, and for good reason: The customer service would be quite poor. We have observed such behavior in many engagements: Planners ignoring policies and having fewer stockouts but at the cost of more inventory. This is a natural reaction to the way planners are instructed: Never, never, never run out of inventory, but don't have too much either! While stockouts due to low inventory can happen daily or weekly, placing turmoil and stress on planners, the issue of too many dollars tied up in inventory is usually reviewed much less frequently, perhaps once a quarter or maybe at the end of each month.

Given the analysis here, the policies in the Supply Corporation ERP system are quite bad. And this is typical because most companies do not have a sophisticated and efficient way to generate optimal policies. Although many texts offer algorithms to generate optimal policies, these must be applied to each item, one at a time. When demand changes, it must be done all over again. This can be a lot of work if not done effectively. No wonder we have seen many ERP systems, years after their implementation, having the same policies as originally set by the implementation team! The analysis shown here makes this much easier. Once the data for item demand levels and replenishment times are in the system, we can generate optimal policies for one item or for 525 items with the same effort.

Deciding where to be on the efficient frontier is a strategic decision that attempts to balance customer service with inventory investment and replenishment capacity. But once we make that decision, we can generate the inventory policy parameters for all 525 items that yield the desired performance. This chapter shows how this is done.

However, before we get into all these details, let's consider a simpler example from Bubba's Lumber.

Bubba's Lumber

Bubba's Lumber uses a simple reorder point/reorder quantity system, and so he must determine the best reorder point (ROP) and reorder quantity (ROQ) for the five inventory items he is managing (Table 5-2). We will describe ROP/ROQ systems in more detail later.

Item	Cost per Unit	Demand per year	Replenishment Time
	($/unit)	(units/yr)	(days)
4X8 3/4" Ash	100	100	60
2X12 10' Pine	250	500	30
2X4 Stud bndl	20	50	100
4X8 1/4" util	10	100	15
1X4 10' Pine	100	1000	30

Table 5-2: Data for Bubba's five inventory items

Because Bubba's ordering capacity is constrained, he does not want to reorder more than 20 times per year. So, that makes the ROQ easy: Just divide the demand by 4. Each item will be ordered 4 times per year, multiplied by 5 items makes 20 orders per year. Bubba also wants to make sure inventory is available when someone wants it. Like the ROQ, he is going to use the demand to set the ROP. By trial and error, he increases the fraction of demand for each ROP from 10% to around 17% and sets ROP = average demand ÷ 6. The result is 20 replenishment orders per year, with only 2% of demands back-ordered. However, resulting investment in inventory is almost $49,000, which is around 5 turns per year (cost of goods sold ÷ inventory investment). See Table 5-3.

Item	Cost per Unit	Demand per year	Replenishment Time	Reorder Quantity	Reorder Point	Order Frequency	Fill Rate	Average Backorder	Inventory Investment
	($/unit)	(units/yr)	(days)	(units)	(units)	(per year)		units	$
4X8 3/4" Ash	100	100	60	25	17	4.0	0.946	0.131	$ 1,319.27
2X12 10' Pine	250	500	30	125	83	4.0	1.000	0.000	$ 26,101.03
2X4 Stud bndl	20	50	100	13	8	4.0	0.540	1.831	$ 47.64
4X8 1/4" util	10	100	15	25	17	4.0	1.000	0.000	$ 253.90
1X4 10' Pine	100	1000	30	250	167	4.0	1.000	0.000	$ 20,980.82
		1750				20.0	98.38%	1.962	$ 48,702.67

Table 5-3: Bubba's current inventory policy and results

Five turns is not terrible but not great. Could Bubba do better? The answer is, yes, much better! Table 5-4 lists Bubba's optimized policies, showing a much lower investment for the same fill rate, while adhering to the ordering constraint.

Item	Cost per Unit ($/unit)	Demand per year (units/yr)	Replenishment Time (days)	Reorder Quantity (units)	Reorder Point (units)	Order Frequency (per year)	Fill Rate	Average Backorder units	Inventory Investment $
4X8 3/4" Ash	100	100	60	42	17	2.4	0.968	0.078	$ 2,163.97
2X12 10' Pine	250	500	30	59	43	8.5	0.971	0.106	$ 7,877.47
2X4 Stud bndl	20	50	100	66	15	0.8	0.986	0.029	$ 686.60
4X8 1/4" util	10	100	15	132	4	0.8	0.993	0.008	$ 658.99
1X4 10' Pine	100	1000	30	132	85	7.6	0.982	0.092	$ 6,890.07
		1750				19.9	97.87%	0.313	$ 18,277.09

Table 5-4: Bubba's optimized inventory policy and results

Next, let's look at the fundamentals for determining those optimal policies.

Basic Inventory Terms and Data

Before we can optimize, we need to model the inventory management system. First, let's examine the usual inventory terms:

cycle stock: the up-and-down level due to order quantity (inventory moving from 5 to 25 in Figure 5-3).

safety stock: the minimum planned level to accommodate variability (inventory at 5 in Figure 5-3).

average inventory: the average number of units on hand available for use or distribution. This is expressed in units, value, and carrying cost.

customer-service measures: a measure of satisfaction of demand. These include fill rate (percent of demands filled from stock), average back-order level, average back-order time when back-ordered, number of stockouts per period, total back-order cost, and others).

replenishment frequency: the number of replenishments per period. This is a measure of the load on the replenishment team (e.g., purchasing).

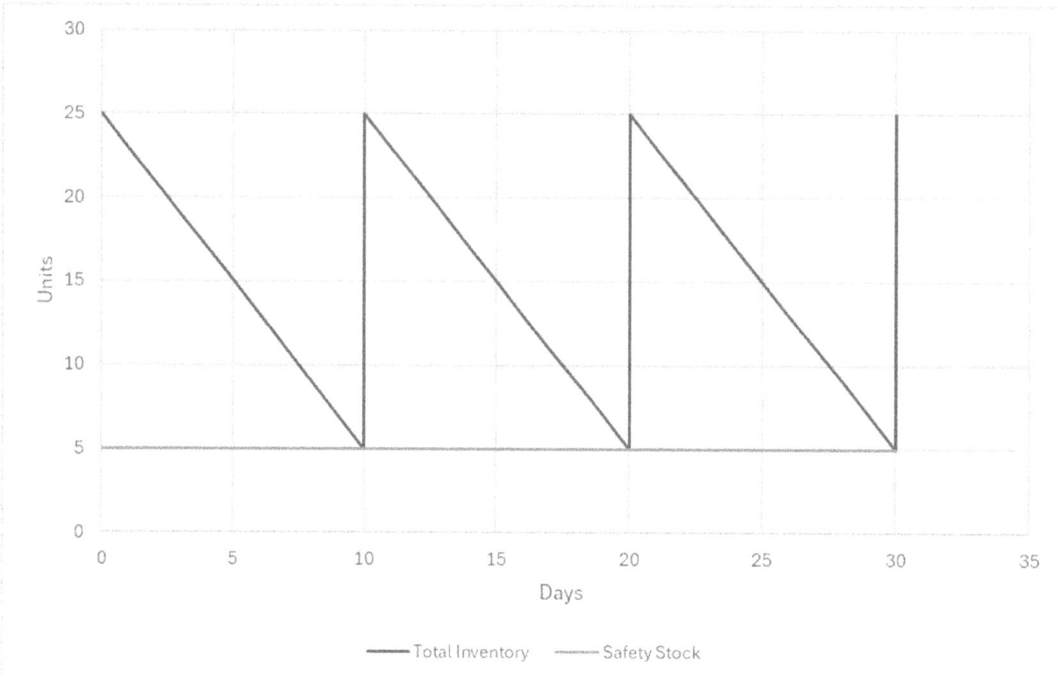

Figure 5-3: Cycle stock and safety stock over time

The state variables that provide the status of a particular item and are used to control the inventory system are as follows:

on-hand inventory, $I(t)$: Number of units on hand at time t.

back-order level, $B(t)$: Number of units on back order at time t. To keep the logic straight, any inventory that is *allocated* for anticipated demand is treated as a back order.

net inventory, $IN(t)$: On hand inventory less back orders. Net inventory is positive when there is stock on hand and negative when there are back orders.

number of units on order, $IO(t)$: The quantity of units with outstanding purchase orders for purchased parts and the WIP level for parts produced.

inventory position, $IP(t)$: Net inventory plus units on order. Unlike net inventory, inventory position should be used to control inventory.

The relationship between inventory position and the other variables is

$$IP(t) = I(t) - B(t) + IO(t)$$

or

$$IP(t) = IN(t) + IO(t)$$

We will see that the inventory position, $IP(t)$, is the most important of these for inventory control. Many people try to use net inventory as a control, but such practice can lead to disastrous results.

Inventory Control Data

Before we describe the data, we need to describe a few more terms. The first is period, as we describe the average demand per period, number of periods in replenishment time, etc. Our choice of period is a week because, unlike months and corporate quarters, a week always has the same number of days, exactly seven.

The other term is replenishment time, which is the time required to replenish the stock, includes the time

1. to realize a replenishment is required
2. to prepare the purchase/work order
3. for the vendor/production facility to receive the order
4. to procure/produce the part
5. to deliver it to the requesting organization, and
6. to make it available for use (including inspection, quarantine, etc.)

Replenishment time should be in terms of the same periods as the demand. If, for convenience, we choose to use a different period, we need to provide a conversion factor. For example, if we have weekly demand buckets and measure replenishment time in days, we will need to convert the days into weeks as $\ell/7$, where ℓ is replenishment time. Even when we only work five days per week, the replenishment times usually follow a seven-day week because most of the time (for suppliers) is outside our facility. We distinguish between lead time and replenishment time as follows: Lead time is typically a planning parameter, a management constant entered into the system to estimate completion or promise dates. In project management, it's often referred to as task duration. While this value may be based on historical data, it is usually applied deterministically, without explicitly accounting for variability.

In contrast, replenishment time reflects the actual time it takes for goods or services to become available and is inherently stochastic, influenced by variability in processing, transportation, and other factors. By modeling replenishment time as a random variable, we adopt a more comprehensive and realistic approach to inventory planning and avoid the problems of the simplistic, deterministic estimates often used in traditional systems.

With these caveats, we now list the data and derived quantities required for effective inventory control:

average demand per period, d: The data will be parts per day, per week, etc. and the period should be consistent with other data, such as the inventory carrying-cost ratio per period described next. To estimate, use the arithmetic mean of the demand, D_j, over n periods,

$$\bar{X} = \frac{\sum_j D_j}{n} \to d$$

variance of demand per period, σ_d^2: Uses the unbiased estimator of the variance,

$$S^2 = \frac{\sum_j (D_j - \bar{X})^2}{n-1} \to \sigma_d^2$$

average replenishment time, ℓ: An estimate from purchasing, etc.

variance of replenishment time, σ_ℓ^2: An estimate. It is easier to estimate σ_ℓ rather than the variance. If it is totally unknown, a reasonable value would be $\sigma_\ell \approx \ell/2$.

unit cost, c: Denotes cost of goods sold.

inventory carrying-cost ratio (ICCR) per period, i, where $0 \leq i \leq 1$: The ICCR is multiplied by the unit cost to obtain the cost per period to carry one unit of inventory. This should be based on actual costs. People tend to underestimate the ICCR, thinking it involves only the cost of capital. But it should include the cost of capital as well as all costs associated with carrying inventory, e.g., the expense of inventory managers and warehouse employees, warehouse rental, and/or maintenance, obsolescence, shrinkage, and so on. A reasonable value for ICCR is between 20% and 35% per year. If we are looking at a weekly period, and our ICCR is 25% per year, then $i = 0.25 \times \frac{7}{365}$. The inventory carrying cost will be ic, where c is the unit cost of the item.

inventory carrying cost per unit, $h = ic$: The cost per period to carry one unit.

back-order cost per unit, b: The cost to have one unit back-ordered one period. This is extremely difficult to estimate, but there are ways to relate it to the desired fill rate and holding cost.

desired fill rate, FR: The percent of time demand is satisfied from inventory, per part or for an entire set of parts. For comparison, a historical analysis of the on hand inventory can provide a good estimate. An easy estimate for a set of parts is the percentage of demand that currently has a typical order quantity on hand (a one-day snapshot).

order frequency, f: A feasible order frequency given current order-management capacity; for instance, the number of orders the purchasing department can manage in the period. One way to do this is to consider the current order frequency by dividing the current demand forecast, d, by the current order sizes, Q, and adding: $f = \sum_j d_j/Q_j$. We know this is feasible because we are currently doing it. Now we can determine what will happen to our inventory investment by increasing or decreasing this frequency.

ordering cost: This may be an actual out-of-pocket (OOP) cost for a particular item, A_j, where we spend a certain amount each time we place an order. An example could be shipping charges that do not depend on the quantity (within a range). But an OOP cost should not include the cost of the labor to process the order; the cost of labor usually does not change if we do 100 orders in a month and then do 101 the next month. Instead, there is a limit, or constraint, on the available capacity to order. Related to that constraint is a cost for ordering, λ, a global constant. λ is not for a particular item because it represents the constraint on the number of total orders across all items. This cost is sometimes called a shadow price, which indicates the amount of inventory carrying cost that would be saved for each extra order placed per period. If we were running a restaurant, we would not want to order our baking supplies (flour, sugar, yeast) one at a time, nor would we want to order a year's worth, either. Indeed, when optimizing order quantities, a constraint on the number of orders is equivalent to having an order placement cost, as we will see.

There are two important derived quantities to understand: the average and variance of replenishment-time demand. We will show why these are important as we get into the models.

average replenishment time demand, θ: The demand that occurs during replenishment time. $\theta = d \times \ell$. It is important that d and ℓ have the same relevant units (e.g., week). But if d is the average demand per week, and ℓ is the average replenishment time in days, we will convert, namely,

$$\theta = d \times \ell/7$$

standard deviation of replenishment time demand, σ_{RTD}: Given by

$$\sigma_{RTD} = \sqrt{\ell\, \sigma_d^2 + d^2 \sigma_\ell^2}$$

This value is a key driver in the cost of optimal policies. It is important to note that σ_{RTD} is the value of the time-inventory buffer, Δ_{TI}, described in Chapter 3. So if σ_{RTD} increases, the average inventory and average back orders, both increase! Moreover, the cost of an optimal policy with inventory carrying costs and back-order costs is a linear multiple of σ_{RTD}.

How Much to Order

The issues are how much to order and when to order. We discuss how much first, since the derivation does not require any probability arguments.

The more frequently we order means smaller quantities are needed with each order, and so we carry less inventory, but we may need more ordering capacity. If we order less frequently, we order larger quantities for each order so that we carry more inventory but will need less ordering capacity. The factors that determine the order quantity are how much demand we have and the cost of the item, along with the capacity to place orders and any costs directly associated with the order. More expensive items should be ordered more frequently to keep inventory costs lower. Bubba did not pay attention to these issues.

We have described all the needed data in the previous section. Now we will derive the cost equation(s) and find the optimal order quantity.

Known Order Frequency

If we have only one item and we have a limit on the frequency of orders, f, and the average demand per day is d, then there is no need to optimize because there is only one reorder quantity, Q, that satisfies the constraint on the order frequency:

$$Q = d/f$$

But, in Bubba's case, we have five items. How does he find the order sizes that result in 20 orders per year and minimize cycle stock inventory? He might try the classic economic order quantity (EOQ) formula, first proposed by Ford Whitman Harris in 1919 which determines the optimal reorder quantity, Q^*,

$$Q^* = \sqrt{\frac{2\,A\,d}{i\,c}} \qquad \textit{eq 1}$$

where A is the ordering cost associated with the constraint of only 20 orders per year, d is the average demand, i is the inventory carrying-cost ratio, and c is the cost of the item. Bubba guesses his ordering cost is $100 per order and the inventory carrying-cost ratio is 35%. If we use Bubba's costs and demand and eq 1, we get the reorder quantities in Table 5-5 which translate to about 35 orders per year—rather more than Bubba's desired 20 orders per year.

Item	Q*	f
Ash	24	4.2
2x12 Pine	34	14.8
Stud bndl	38	1.3
Util	76	1.3
1x4 Pine	76	13.2
		34.8

Table 5-5: Bubba's reorder quantities calculated using the EOQ

So guessing the ordering cost might not be effective. Instead, we should solve a different problem: *Minimize inventory carrying cost subject to an order frequency that is less than or equal to 20.*

Looking at Figure 5-3, average on-hand inventory units will be $\frac{Q_j}{2} + s_j$ so we can write down the inventory carrying cost and the constraint,

$$\min Z = \sum_j i\, c_j \left(\frac{Q_j}{2} + s_j\right) \qquad \textit{eq 2}$$

where s_j is safety stock for item *j*, subject to

$$\sum_j \frac{d_j}{Q_j} - 20 \leq 0$$

We can rewrite the constrained problem as an unconstrained problem with an extra variable for the constraint. In this case it would be

$$\mathcal{L} = \sum_j i\, c_j \left(\frac{Q_j}{2} + s_j\right) + \lambda\left(\sum_j \frac{d_j}{Q_j} - 20\right)$$

where λ is the Lagrange multiplier of the constraint.[i]

Since the minimum of a function always occurs where the function is flat, we take the derivative with respect to all the variables, namely, Q_j and λ, and set them equal to 0. The general equations for any order frequency, f, are

$$\frac{d\,\mathcal{L}}{d\,Q_j} = \frac{i\, c_j}{2} - \frac{\lambda\, d_j}{Q_j^2} = 0 \qquad \textit{eq 3}$$

$$\frac{d\,\mathcal{L}}{d\,\lambda} = \sum_j \frac{d_j}{Q_j} - f = 0 \qquad \textit{eq 4}$$

eq 3 yields

$$Q_j^* = \sqrt{\frac{2\,\lambda\, d_j}{i\, c_j}} \qquad \textit{eq 5}$$

which is the classic EOQ but with a catch: eq 4 forces the order frequency to be equal to f. To solve it, we substitute Q_j^* into the second constraint, eq 4, and solve for λ.

$$\lambda = \frac{i}{2\,f^2}\left(\sum_j \sqrt{c_j d_j}\right)^2 \qquad \textit{eq 6}$$

Using $f = 20$ yields $\lambda = 303.60$. So, in this case, $\lambda = A = 303.6$. Plugging this into eq 5 yields the same optimal values for Q as in Table 5-4 (i.e., 42, 59, 66, 132, 132). Note that as the value of λ goes up, the number of replenishments per year goes down.

The takeaway here is that setting a limit on the number of replenishments determines the effective order cost and vice versa. In general, the Lagrange multiplier represents the shadow price of the constraint. In other words, it shows how much the objective function (eq 2) could be reduced (for minimization or increased for maximization) if we relax the constraint by one unit.

Known OOP Cost

There may be situations when the true OOP costs provide a feasible order frequency. For instance, there may be a fixed price for shipping that is the same for 1 unit or for 100 units. Or there could be a fee to expedite, and so on. In such cases we could use the EOQ straight away with A equal to the OOP order cost,

$$Q_j^* = \sqrt{\frac{2\,A_j d_j}{i\, c_j}} \qquad \textit{eq 7}$$

but only if we do not need to consider a capacity constraint.

Likewise, in the next chapter, when we look at production batching, we find production situations where there are true OOP costs. In situations where we scrap a number of items when we change over or do a setup, there are real OOP costs. But, again, we should not consider a cost of capacity since we are solving for a situation with a given capacity and therefore have a constraint on capacity. This means that there are situations where we have both OOP costs and a capacity constraint.

Optimal Quantity with OOP Costs and an Order-Frequency Constraint

Suppose we had actual OOP order costs along with a constraint on the number of orders per period. In this case, for n products, we would have n OOP order costs, A_j, along with one order cost for the constraint, λ. The EOQ becomes

$$Q_j^* = \sqrt{\frac{2\,(A_j + \lambda) d_j}{i\, c_j}} \qquad \textit{eq 8}$$

And we substitute as before:

$$f(\lambda) = \sum_j \sqrt{\frac{d_j \, i \, c_j}{2\,(A_j + \lambda)}} \qquad \text{eq 9}$$

Unfortunately, eq 9 cannot be solved for the constraint order cost, λ, in closed form and so must be solved numerically. For instance, we could plot $f(\lambda)$ given above and see what value of λ yields the desired frequency.

As mentioned in the previous endnote, it is possible for the solution for λ to be 0. In this case, the A_j values are large enough to force the order frequency below the constraint. If this happens, we set the reorder quantities equal to the original EOQ formula.

For example, suppose there are OOP costs for Bubba's situation, and they are all $A_j = \$350$. If we use this value in eq 7, we get reorder quantities yielding an order frequency of 18.6, well below the requirement of 20 (see Table 5-6). So, there is no need for another order cost and $\lambda = 0$.

	c	d	A	Q*	f
Ash	100	100	350	44.7	2.2
2x12 Pine	250	500	350	63.2	7.9
Stud bndl	20	50	350	70.7	0.7
Util	10	100	350	141.4	0.7
1x4 Pine	100	1000	350	141.4	7.1
					18.6

Table 5-6: Optimal order frequency when A = \$350

However, if all the A_j values were \$250, using eq 9 and $\lambda = 0$, the order frequency would be 22.0 (see Table 5-7, Lamda1), which is not feasible. In the case without OOP costs, we found the single $\lambda = A = 303.6$ (see eq 6). So, in this case, since $A = 250$, $\lambda = 303.6 - 250 = 53.6$. If we recompute the order frequency for each item using eq 9 with this λ value, we get the desired order frequency of 20 (see Table 5-7, Lamda2).

		i	0.35			Lamda1	Lamda2
						0	53.6
Item		c	d		A	F(Lamda1)	F(Lamda2)
Ash	\$	100	100	\$	250	2.6	2.4
2x12 Pine	\$	250	500	\$	250	9.4	8.5
Stud bndl	\$	20	50	\$	250	0.8	0.8
Util	\$	10	100	\$	250	0.8	0.8
1x4 Pine	\$	100	1000	\$	250	8.4	7.6
						22.0	20.0

Table 5-7: Order frequencies when A= \$250

The Incredible Robustness of the EOQ Formula

Suppose we do not know the exact values of the parameters that go into the EOQ equation. The good news is that it does not affect the performance much. For example, if have an item that costs \$250 and has an inventory carrying cost of 35% per year. Each time we order, the shipping cost is \$20 for any order from 1 to 100 items. Our current forecast is for 150 per week, but suppose it is actually 75 per week. The holding cost per unit per week will be

$$h = 0.35 \times \frac{7}{365} \times \$250 = \$1.678$$

The true optimal quantity will be

$$Q^* = \sqrt{\frac{2 \times 20 \times 75}{1.678}} = 42.28 \approx 42$$

The true optimal cost will be

$$\begin{aligned} TC &= h\frac{Q}{2} + \lambda\frac{d}{Q} \\ &= 1.678 \times \frac{42}{2} + 20 \times \frac{75}{42} \\ &= \$70.95 \text{ /wk} \end{aligned}$$

But our forecast was 150 per week, making the order quantity

$$Q_f = \sqrt{\frac{2 \times 20 \times 150}{1.678}} = 59.80 \approx 60$$

The cost with this order quantity will be

$$\begin{aligned} TC_f &= 1.678 \times \frac{60}{2} + 20 \times \frac{75}{60} \\ &= \$75.34 \text{ /wk} \end{aligned}$$

which is 6% above the optimal. So, we made a 100% error in our forecast that resulted in costs that are only 6% above the optimal—pretty robust!

Now that we know how much to order, we need to know when to order. We don't want to order too early because that increases inventory but also not too late, for that will decrease customer service.

Determination of Optimal Inventory Investment

After seeing how to determine optimal order quantities in the previous section, the reader may be enthusiastic about optimizing inventory investment with optimal order quantities. However, *optimal inventory investment cannot be determined by reorder quantity alone*. We will cover the details shortly, but it is important to understand that optimal inventory investment is an interplay between how much to order and when to order.

When to Order

Notice that the safety stock, s, did not appear in the above analysis of *how much* to order. It does, however, have everything to do with *when to order*. Before we get to that, we need to describe the timing of replenishment orders.

We note that the primary driver of safety stock is variability in demand during replenishment time and that the reorder quantity and safety stock are interdependent. However, given the robustness of the EOQ, we can treat them independently and get good results. But if we really want to optimize, we need to consider ROQ and ROP simultaneously. This is beyond the scope of this book, but we can recommend Zipkin (2000), Section 6.5.3[ii], for an excellent algorithm that is truly optimal.

It is important to note that we do not monitor net inventory, $IN(t)$, to determine when to order. Nor do we monitor back orders, $B(t)$. Instead, we monitor inventory position, $IP(t)$, which is the on-hand inventory minus the back orders plus what is on order, $IO(t)$.

For external replenishment, on order includes all those purchase orders that are currently outstanding. But it is more than that. It includes all the purchase orders that have not yet been sent but that need to be sent due to the current inventory position (e.g., inventory position below the ROP but no one has yet put in a purchase order). It also includes all the material that has been received but is not yet ready for use (e.g., items not yet on the shelves, still in inspection, etc.). For internal replenishment, one order includes all the current work in process and anything finished but not yet available for use. We repeat the relationship between all of these:

$$
\begin{aligned}
IP(t) &= I(t) - B(t) + IO(t) \\
&= IN(t) + IO(t)
\end{aligned}
$$

We denote the average of any of these variables as the same symbol without the time dependency, namely, P, I, B, IO. Since the sum of an average is the average of the sums, the following average relationship also holds,

$$IP = IN + IO$$

The ROP is the level of the inventory position that generates the next order. Whenever the inventory position falls at or below r, we must place an order to bring it back to

$$r + 1 \le IP(t) \le r + Q$$

In daily execution, an organization's actual inventory position often deviates outside of inventory position limits. Nonetheless, the operative inventory position is completely under the control of the planner who is executing it. If the actual inventory position is below $r + 1$, place an order. If the actual inventory position is above $r + Q$, there should have been a good reason for this being the case. An example of a good reason occurred in 2025 when Apple chartered six cargo jets to ship 1.5 million iPhones from their production facilities in India to avoid tariffs. It is estimated they saved over $100 million dollars in tariffs by doing so.[iii] This was a reasonable approach given expectations of impending tariffs even though it probably put their iPhone inventory position far above the planned $r + Q$ level.

The system in Figure 5-4 and Figure 5-5Figure is executing according to the policy set, and we see that the inventory position never falls below $r + 1 = 16$ nor goes above $r + Q = 19$. So, even though we have back orders around three days into it, the system is under control, meaning that it is doing what the policy calls for. If there are too many back orders, then we need to modify the policy, not ignore the policy and do something different.

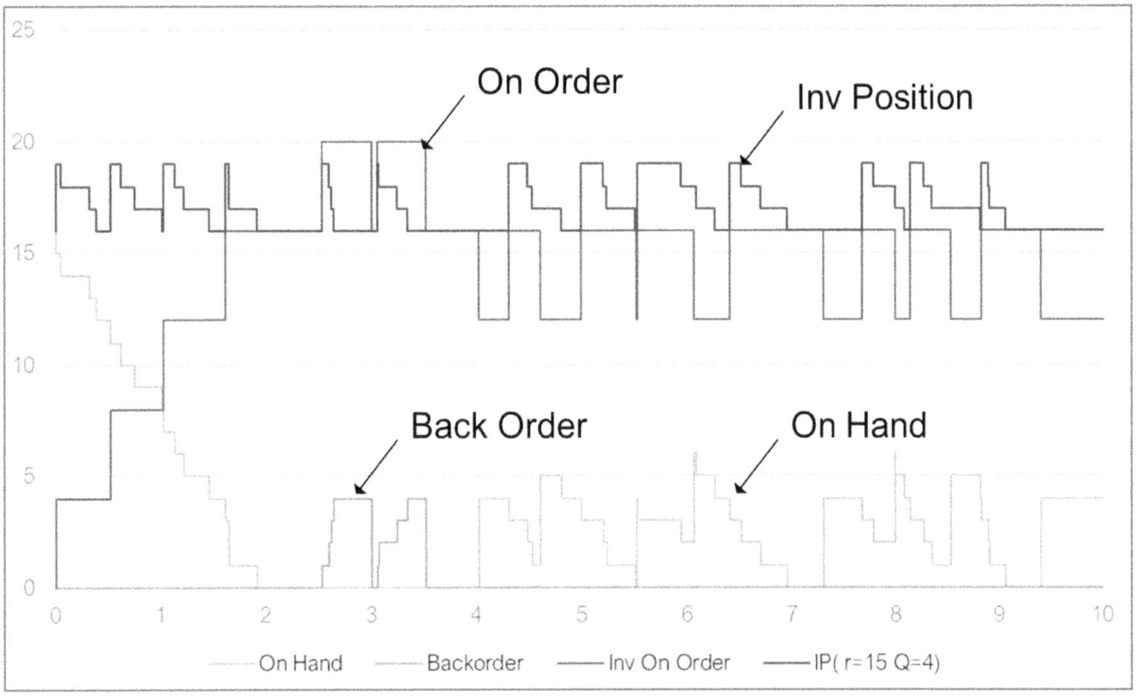

Figure 5-4: Inventory measures versus time (days)

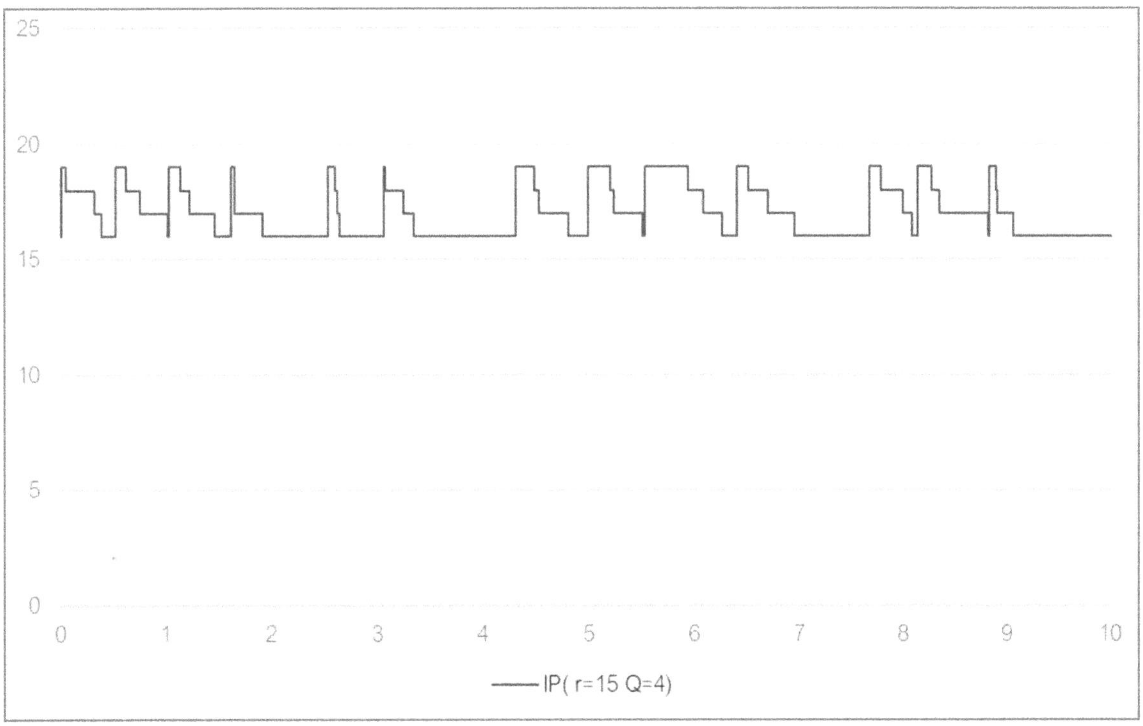

Figure 5-5: Inventory position by day

A special case of an ROP system called a base stock system, with $Q = 1$ and $S = r + 1$ so that $IP(t) = S$ for all t. Another way to look at the base stock system is for internal production where the on-order inventory is WIP, $w(t)$. Again, we keep the inventory position at a constant value, S:

$$S = IP(t) = IN(t) + w(t)$$

If the average demand is d, then from Little's law, we get the average time in inventory and time back ordered:

$$T_I = I/d$$

$$T_B = B/d$$

While the base stock system is simple, ordering one at a time can be tedious and time consuming. Thus, many base stock systems have a standard order quantity that is greater than one. The inventory position and on order are measured a multiple of these standard quantities. Kanban is a good example of a base stock system with a standard quantity greater than one.

If there is no variability, things work out very well. We can keep 0 inventory with a fill rate of 100%. For example, suppose the demand is always 15 items per day, and the replenishment time is always exactly 10 days. Then the average demand during the replenishment time will be 150 items (10 days × 15 items per day). If we divide by demand to obtain a plot of back order days and inventory days versus the base stock level, we get something like what we see in Figure 5-6. The optimal inventory base stock level is 150 whereby we carry no inventory and have 0 stockouts as well.

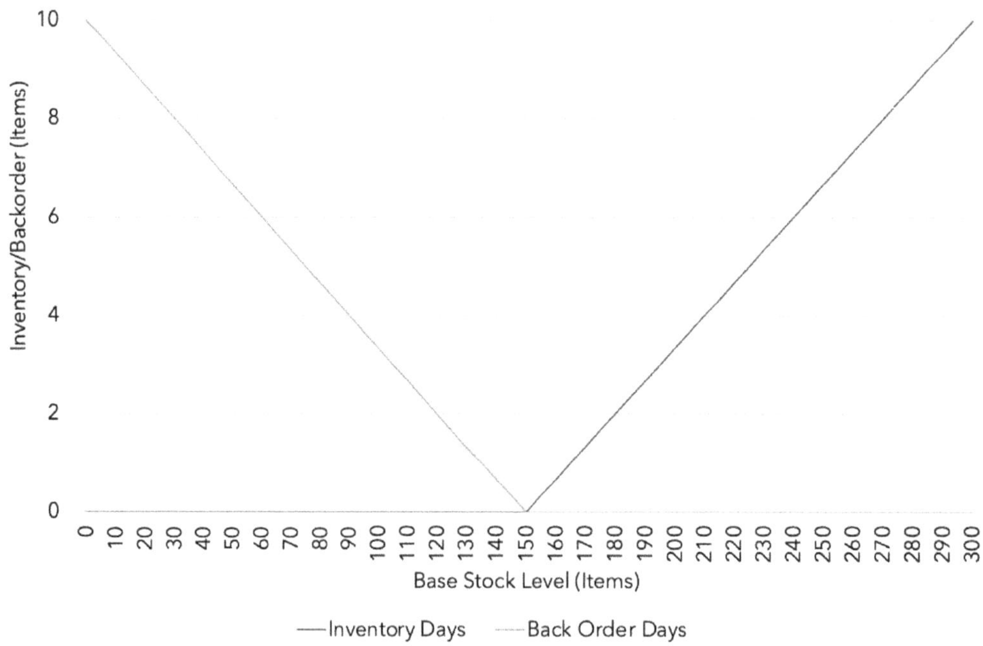

Figure 5-6: Inventory and back order days in a base stock system

But this is not very realistic. Here we have exactly 15 items of demand *every day* and the replenishment time is *always* 10 days. What if there were variability in demand and in the time to replenish? In such a case our graph looks very different, as is seen in Figure 5-7. Now the amount of base stock is not so clear. If we choose 150, we have around 1.3 inventory days and 1.3 back-order days. If we want to keep back-order days below one-half day so as to ship almost every day, we need a base stock level of 185, yielding about 3 days of inventory.

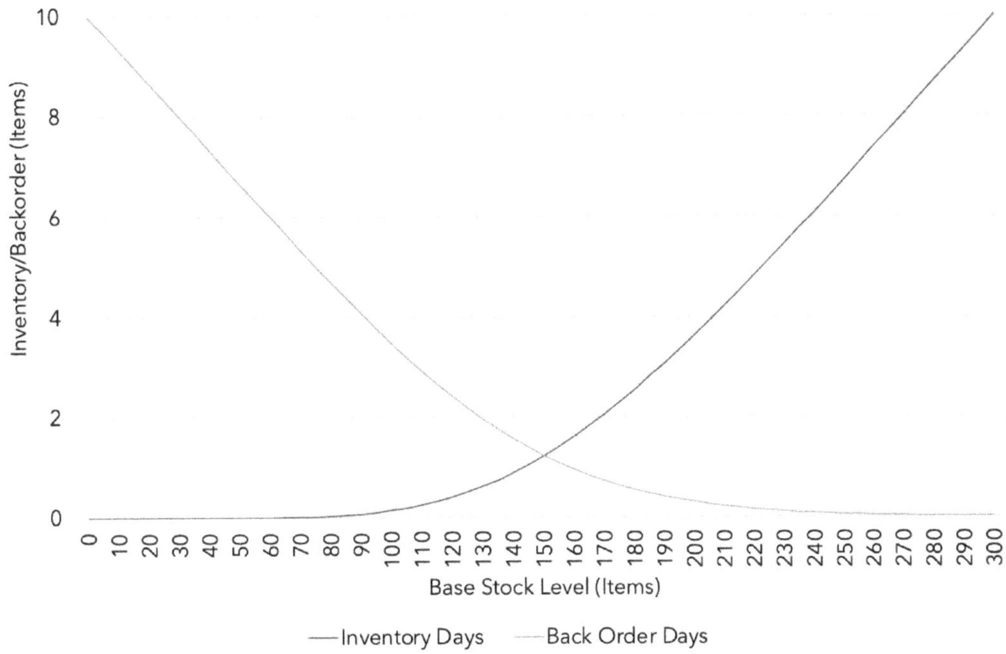

Figure 5-7: Base stock system with variability

So to preserve customer service we need to set the base stock level above the average replenishment-time demand. And this is the definition of safety stock,

$$s = r - d\,\ell$$

where r is ROP, d is demand, ℓ is average replenishment time for the item, and s is safety stock.

Reorder Points and Safety Stock

Some people on the internet use the terms *ROP* and *safety stock* interchangeably, which creates a great deal of confusion. And there is even more confusion about calculating safety-stock levels to guarantee a given customer-service level.

In Figure 5-3, we see that the average demand is 2 items per day (from 25 to 5 in 10 days, with 0 variability) and the safety stock is 5 items. If the replenishment time had an average of 5 days, the ROP would be $2 \times 5 + 5 = 15$.

Now consider another example that has variability in demand with a larger ROQ, as shown in Figure 5-8. Here we have random demand with a mean of 2 and a variance of 2 per day, while the replenishment time is always 5 days. The probability of having no stockouts can be calculated and is around 95%. Here, out of 10 cycles, we see only 1 cycle had any stockout, which would be 90% of the cycles. This is reasonable given the randomness in demand. Another 10 cycles might see no stockouts.

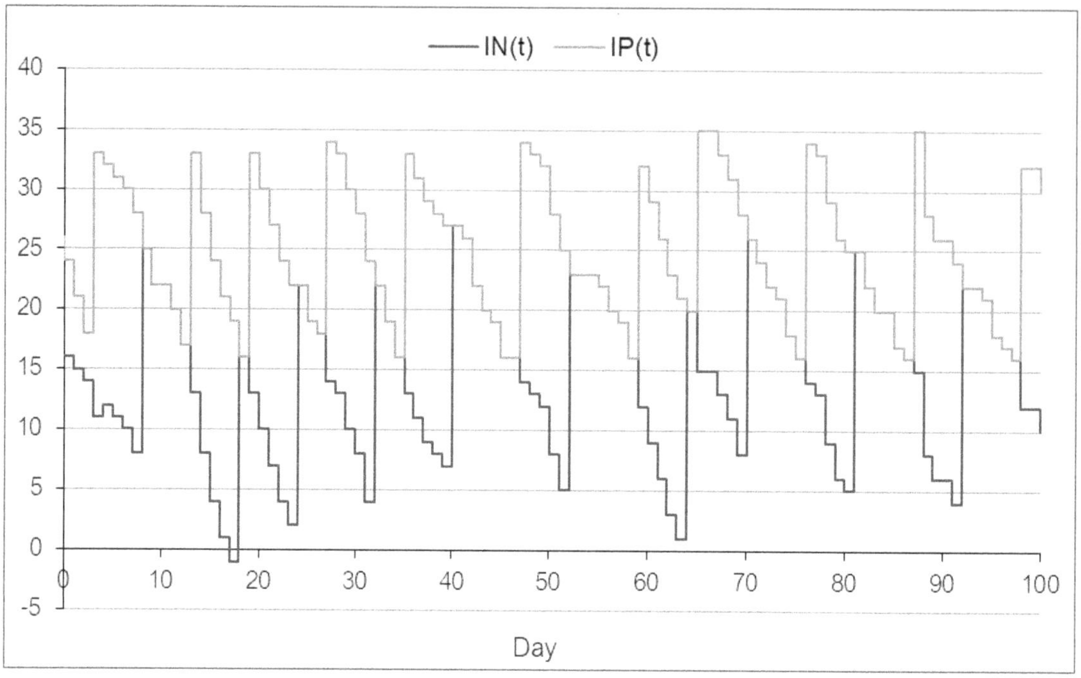

Figure 5-8: Inventory plot with r = 15, Q =20

But we should not confuse the percentage of cycles with no stockouts with the fill rate. The fill rate is what the customer sees, not what the planner sees. And the customer would see one stockout out of around 200 demands, a 99.5% customer fill rate compared to a 95% cycle fill rate.

There are many websites that compute safety-stock levels for a given *cycle* fill rate and present it as the *customer* fill rate. A popular safety-stock formula that appears on the web is

$$s = z_\alpha \, \sigma_{RTD} \, \{WRONG \; IF \; Q > 1!\} \qquad \text{eq 8}$$

where z_α is value that has a probability of $1 - \alpha$ of being greater than a standard normal random variable. So if the replenishment-time demand is normally distributed with a mean of θ and a standard deviation of σ_{RTD}, then the fill rate will also be $1 - \alpha$. But this is true only when the ROQ is 1, i.e., for a base stock system. If the ROQ is greater than one, then this level of safety stock will have a fill rate greater than $1 - \alpha$ along with more inventory than needed to achieve the desired fill rate.

Indeed, if you find a formula for safety stock that provides a given fill rate and a ROQ greater than 1, it is wrong! Why? Because there is no formula—only an algorithm as we describe as follows.

The reason an algorithm is required is that safety stock is dependent not only on the level of variability in demand and replenishment time but also on reorder size. For the same demand and replenishment processes and the same fill-rate requirement but with different order quantities, the safety stock is less for the larger order quantity.

A good check is to observe how often one stocks out. If you want a 95% fill rate, then about 1 out of every 20 orders should stock out and be back-ordered. This assumes demand is one at a time. If demand comes in quanta larger than 1, then the computed fill rate is only approximate and depends on if partial orders are fulfilled. Nonetheless, if you are *not* stocking out around 1 out of every 20, your fill rate is not

95% but something much higher. This is one reason why many companies have way too much inventory—they use the safety-stock formula for a ROQ of 1 but also use reorder sizes larger than 1.

Safety Stock and Safety Lead Time

There is also much confusion regarding safety stock and safety lead time. Safety lead time is extra time added to the average replenishment time, ℓ. Remember, the replenishment time includes the duration from the time the system hits the ROP until the replenished items are available for use (including time to recognize the need, prepare purchase orders, receive, inspect, etc.). From Little's law, we see that adding extra time, δ, to the average will increase on-hand inventory, making the effective safety stock, s_e, go up. If the average demand is d, then the effective safety stock will be

$$s_e = s + d \times \delta$$

Consequently, if we have safety stock, do not add safety lead time and vice versa. However, safety stock is more effective for variability in demand, while safety lead time is more effective for dealing with variability in delivery. But if we use both, we get more inventory than we had planned for.

Base Stock Systems

The base stock system has a ROQ of 1 unit (or standard container), and so the inventory position should always be at the same level, S. Consequently, we do not consider the ordering cost because we always order whenever we consume the standard quantity (which could be 1). Consider a base stock system with the following parameters:

$$d = 10$$

$$\sigma_D^2 = 10$$

$$\ell = 5$$

$$\sigma_\ell^2 = 1$$

What should our base stock level, S, be in order to have a fill rate equal to 95%? To answer this, we realize that we do not want our net inventory to fall below 0 with probability 0.95. To find this level, we first compute the mean and standard deviation of the replenishment-time demand:

$$\theta = d\,\ell = 10 \times 5 = 50$$

$$\sigma_{RTD} = \sqrt{\ell\,\sigma_D^2 + d^2\sigma_\ell^2}$$
$$= \sqrt{5 \times 10 + 100 \times 1}$$
$$= 12.25$$

The standard deviation is small enough that we can use a normal distribution. So we need to find S such that

$$FR = P\{D(T) \leq S\} = 0.95$$

$$P\left\{Z \leq \frac{S - \theta}{\sigma_{RTD}}\right\} = 0.95$$

where Z is the standard normal random variable with mean of 0 and standard deviation of one. We can look up the value for 0.95 (or use a spreadsheet function),

$$z_{0.95} = 1.645 = \frac{S - 50}{12.25}$$

Rounding yields $S \approx 70$ with a fill rate of 94.88%.

The average inventory and average back orders are a bit more complex. Let x be a random variable representing the replenishment-time demand with a mean of θ and a variance of σ^2_{RTD}. Back orders occur only when $x > S$. So to obtain the average B we multiply by the probability density function and integrate from S to infinity,

$$B = \int_S^\infty (x - S)\ f(x)dx$$

where $f(x)$ is the probability density function of the replenishment-time demand. If the standard deviation of replenishment-time demand is not too large compared to the mean, the normal distribution can be used. In this case, the average back orders is

$$B = \sigma_{RTD}[\phi(z) - z(1 - \Phi(z))]$$

where $z = (S - \theta)/\sigma_{RTD}$, $\phi(z)$ is the probability density function, and $\Phi(z)$ is the cumulative distribution of the standard normal distribution. Both ϕ and Φ are typically available in most spreadsheet applications such as Excel. The quantity in the brackets is $\Phi^1(z)$, the standard normal first-order loss function.

Once we have B, it is easy to compute the average inventory, since

$$\begin{aligned} S &= I - B + IO \\ &= I - B + \theta \end{aligned}$$

Solving for I yields

$$I = S - \theta + B$$

If we have the inventory carrying and back-order costs, we can write an equation for the total cost:

$$TC = h\,\bar{I} + b\,\bar{B}$$

The optimal back-order level is found by finding the value of z^* from

$$z^* = \Phi^{-1}\left(\frac{b}{b + h}\right)$$

where $\Phi^{-1}(p)$ the inverse of the standard normal CDF. Then the optimal base stock level will be the rounded value of

$$S^* = \theta + z^*\sigma$$

For example, suppose the carrying cost is \$5 per unit per day and the back-order cost is \$95 per unit per day. Then z^* will be

$$z^* = \Phi^{-1}\left(\frac{b}{b + h}\right) = \frac{95}{100} = \Phi^{-1}(0.95) = 1.645$$

which is the same as the previous example when we were given the desired fill rate, where $S = 70$. The average back orders are

$$\bar{B}(1.645) = 12.25[\phi(1.645) - 1.645(1 - \Phi(1.645))] = 0.256$$

And average inventory

$$\begin{aligned}\bar{I} &= s - \theta + \bar{B} \\ &= 70 - 50 + 0.256 = 20.256\end{aligned}$$

Then, the rounded optimal total cost will be,

$$TC^* = 5 \times 20.256 + 95 \times 0.256 = \$125.58$$

The normal distribution is interesting because the cost of the optimal solution can be written without first finding the optimal base stock level and is a function of only the cost parameters and σ_{RTD}:

$$TC^* = (b + h)\phi(z^*)\sigma_{RTD}$$

Using the same parameters as above, we get

$$\begin{aligned}TC^* &= (95 + 5)\phi(1.645) \times 12.25 \\ &= \$126.31\end{aligned}$$

And so the rounded solution is very close to the optimal,.

Now that we have found the optimal policy for base stock systems with a ROQ of 1, let us move to more general systems where the ROQ can be any integer.

Q, r Models

The base stock system places a replenishment order every time there is a demand. Obviously, this can be very inefficient. Bubba used a Q, r system that places an order for Q items whenever the inventory position hits or falls below the ROP, r, in order to bring the inventory position back to its desired level. Unlike the base stock system, which keeps the inventory position at a constant value, we showed that the Q, r system keeps the inventory position within a range of

$$r + 1 \leq IP(t) \leq r + Q$$

For purchased parts, we may obtain the value for Q using the EOQ method described previously.

So now we need to describe methods for computing the fill rate, average back orders, and average inventory for a Q, r system. Actually, it is quite easy once we have the formula for the base stock system. For the Q, r system we simply compute the value for each of the possible inventory positions and average. The possible inventory positions are: $r + 1, r + 2, \ldots, r + Q$. So for a Q, r system with normal replenishment-time demands whose average is θ and standard deviation is σ_{RTD}, we use

$$FR(Q, r) = \frac{1}{Q} \sum_{s=r+1}^{r+Q} P\left\{Z \leq \frac{s - \theta}{\sigma_{RTD}}\right\}$$

$$= \frac{1}{Q} \sum_{s=r+1}^{r+Q} \Phi\left(\frac{s-\theta}{\sigma_{RTD}}\right) \qquad \textit{eq 9}$$

In order to make the mathematics less complex, we consider normally distributed replenishment-time demands. This may not be a good assumption if the coefficient of variation of RTD, σ_{RTD}/θ is greater than 0.5. For such cases, we would need more complex first- and second-order loss functions. We can eliminate the sums in eq 9 by using the relationship:

$$FR(Q,r) = \left(\frac{\sigma_{RTD}}{Q}\right)\left[\Phi^1(z_r) - \Phi^1(z_{r+Q})\right] \qquad \textit{eq 10}$$

where $z_r = (r - \theta)/\sigma_{RTD}$ and $z_{r+Q} = (r + Q - \theta)/\sigma_{RTD}$. The quantity in the brackets is the standard normal first-order loss function.

Like the base stock system, we cannot simply write a formula to determine the ROP when we know the ROQ for a given fill rate. The only way to do this, in general, is to create an algorithm to search for the minimum value of r that meets or exceeds the desired fill rate. We present such an algorithm at the end of this section.

Recall that the safety stock, *s*, is given by

$$s = r - \theta$$

So if there is no formula to determine the ROP for a given fill rate then, obviously, there will be no formula to determine the safety stock for a given fill rate. If you see one on the Internet for $Q > 1$, it is wrong. To see this, consider an example with an average replenishment-time demand of 10 with a standard deviation of 3 and ROQ of 12 with a desired fill rate of 95%. The Internet formula gives

$$s = 1.645 \times 3 = 4.93$$

or $r = 15 \approx 10 + 4.93$, which results in a fill rate of 99.5% with 11.0 items of inventory, on average. But if we search for the minimum value of r such that

$$\frac{1}{12} \sum_{s=r+1}^{r+12} \Phi\left(\frac{s-10}{3}\right) \geq 95\%$$

we find that $r = 11$ results in a fill rate of 95.1% with an average of 7.58 items of inventory. The Internet formula (eq 8) results in 45% more inventory than is needed.

Why is this so common? We think the answer is that no one sounds the alarm when the fill rate goes up. With a fill rate of 95%, there should be one stockout for every 20 or so demands. No one complains when performance is better than this. But there is much consternation at the end of the quarter when inventory is too large. For anyone looking for predictable control leading to credibility, success, and pay raises, it is vital to understand the operations science of inventory. Understanding the expected results for a given performance envelope[iv] avoids both the daily beating that is received when there is not enough inventory and the monthly or quarterly beating that occurs when there is too much inventory. Operations science doesn't just improve planning; it protects people from fire drills and finger pointing. And it builds a foundation for sustained successful performance.

A quick sanity check is when the ROP is more than two standard deviations above the average replenishment-time demand, i.e.,

$$r > \theta + 2\,\sigma_{RTD}$$

Then it is probably too high, especially when the ROQ is much greater than one. A good way to find the optimal ROP is to start with $r = S^*$, where S^* is the base stock level that yields the desired fill rate and work down. The smallest value of r that meets or exceeds the desired fill rate is optimal.

Knowing the value of r that yields the desired fill rate, we find average back orders and average inventory by doing the same type of averaging. In the case for normal replenishment times, we use

$$B(Q,r) = \frac{1}{Q}\sum_{s=r+1}^{r+Q} \sigma_{RTD}[\phi(z(s)) - z(s)(1 - \Phi(z(s)))]$$

where $z(s) = (s - \theta)/\sigma$, we can make use of the standard normal second-order loss function,

$$\Phi^{(2)}(z) = \frac{1}{2}[(z^2 + 1)\Phi^0(z) - z\phi(z)]$$

Then

$$B(Q,r) = \left(\frac{\sigma_{RTD}}{Q}\right)\left[\Phi^{(2)}(z_r) - \Phi^{(2)}\left(z_{r+Q}\right)\right]$$

where $\Phi^{(2)}(z)$ is the standard normal second-order loss function. To avoid confusion with $\Phi^2(z)$ we put the superscript 2 in parentheses. Once we have $B(Q,r)$ we can obtain the average inventory:

$$I(Q,r) = \frac{Q+1}{2} + r - \theta + B(Q,r)$$

At the beginning of this chapter, Bubba wanted to use a Q, r policy with a target of 20 orders per year and an aggregate fill rate of 98%. The word *aggregate* is important because not all items will need a 98% fill rate. Nonetheless, we typically maintain a lower bound on the fill rate for any particular item. But allowing some items to have lower fill rates (but not too low) allows us to reduce the amount of money tied up in inventory. As it is, although Bubba's aggregate fill rate is 98%, individual items have fill rates that range from 54% to 100%, much like those in the Supply Corporation example.

For more clarity on the meaning and calculation of aggregate fill rate, we return to Bubba's initial efforts. In Table 5-8, we show each item's demand and fill rate for Bubba's initial policies.

Item	Demand (units/yr)	Fill Rate (%)
4X8 3/4" Ash	100	95%
2X12 10' Pine	500	100%
2X4 Stud bndl	50	54%
4X8 1/4" util	100	100%
1X4 10' Pine	1000	100%
	1750	98.38%

Table 5-8: Fill rate data

Note that taking the average of the fill rates listed will yield a fill rate of 90%, not 98.38%. This provides undue influence to 2×4 Stud bndl, which has the lowest fill rate but also the lowest demand. Instead, the aggregate fill rate is demand weighted and computed as

$$\text{aggregate fill rate} = \frac{\sum_j d_j FR_j}{\sum_j d_j}$$

Another consideration for fill rate is for assembly of items from inventory components. Consider that Bubba's items in Table 5-2 are all required for construction of some structure. They all must be on hand for the structure to be built. The assembly fill rate (AFR) is the product of the components' fill rate (CFR):

$$\text{AFR} = \prod_{1}^{n} \text{CFR}_n$$

where n is the number of components required for the assembly. In Bubba's case, the AFR = 0.95 × 1 × 0.54 × 1 × 1 = 51.3%. The AFR is the probability of having all required components available to complete the assembly.

Likewise, if a desired fill rate for an assembly of components is specified the required fill rate for each component should be

$$CFR = ASR^{\left(\frac{1}{n}\right)}$$

where n is the number of components required for the assembly. If Bubba wants a 90% AFR, the CFR = $0.9^{\left(\frac{1}{5}\right)} = 0.98 \text{ } or \text{ } 98\%$.

In these cases, consideration of component back-order time is key for buffering appropriately. It may be possible to allow much less than 100% fill rate for components if the back-order times are acceptably short. Suppose 3 days are allowed for kitting prior to assembly. Then, if back-order times are less than 3 days for every component, one would achieve 100% availability of components to start assembly.

The optimization of the base stock system suggests a method to optimize a Q, r system. We first find the optimal ROQ, using the methods described previously, with capacity constraints and, possibly, OOP costs. Then we find the ROP values using an equation like the base stock system, i.e.,

$$r_j = \theta_j + z\,\sigma_j$$

where σ_j is the standard deviation of the replenishment-time demand. We present an algorithm that searches for the best value of z that satisfies the fill rate constraints.

Optimal ROP Algorithm

Set FR_0 = target fill rate

Set $z = 3, z_{hi} = 6, z_{lo} = 0$, epsilon (eps) = 0.01

Compute θ_j, σ_j FOR ALL j

DO WHILE $z_{hi} - z_{lo} >$ eps

DO FOR ALL j

$r_j = \theta_j + z\,\sigma_j$

$FR_j = \left(\frac{\sigma_j}{Q_j}\right)\left[\Phi^1(z) - \Phi^1(z + Q_j)\right]$

NEXT j

$FR = 0$, SUMFR= 0

DO FOR ALL j

$FR = FR + d_j FR_j$

SUMFR = SUMFR $+d_j$

$FR = FR/SUMFR$

IF $FR < FR_0$ THEN

$z_{hi} = z$

ELSE

$z_{lo} = z$

END IF

$z = (z_{lo} + z_{hi})/2$

LOOP

We apply this algorithm to Bubba's data, and after 10 iterations we converge on $z = 0.366$ with a fill rate of 98.1% and a total inventory investment of \$18,483. These are the results shown at the beginning of this chapter.

Summary of Formulas

Here we summarize the different formulas for the different situations. First, the state variables (all are function of time, t):

$I(t)$ = on-hand inventory

$B(t)$ = back orders

$IO(t)$ = amount on orders

$IN(t)$ = net inventory = $I(t) - B(t)$

$IP(t)$ = inventory position = $IN(t) + IO(t)$

When the state variables are written as a function of the parameter instead of time are assumed to be the mean of the variable, e.g., B is average back-order level. Because fill rate is an average itself, we will never see an equation involving $FR(t)$.

Next, the input data:

d = average demand per period

σ_D^2 = variance of demand

ℓ = average replenishment time
σ_ℓ = standard deviation of replenishment time in periods
θ = the average replenishment-time demand = $d \times \ell$
σ_{RTD}^2 = variance of replenishment-time demand, = $\ell\sigma_D^2 + d^2\sigma_\ell^2$
i = inventory carrying-cost ratio per period

c = unit cost of the item
$h = i\,c$ = the holding cost of the item per period

b = back-order cost per item per period
A = OOP ordering cost

f = desired order frequency per period
λ is the shadow price of the order-frequency constraint

Next, the control variables that we will seek to optimize:

S = base stock level
Q = ROQ

r = ROP

s = safety stock = $r - d\,\ell$

Finally, we define some useful functions:
$\Phi(z)$ = the standard normal cumulative distribution function,
$\phi(z)$ = the standard normal probability distribution function,
$\Phi^1(z)$ = the standard normal first-order loss function = $\phi(z) - z(1 - \Phi(z))$
$\Phi^{(2)}(z)$ = the standard normal second-order loss function = $\frac{1}{2}[(z^2 + 1)\Phi^0(z) - z\phi(z)]$

Now for the formulas for different situations.

1. Order quantity when there is a constraint on the number of orders per period and no OOP costs.

   For one item:

$$Q = d/f$$

   For more than one item:

$$Q_j = \sqrt{\frac{2\,A\,d_j}{i\,c_j}}$$

where

$$A = \frac{i}{2\,f^2}\sum_j \sqrt{c_j\, d_j}$$

2. Optimal base stock level, given h and b

$$S^* = \theta + z^*\sigma_{RTD}$$

where $\Phi(z^*) = b/(b+h)$.

3. Fill rate for normal replenishment-time demand in a base stock system with base stock level, S:

$$FR = \Phi\left(\frac{S-\theta}{\sigma_{RTD}}\right)$$

4. Optimal fill rate, FR^*, given h and b:

$$FR^* = \frac{b}{b+h}$$

5. Average back-order level for normal replenishment-time demand in a base stock system with base stock level, S:

$$B = \sigma_{RTD}\,\Phi^1\left(\frac{S-\theta}{\sigma_{RTD}}\right)$$

6. Average inventory level for normal replenishment-time demand in a base stock system with base stock level, S:

$$I = S - \theta + B$$

7. Fill rate for normal replenishment-time demand in a Q, r system

$$FR(Q,r) = \left(\frac{\sigma_{RTD}}{Q}\right)\left[\Phi^1(z_r) - \Phi^1(z_{r+Q})\right]$$

where $z_r = \frac{r-\theta}{\sigma_{RTD}}$, $z_{r+Q} = \frac{r+Q-\theta}{\sigma_{RTD}}$.

8. Average back-order level for normal replenishment-time demand in a Q, r system

$$B(Q,r) = \left(\frac{\sigma_{RTD}}{Q}\right)\left[\Phi^{(2)}(z_r) - \Phi^{(2)}(z_{r+Q})\right]$$

9. Average inventory level for normal replenishment-time demand in a Q, r system

$$I(Q,r) = \frac{Q+1}{2} + r - \theta + B(Q,r)$$

Note that there is no formula for the ROP given a ROQ > 1 and a desired fill rate. But one can use the optimal ROP algorithm to determine the optimal r value(s).

Execution

This section will describe how to convert ROP and ROQ values to traditional MRP execution parameters that reside on most ERP systems. However, there are conditions that can be handled using MRP data and a simple spreadsheet. We discuss both situations in the following paragraphs.

Before we get into the execution, we should mention one important and dangerous artificial restriction that we have found in MRP systems: The order quantity must be at least the average demand during the lead time. In our notation,

$$Q \geq \ell \times d$$

But this assumes that we can have only one outstanding order, which is not true. This is important for long lead-time items. There can be multiple outstanding orders, so we do not need to set our order quantity to satisfy the previous equation. Indeed, the minimum order quantity should be the smallest order quantity that allows us to pay the negotiated price.

With this caveat in mind, we now consider execution procedures for two different situations regarding visibility to customer demand.

Execution with No Visible Customer Demand

If we are in a situation where we have no visibility to actual customer orders (e.g., we do not quote delivery dates but do expect to have inventory on hand when the customer asks for it), the implication is that we are working from a forecast only. Under these conditions, here is a recommended way to execute the optimal policy.

In this case we have a forecast for each item. The forecast may be a constant, or it could change over time. It is important to note that forecasts should be done at the replenishment time of the part. This is a common mistake in forecasting software. It is of limited utility to forecast and measure forecast accuracy 3 or 9 months out if the replenishment time of the item is 6 months. Whether or not visibility to actual orders is available, the average of the forecast per week for d is computed or estimated. For the variance, we estimate the mean, $\bar{X}$, and variance, S^2, of the historical demand. In addition, demand forecasting software packages often do not provide variance projections. We can estimate the variance of future demands by assuming that the variance-to-mean ratio stays fairly constant, hence

$$\sigma_D^2 = d\frac{S^2}{\bar{X}}$$

We use the MRP lead time for ℓ and and estimate σ_ℓ, depending on experience with the supplier. From the MRP system, we also obtain minimum order quantities, Q_{min}, and an order multiple, Q_{mult}. With these parameters, we compute the r and Q for all of the items using the optimal ROP algorithm.

Using MRP status data, we determine the on-hand inventory, $I(t_0)$, any back orders or allocations, $B(t_0)$, and all that is on order, $IO(t_0)$, where t_0 is today. Then we compute $IP(t_0) = I(t_0) - B(t_0) + IO(t_0)$. Then, if for any item,

$$r + 1 \leq IP(t_0) \leq r + Q$$

nothing needs to be done. But if the IP is below $r + 1$, we need to place an order. The order size should be

$$Q = \min_n\{Q_{min} + n\, Q_{mult} > r + 1\}$$

So the entire process can be converted to a single spreadsheet sorted by $IP(t_0) - r - 1$ and action taken for all items with a negative value. The execution is extremely simple and straightforward.

We also monitor the amount on hand and the percentage of demands filled from inventory and the fill rate, and then compare these to what is predicted by the model. If these are consistently far apart, the statistical analysis of the demand and replenishment times needs to be redone.

Execution with Visible Customer Demands

In this case we need to use the MRP feature in the ERP system. We will still compute the optimal Q and r but translate these into typical MRP parameters.

If the MRP system has a ROQ, ROP (i.e., Q, r) process, then populate these with the optimal parameters adjusted by the Q_{min} and Q_{mult} parameters. For example, suppose the optimal $Q = 333$, and the $Q_{min} = 100$ and $Q_{mult} = 25$. The MRP order quantity will be

$$Q_{MRP} = \min_{n}\{Q_{min} + n\, Q_{mult} \geq Q\}$$

and the result would be, $Q_{MRP} = 350$. Of course, this can be modified if desired. For instance, we might make the minimum quantity be $p\, Q$, where p is a number close to 1. That way we are not rounding up when we are very close.

If the MRP system uses safety stock, then compute the optimal safety stock as

$$s = r - d \times \ell$$

and round up.

Some MRP systems use something like days of supply (DoS) to determine the order quantity. For example, if DoS = 10 and the first uncovered[v] demand occurs 14 days out, then the MRP system will sum all the actual demand and/or forecast from day 14 to day 23 to create the replenishment order (subject to minimum and multiple requirements). To obtain the optimal DoS, we use

$$DoS = 7\frac{Q}{d}$$

and round up. This assumes d is average demand per week.

If the MRP system is using something like a min-max policy, we recommend caution. Most min-max policies look at on-hand inventory not, inventory position. If it is using inventory position, then the min will be $r + 1$ and the max will be $r + Q$, but it is often difficult to tell what the MRP system is doing by looking at the documentation or by asking the installation consultants because of the rampant confusion of the terms. If it is unclear what the min-max policy is doing, switch to a policy where the MRP mechanics are clearly understood.

Conclusion

In Chapter 3, we showed that there were only three buffers: time, inventory, and capacity. The relationship between two of these, time and inventory, is the subject of the current chapter, which focuses on minimizing the inventory buffer for a given time buffer. There are several possible situations.

Known Fill Rate and Order Quantity

If we know the desired fill rate, FR^*, and the order quantity, Q, then all we must do is find the minimum ROP that satisfies the fill rate:

$$\min\{r | FR(Q, r) \geq FR^*\}$$

The optimal ROP algorithm accomplishes this.

Known Fill Rate, OOP Cost, and Holding Cost

The total cost can be written as

$$TC(Q,r) = A\frac{d}{Q} + h\left(\frac{Q+1}{2} + r - \theta\right)$$

subject to

$$FR(Q,r) \geq FR_0$$

A heuristic solution is to use the EOQ formula, $Q^* = \sqrt{2\,A\,d/(h\,c)}$, to obtain the ROQ and then use the optimal ROP algorithm to find r.

Known Order Cost, Holding Cost, and Back-order Cost

In this case, the total cost has an additional term.

$$TC(Q,r) = A\frac{d}{Q} + h\left(\frac{Q+1}{2} + r - \theta\right) + (h+b)\left(\frac{\sigma_{RTD}}{Q}\right)\left[\Phi^{(2)}(z_r) - \Phi^{(2)}(z_{r+Q})\right]$$

where $z_r = (r-\theta)/\sigma_{RTD}$ and $z_{r+Q} = (r+Q-\theta)/\sigma_{RTD}$.

A heuristic would be to use the EOQ formula for Q^* and then use the optimal ROP algorithm to find r that gives a fill rate of $b/(b+h)$. This yields an approximate solution.

The previously mentioned methods and algorithms may seem daunting but once programmed into something like Excel using VBA to create the necessary functions, it is quite routine.

Having done so, one can create the efficient frontier plots described in Chapter 3 and shown in Figure 5-2. We use the EOQ formula to compute the ROQ value by searching for A that results in the desired order frequency. Then we use a range of z values for the plot. For each z value we compute the r_j values, which, along with the Q_j values, yield the total inventory investment (y value) and the resulting fill rate (x value).

In the next chapter, we will pull together the ideas from this and the previous two chapters to find policies that minimize the carrying cost of WIP and inventory, along with OOP costs, while meeting a given customer fill rate. This will take us into production batching which, though similar, is different from inventory batching.

---

[i] Both classical physics and optimization theory use Lagrange multipliers. In optimization theory, we always set the right-hand side (RHS) of the constraints to be 0 by subtracting whatever may have been the RHS before, e.g., something like, $x \leq 5$ becomes $x - 5 \leq 0$. For minimization problems, the constraints should be greater than or equal to 0, and for maximization problems this is reversed. The Lagrange multiplier is always positive or 0 and is subtracted from the objective function. In the case here, we have a minimization problem that starts with a less-than-or-equal-to constraint. To make it a greater than or equal to constraint, we multiply both sizes by -1, which then makes the λ coefficient in the new objective positive instead of the usual negative. If the constraint is tight (i.e., equal to 0), then the multiplier will be positive (or 0). But if the constraint is not tight, the multiplier will be 0. Thus, the product of the constraint value and the multiplier will always be 0.

[ii] Zipkin, P.H, [2000]. *Foundations of Inventory Management,* McGraw-Hill, Boston

[iii] Doloresz Katanich. "How Apple Is Airlifting Tonnes of iPhones to Avoid Trump's Tariffs." *Yahoo Finance*, accessed August 11, 2025, https://sg.finance.yahoo.com/news/apple-airlifting-tonnes-iphones-avoid-110211789.html.

[iv] A performance envelope is the expected average and variance of demand and replenishment time for items or services when determining operations strategy. We will have much more on this in subsequent chapters.

[v] *Uncovered* means demand in the system, either forecast or actual orders, that has no supply as on-hand or purchase orders to satisfy, or cover, it.

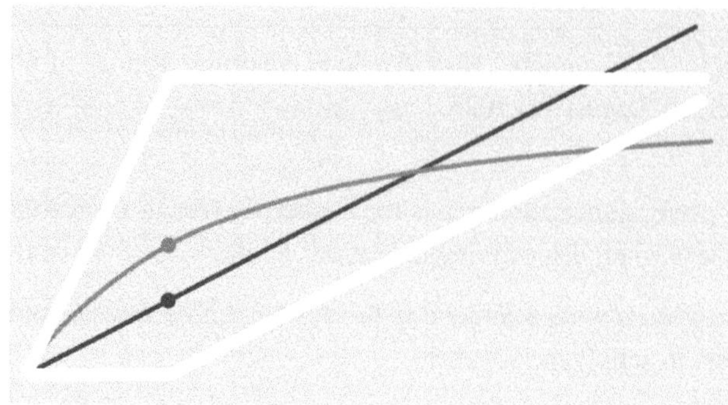

6

Pulling It All Together

Introduction

Chapters 2 through 5 provide a powerful, practical foundation of the science for managing operations and organizations in general. There is tremendous benefit to understanding the major operations science relationships on their own. However, we have spent decades in research and application of the concepts in practice and have seen the power of their use in a comprehensive framework. We now take a first principles approach to unite these concepts with a framework that can be applied in any operations to achieve repeatable, sustainable success.

In this chapter we tie together all the elements of an operation, as illustrated in Figure 6-1.

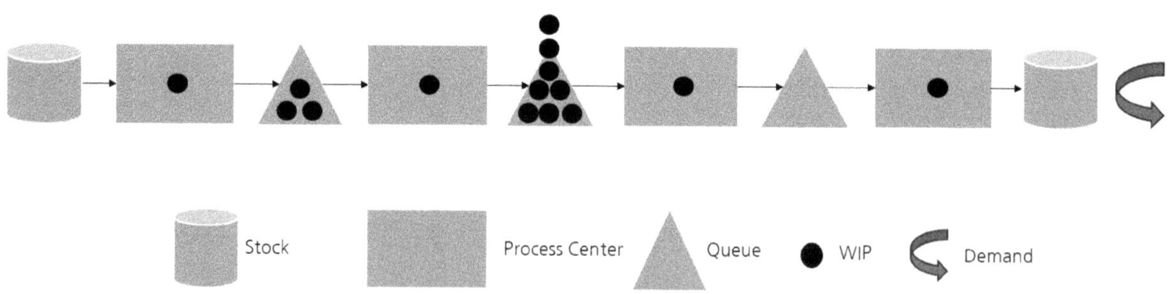

Figure 6-1: All components of an operation

The chapter provides an initial discussion about the importance of starting with first principles, especially in contemporary times, where AI holds great promise but also significant danger by atrophying independent thought. The first principles of operations science are bedrock concepts required for any predictable, sustainable approach to managing operations. The concepts should be embedded accurately in any AI model and, more importantly, embedded in managers' thinking so they know when AI is hallucinating or just blathering information. Since we are focused on operations, we then dig into the requirements for building a great operations plan based on strategic choices for the capacity, inventory, and time buffers to mitigate the inherent variability described in previous chapters. We demonstrate the practices of operations science planning and execution. A central technical concept connected to all elements of operations is batch sizing, so we provide the math and science of optimal batch sizing. We then provide a detailed example to tie all the buffers together in a total cost calculation and end with an operations science vignette in conclusion. This completes the foundational technical discussion and sets the reader up to move into Chapters 7 through 13 for more particular operations science application guidance.

First Principles

First principles are basic knowledge assumed to be true,[i] upon which all other knowledge is built. For example, the statement that inertial mass and gravitational mass are equal is a basic concept known as the Equivalence Principle and has been tested to the point where it is now assumed to be true. "Be lean and reduce waste" is not a first principle of operations, though we have seen many organizations take that as their first principle for improving performance and wander from one effort to the next, often with little success.

AI can intensify wandering through pseudo principles as people delegate their knowledge to AI. In a June 2025 *Wall Street Journal* article,[ii] Allysia Finley writes, "The brain continues to develop and mature into one's mid-20s, but like a muscle it needs to be exercised, stimulated, and challenged to grow stronger. Technology and especially AI can stunt this development by doing the mental work that builds the brain's version of a computer cloud—a phenomenon called cognitive offloading." Further, AI is fallible: "A professor at a large research university tells me that a student paper on the Bosnian War declared that 'brave Nazi soldiers were raped in enormous numbers.' Chatbots can say the darnedest things."

AI is a tool. Digital technology such as ERP systems and Manufacturing Execution Systems (MES) are tools. In another June 2025 article, [iii] Dennis Scimeca of *Industry Week* reports on the results of Rockwell Automation's 2025 "State of Smart Manufacturing" report. He writes, "Most manufacturers (81%) say they are accelerating their digital transformation considering current challenges, and 'deploying and integrating new technology' took the top spot for internal obstacles..." Organizations continue to grapple with so-called technology solutions. The report results also provide evidence that organizations recognize this disconnect between people and technology: "Most (83%) respondents in the 2025 report cite analytical thinking and communications/teamwork as the most important skills for new workers." Without a first principles understanding of operations, tools are often used to poor effect.

The first principles approach to operations provided here creates a practical, powerful method for establishing and sustaining high levels of performance for any organization. It is a Goldilocks[iv] approach that provides just the right amount of math and science to be accessible to anyone. Thus, no more than a high school education is needed for grasping the basic concepts. But for some of the technical discussion, one has seen (and will continue to see) the use of elementary calculus, basic statistics, and even a bit of renewal theory in order to provide a sound basis for the general methods we discuss in Chapters 7 through 13. Moreover, this approach provides the mechanics of operations science for those who desire to dig deeper as they implement these ideas in their organization.

Next, we describe a first principles approach for creating and executing a great operations plan.

Creating Great Operations Plans

We have introduced the power and comprehensiveness of operations science given its foundation in first principles. As we discussed in Chapter 3, the primary first principles of operations are

1. the operations science fundamental relationship (OSFR) between variability and the buffers
2. Little's law relating to throughput, cycle time, and WIP
3. the VUT equation relating queue time, utilization, and process time
4. variance of replenishment time demand and its effect upon inventory requirements
5. batch sizing and its relationship to cycle time and utilization

Before we start on developing a great operations plan, a few words to put the operations plan into perspective. As mentioned, business strategy drives operations strategy. The basic elements of any business management system are as follows:

Strategy: Plans

Tactics: Policies or actions

Controls: Methods or systems

We will explore these elements in more detail in Chapter 8.

In developing great operations plans, we review our discussion in Chapter 3 regarding the OSFR. Then we will provide descriptions of different elements belonging to a great operations plan. The essence of a great operations plan is to balance the buffers of capacity, inventory, and time in the face of variability to drive desired results.

Because of variability in both transformation and demand, the two will never be perfectly synchronized. Either the transformation will occur first (creating an inventory buffer), or the demand happens first and must wait (creating a time buffer). But, from what we learned in the last chapter, we can balance these two buffers by choosing a target inventory level such as a base stock level or a reorder point.

Where an inventory buffer is not available, e.g., in services such as health care or construction, a response time can be set for a given demand. Since inventory is not an option for services, of significant utility is the ability to set a capacity trigger as an alert to the need for additional capacity or, conversely, when demand is low, an alert that capacity is underutilized. We will discuss this in more detail shortly in the section titled "Operations Science Planning and Execution."

Returning to target inventory levels, a high target creates more inventory and fewer stockouts and vice versa. So, there are not two buffers here but one time-inventory buffer. And this buffer is represented by the standard deviation of the net inventory in a base stock system (e.g., the height of the "squiggles" in the following graphs). The value of this buffer is also equal to the standard deviation of the replenishment-time demand, $\Delta_{TI} = \sigma_{RTD}$.

Now consider two very simple but different systems, composed of a single resource producing a single item to stock using a base stock policy. Figure 6-2 and Figure 6-3 present plots of the net inventory for systems A and B, respectively. Systems A and B must satisfy an average of 10 units per day for demand. System A has a capacity of 12 units per day and the SCV of both demand and production is 1 while the capacity of System B is only 10.5 units per day, with a SCV of demand and production of 0.267. Both systems have a base stock level of 20.

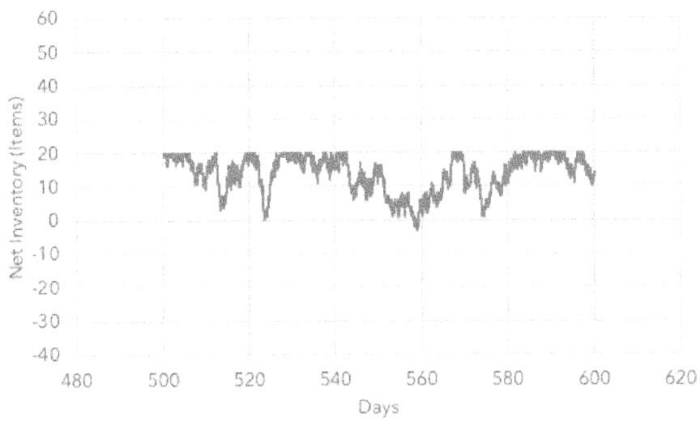

Figure 6-2: Base Stock System A with demand =10, capacity = 12, SCV = 1

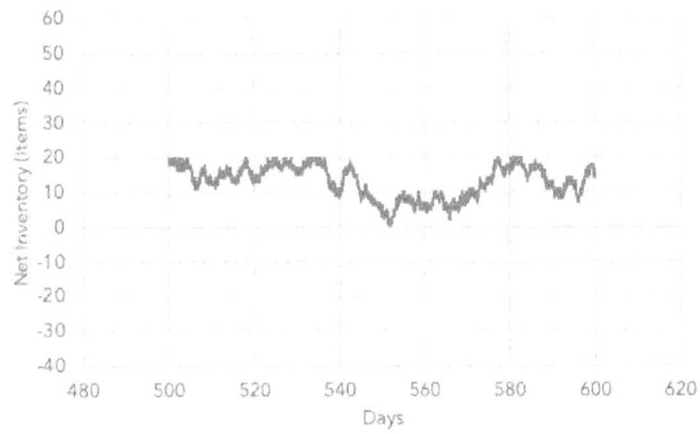

Figure 6-3: Base Stock System B with demand =10, capacity = 10.5, SCV = 0.267

Even though the graphs are not the same, they exhibit similar characteristics. Both stay close to the base stock level with significant dips at random times. Indeed, the graphs for average on hand and average backorders will be nearly identical (see Figure 6-4 and Figure 6-5) because they both have the same time-inventory buffer, i.e., the same standard deviation of replenishment time demand.

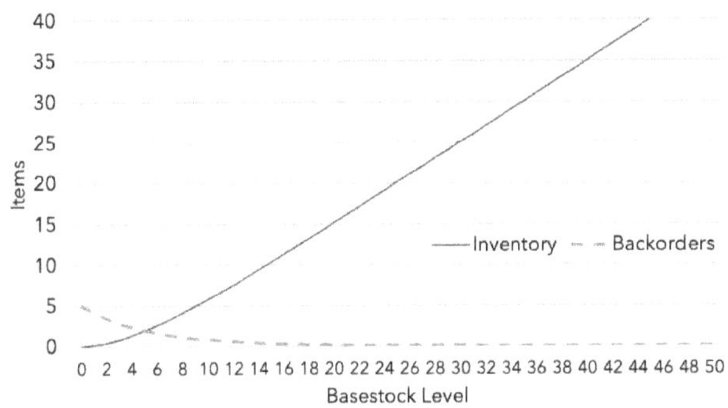

Figure 6-4: Average net inventory and average backorders versus base stock level for System A

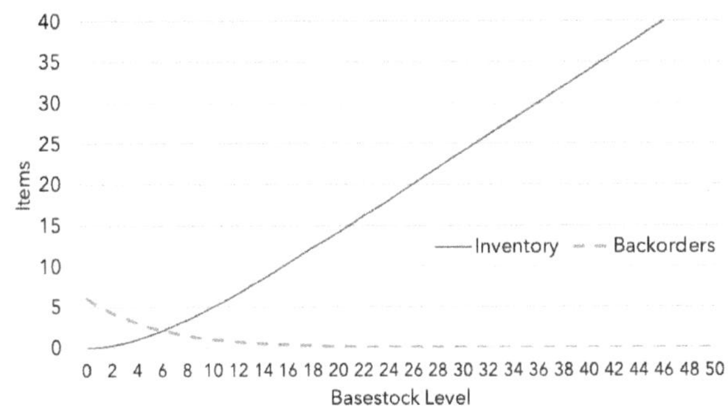

Figure 6-5: Average net inventory and average backorders versus base stock level for System B

This illustrates the Operations Science Fundamental Relationship (OSFR) that we can mitigate the effects of variability and randomness with extra capacity. So there is a different buffer, the capacity buffer, which is defined as the installed maximum capacity minus the average demand, $\Delta_r = r - d$. These two buffers are related to a measure of the average variability by the OSFR, namely,

$$\Delta_{TI} \times \Delta_r \approx \bar{V} = \frac{c_a^2 + c_e^2}{2} d \qquad \textit{eq 1}$$

The expression is approximate and is most accurate at high utilization when the capacity is not much greater than average demand. Recall that $\Delta_{TI} = \sigma_{RTD}$ is a dimensionless quantity, while Δ_r has dimensions of $1/t$ (e.g., units per week) as does $\bar{V}$. The expression for σ_{RTD} was presented in the last chapter and is messy, but we can approximate it using

$$\varDelta_{TI} \times \varDelta_r \approx \bar{V} = \frac{c_a^2 + c_e^2}{2}\, d \qquad eq\ 1$$

:

$$\sigma_{RTD} = \Delta_{TI} \approx \frac{\bar{V}}{\Delta_r}$$

If a_1 is the cost per unit of the time-inventory buffer (related to inventory carrying cost and backorder cost), and a_2 is the cost per unit of the capacity buffer, then the total cost can be written as

$$Z(\Delta_r) = a_1 \frac{\bar{V}}{\Delta_r} + a_2 \Delta_r$$

Taking the derivative with respect to Δ_r, setting the result equal to zero, and solving for Δ_r yields the optimal value of the capacity buffer, Δ_r^*:

$$\Delta_r^* = \sqrt{\frac{a_1}{a_2} \frac{(c_a^2 + c_e^2)\, d}{2}}$$

Once we have the optimal capacity buffer, we simply add the average demand to obtain the optimal capacity, $r^* = d + \Delta_r^*$,

$$r^* = d + \sqrt{\frac{a_1}{a_2}\, \bar{V}}$$

When we substitute into the total cost equation we find

$$Z(\Delta_r^*) = \sqrt{a_1\, a_2\, V_d} + \sqrt{a_1\, a_2\, V_d}$$

where the first term is the optimal time-inventory cost and the second, the cost of extra capacity. Since the two terms are equal, this has a significant impact on what it means to be efficient. A common misconception is that higher utilization is always better, but this shows there needs to be a balance between high utilization and the time-inventory cost.

Recall that if there were no variability, one could set the takt time equal to the reciprocal of demand and have no inventory, no backorders, and no extra capacity. Capacity would be exactly equal to demand. Thus, the cost of variability is equal to the cost of the extra capacity needed to achieve optimal operations. And when there is variability for simple systems, we should see the optimal inventory backorder cost approximately equal to the cost of the extra capacity.

Unfortunately in the real world, things are not so simple. We seldom have a single product made on a single resource. Moreover, the utilization of the resources depends on more than just the speed of the resource. Recall from Chapter 4 that utilization also depends on setup times and batch sizes. As an example of the impact of complexity, when setting production batch sizes, one must consider the utilization of one or more resources, and so it is a bit trickier than setting reorder quantities for an inventory system where there is a simple reorder frequency constraint. Nonetheless, we can use much of the mechanics developed in Chapter 4 to address process batch sizing as we do later in this chapter. All this said, a key insight of operations science is that balancing the buffers of capacity, inventory, and time in the face of variability drives desired results. In complex cases, the cost of the time-inventory buffer and capacity buffers may not be equal, but the science shows that a little buffer goes a long way. Attempting to eliminate one buffer entirely (e.g., zero capacity buffer by attempting 100% utilization) can lead to disastrous results.

As we discussed in Chapter 2, there is an extensive history of management best practices, continuous improvement efforts, and academic models to master operations behavior in all types of businesses from manufacturing to project management and services. Many become recipes that are used for all situations whether they are appropriate or not (remember Boeing's moving assembly line). But by understanding basic operations science, much of the confusion and wasted effort could have been avoided. For example, Toyota could have compressed development of its famed production system from 30 years to about 5 years with an operations science approach.

Great operations plans start from a foundation of first principles. Of course, more than first principles are required for a great operations plan. A list of additional required elements is provided as the Operations Science Application Summary Table (Table). More detail on the elements is provided in Chapters 8 through 13 and a non-exhaustive list of these elements is included in summary tables for the application examples in Chapters 15 through 20. Next, we will explore these elements in the context of Toyota's effort at developing its own business management system.

Operations Science Application Summary Table

ELEMENT	DESCRIPTION
BUSINESS STRATEGY	The organization's plans for products, services, projects and their respective markets
OPERATIONS STRATEGY	
DEMAND	Segmentation, average, variance, cost, price
TRANSFORMATION	What, how, who
VARIABILTY	Sources, level
BUFFERING OPTIONS: CAPACITY	How much, configuration, control methods
BUFFERING OPTIONS: INVENTORY	Order frequency, desired fill rates, turns
BUFFERING OPTIONS: TIME	% on time, average backorder time when backordered, customer expectations
TOOLS AND APPLICATIONS	Utilization and capacity buffer size Batch sizing—process batches and move batches WIP control *Q,r* or time-phased inventory policy Inventory policy Virtual queues and due date quoting

Table 6-1: Elements of great operations plans

What if Toyota Had Started from First Principles?

The history of the Toyota production system (TPS) and Lean provide a great example of the power of first principles. Taiichi Ohno, in his book, *Toyota Production System: Beyond Large-Scale Production* (1978 in Japanese), describes the development from 1947 to 1971[v] of Toyota's system using trial and error to develop the essential tactics and methods. In 1982, Richard Schonberger published *Japanese Manufacturing Techniques*, one of the first books in English describing Toyota's system. Since that time, many more books have been written, with one of the more popular being *The Machine That Changed the World* by Womack and Jones (1990). In a later book, *Lean Thinking,* (1996) Womack and Jones popularized the term *Lean*. Although Lean is a more general concept than the TPS, it has many of the same features and requirements. What if Ohno had been able to start with a first principles approach, as opposed to using trial and error for 25 years? What if Lean development from the TPS had been based on first principles instead of on initiative by imitation?

Starting from First Principles

If Taiichi Ohno had understood the first principles of operations science, it would not have taken him 24 years to develop his system. This lack of understanding may be why the vast majority of subsequent Lean initiatives have failed (greater than 70%, according to the Lean Enterprise Institute[vi]) and would have succeeded with a first principles approach. Indeed, what we often hear from clients is usually something like "We are not like Toyota;

- we are a high-mix, low-volume operation with thousands of unique SKUs, or
- we treat people, we don't make parts, or
- we build massive one-off structures like buildings or bridges,

moving assembly lines are out of the question!"

So, why are the majority of Lean implementations not successful? The Internet is filled with explanations from lack of leadership to dumping the consultants too soon (per a consultant's website). We think there is a simpler explanation: If you do not understand how operations work, you will not be able to design and control a good operation. The same principle applies to designing cars, health care services, restaurant services, cameras, and nuclear reactors. And given the experience we described earlier in this book (Chapter 2, Figure 2) in which large groups of seasoned managers could not describe the basic relationship between WIP and cycle time, we know that most managers do not understand the basic workings of operations.

Let's apply operations science principles to TPS and see if we can improve its efficiency for moving assembly lines and increase its applicability to almost any operation including hospitals, large capital construction projects, maintenance repair and overhaul (MRO), and other services. This has been the goal of Lean, but, as we see from the track record, the application of Lean to a broader scope of operations has had limited success. We first consider means to generalize and improve manufacturing operations.

The Virtual Queue

A key concept of TPS is the use of takt time production. The takt time is the time per station for the moving assembly line and is, essentially, the reciprocal of the demand. So if the plant is scheduled for 310 production days per year, with each day comprised of two 10-hour shifts, there are 22,320,000 seconds available per year. If the demand forecast is 540,000 units per year, the takt time is 22,320,000 / 540,000 ≈ 41 seconds.

From an operations science perspective, we see that, by using takt time to release into the line, Toyota has prevented all variability of demand from coming into the line. In other words, the SCV of arrivals (c_a^2) is zero for all the work stations resulting in minimal queueing. But where did the variability go? It must go somewhere because it does not go away. The somewhere is what we are calling a virtual queue (VQ). See Figure 6-6.

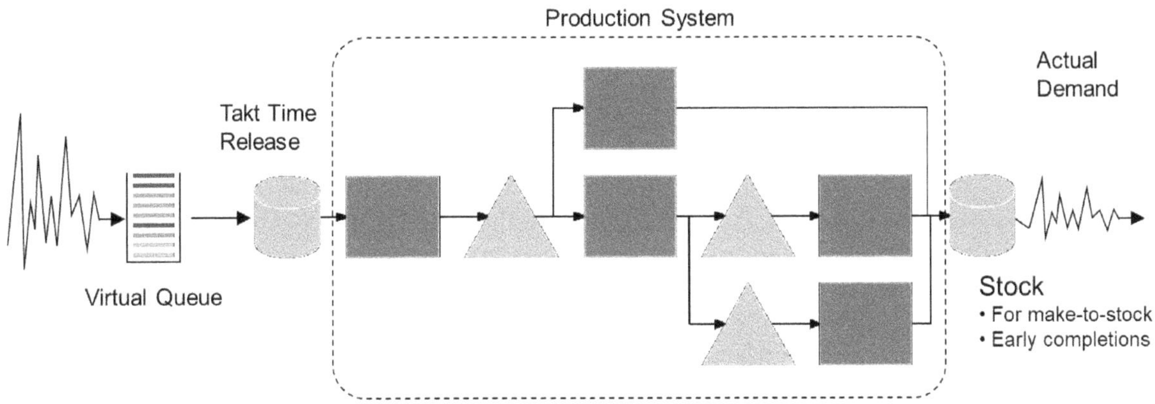

Figure 6-6: The TPS virtual queue

Although the virtual queue is never mentioned in either TPS or Lean literature, it must exist because the average demand is never exactly equal to the reciprocal of the takt time. If the time between demands is shorter than the takt time, the virtual queue increases; if longer, the virtual queue decreases. By monitoring the virtual queue, Toyota can tell when they need to rebalance the line, i.e., reset the takt time. They either make it shorter to handle more demand or longer to handle less. The virtual queue concept is powerful and can be used to predictively determine when items and tasks will be completed. It also helps monitor and manage the impact of capacity on time.

It is vital to note that this also applies in all types of operations including services, such as health care and MRO, and in project management.

CONWIP, Not Kanban

Instead of kanban, which requires a "card" for every item at every work station, we use CONWIP, which keeps a more or less CONstant a generic measure of WIP (CONWIP) for an entire flow (i.e., multiple parts over multiple work stations). Thus, when the product mix changes, we do not need to change the numerous production cards for each part at every station: We need only change the aggregate maximum WIP level.

Moreover, CONWIP flows are always more efficient than a moving assembly line. Why? Because the bottleneck of a moving assembly line is not the slowest work station but the takt time itself. To see this, consider what would happen if the takt time were shorter than the average process time of any work station. Work would come into that station at a rate faster than the capacity of the station, and the line would have to stop for that work station to catch up. And, if the line stops for one station, it stops for all the stations until it is restarted. The CONWIP line will naturally run near the speed of its bottleneck because any WIP above the critical WIP will, according to operations science, tend to accumulate in front of the bottleneck.

Few-Piece Flow

Unlike the TPS, a CONWIP line does not need to have one piece flow where there is one piece at each work station. One piece flow is good because it reduces WIP, which reduces cycle time. However, with real-world variability, the danger is that WIP is reduced so low as to constantly starve the bottleneck and decrease throughput precipitously. This reduces efficiency and can impact customer service. Back in Chapter 4, we demonstrated that a few-piece flow is as good as one piece flow for reducing cycle time, while having higher throughput than one piece flow. Indeed, a few-piece flow line achieves the desired short cycle times with less installed capacity than a one piece flow line.

The goal of one piece flow is to have batch sizes of one. To facilitate this, Toyota developed SMED—single-minute exchange of die. The dye is in a punch press and the *single minute* refers to single digit minutes (i.e., less than 10 minutes), not strictly one minute changeovers. So, if short enough changeovers (or setups) can be achieved, batch sizes of one are feasible and, with them, near perfect flow.

But as with most absolutes, there are exceptions. First, not all processes are stamping presses with dies. Some are actually batch processes, such as a heat treat operation, where many parts are loaded into an oven at one time. And, even with sequential processes, it is not always possible to have a short changeover, such as when it takes a number of adjustments before the first part is made to specification. Another example is when lab calibration runs are required. Other cases include labs that require test batches for certification or, in the case of a hot press, when the tool is too hot for an immediate changeover. Nonetheless, changeover times should be reduced as much as possible.

But with operations science, we can optimize production batch sizes in much the same way as we optimized reorder quantities but considering available capacity. Later in this chapter, we present several of these methods to find the optimal batch size, which may or may not be one item.

Less Capacity Buffer

The TPS relies on "two-shift scheduling" and a moving assembly line. This combination results in a fairly large capacity buffer. In Schonberger's book, *World Class Manufacturing*[vii], he describes how Toyota and Japanese manufacturers went to two-shift production work schedules of 12 hours per shift with 8 hours of work scheduled and 4 hours of time available *in each shift* for preventative maintenance and process improvement. This is a 33% capacity buffer! Not all manufacturers use the same schedule and shift schedules vary within companies, but let's assume Toyota scheduled 10 hours of work in a 12 hour shift. According to operations science, this utilization is

$$u = \frac{t_{used}}{t_{avai}} = \frac{10}{12} = 0.833$$

But the takt time must be somewhat longer than the longest average task time at any of the work stations. If the line were perfectly balanced (i.e., all the stations had the same average time), the takt time would still need to be longer than that average.

Suppose we have a 10 station line and need to have a 95% probability that all the process times are less than or equal to the takt. Then the probability for any one station must be

$$0.95 = p^{10}$$

and $p = 0.995$. If the times are normally distributed, $z_{0.995} = 2.58$. If the average were 45 seconds with a standard deviation of 2 seconds, and we need 2.58 sigmas of extra time, the takt time becomes

$$tt = 45 + 2.58 \times 2 \approx 50 \text{ sec}$$

Then the utilization would be 45 / 50 = 0.9. The overall utilization will be the product of the scheduled utilization and the station utilization:

$$u = 0.833 \times 0.9 = 0.75$$

which is quite low. We typically suggest utilizations of between 0.85 and 0.90, and in CONWIP lines, where the WIP cannot "blow up," we can live with even higher utilizations. However, it is situation dependent, such as when first installing an additional, large-capacity resource in order to meet increasing demand.

The remarkable thing about the TPS is that it has very low inventory and is always on time. But Toyota pays for this low time-inventory buffer with a very large capacity buffer. If the cost to make a unit is around \$15,000, and 20% is labor, with another 10% for tool depreciation, then the unit capacity cost would be \$4,500. With a utilization of 0.75, the cost of the capacity buffer will be

$$c_r\Delta_r = \$4500 \times 1440\frac{\text{units}}{\text{day}}(1 - 0.75) = \$1.620 \text{ million/day}$$

If using CONWIP allowed a 90% utilization, the cost would be

$$c_r\Delta_r = \$4500 \times 1440\frac{\text{units}}{\text{day}}(1 - 0.9) = \$0.648 \text{ million/day}$$

That's a savings of almost \$1 million per day.

Operations Science Planning and Execution

The previous review of the key elements of operations science explaining the TPS and ways to implement Lean more efficiently and in more general environments demonstrates that we have all the components needed for successful planning and execution. We now pull it all together as an operations science planning and execution framework that applies generally across manufacturing, service, and project management settings. Note that we do not mention scheduling: We have replaced it with a set of execution rules that have feedback mechanisms. We do this because optimal scheduling algorithms do not exist for realistic situations.

Execution Without Detailed Scheduling

While many ERP vendors offer some sort of detailed planning and scheduling module, they all suffer from the same two issues. First, they purport to provide an optimal schedule. But this is not possible, given the number of variables and complexity of the problem. Such problems are said to be NP hard meaning that no so-called polynomial solution (i.e., one that has reasonable computation times) exists for the problem.[viii] The second issue is variability and randomness. Almost all commercial scheduling solutions assume fixed process times with no randomness. And we all know that if there is no variability, then, by the VUT equation, there is no queueing at all so long as the utilization is less than one (100%). And if the utilization is above one, the average queue continues to grow until additional capacity is added. So, the idea that we can find an optimal schedule for a realistic situation with random demands and random process times in a reasonable time is not possible.

Instead, we will use operations science concepts to create near optimal situations and use a set of execution rules to control the system. We demonstrate this idea by further describing the CONWIP system that controls WIP and determines when to start new work orders.

Constant Work in Process (CONWIP)

CONWIP is an operations control system that maintains a, more or less, constant level of WIP in the process flow. In CONWIP we refer to incoming work as jobs regardless of whether they are for production, construction, or services. The measure of a job is the time required to perform it at the bottleneck of the system. That way we can control systems that have jobs with widely disparate process times.

The CONWIP level is the maximum amount of WIP that allows new jobs to be released. It should be set at a level that will keep the bottleneck process busy most of the time (e.g., a 90%–95% utilization of the bottleneck). If the WIP is currently above the CONWIP level, no more work can be released, and any new jobs will go into the VQ. Figure 6-7 presents the basic components.

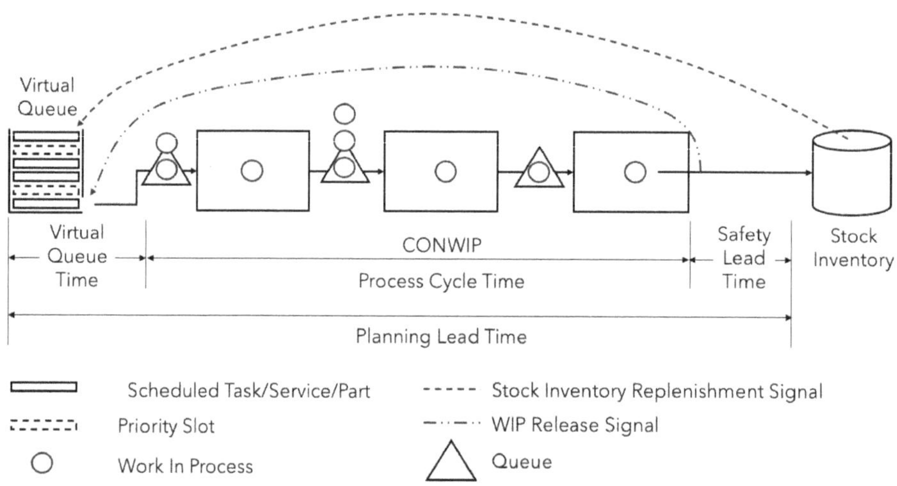

Figure 6-7: A CONWIP operation

As can be seen from the figure, a CONWIP process line is not dissimilar to a moving assembly line. The difference is that there is no moving conveyance, and so the WIP moves from station to station once it completes a process. This is important because the WIP will naturally accumulate in front of the bottleneck of the line, thereby providing protection from starvation and loss of throughput. Consequently, a CONWIP line can run at a higher utilization than a push system because there is no chance of having a WIP explosion since the WIP is capped. Moreover, if the product mix changes and a different work station becomes the bottleneck, the WIP naturally will migrate to the new bottleneck. All this happens because of the natural behavior of the system and not because of any scheduling.

So, instead of blowing up the bottleneck, a burst in demand might blow up the virtual queue. But that is what the virtual queue is for—to monitor demand and determine whether more or less capacity is needed. So, instead of discovering that a particular batch will be late at the last minute, the virtual queue shows that there is a problem while there is still time to remedy it.

New jobs arrive in a virtual queue whether they are generated by customer orders, new service requests, the master production schedule, or a construction schedule. The queue is called virtual because it does not contain physical entities. This is convenient because the VQ can be used for demand from customer orders along with orders to replace stock (e.g., safety stock orders). The VQ will also be used to signal when there is a need to change existing capacity—a capacity trigger. It is best to have some flexible capacity, such as an extra shift or extra workers, when needed. Then, if the VQ goes beyond a given limit as indicated by the quoted due dates, we can add this flexible capacity. If the VQ continually runs out of work, we may need to reduce capacity.

The planning lead time shown in Figure 6-7 is composed of the virtual queue time, operations cycle time, and safety lead time. The average operations cycle time is, from Little's law, the CONWIP level divided by the throughput of the line. Variability in the process is dealt with using both the virtual queue time and safety lead time. For inventory stocking orders, a planning lead time (PLT) is used and is given by,

$$PLT_{stock} = \frac{CONWIP + SS}{d}$$

where CONWIP is the maximum WIP level and SS is the safety stock as discussed in the previous chapter. As before, SS/d represents the safety lead time.

A due date quoting mechanism is provided to ensure that specific jobs finish within the quoted lead time and is described more fully in the following section.

In order to fill orders from demand with a CONWIP line and a VQ, we must decide upon three parameters: the level of service desired (i.e., the probability of jobs being finished on or before their due date), the CONWIP level, and the capacity trigger. The capacity trigger is the indicator that extra capacity will be needed in order to finish jobs on time. The trigger can be simply when the level of the VQ is exceeded or when the quoted lead time creates a due date too far into the future.

Having a capacity trigger is not that dissimilar from Toyota's two-shift scheduling, where they schedule 10 hours of work for every 12 hours. The extra 2 hours are used to make up for what was not done during the time the line was shut down for quality reasons. That way they always make the same number of cars in every 12-hour period. But, if no problem is detected, the 2 hours are still available but not needed. That does not happen when using a VQ with a capacity trigger. If the extra capacity is not needed, the extra period is not used.

Now, consider a four-station line with average process times of 2, 2.5, 1.5, and 1.5 hours with standard deviations equal to their means. Thus, the RPT = 7.5 hours and the BNR = 0.4. We must meet an average demand of 0.36 units per hour for a utilization of 90% (0.36 / 0.4). Using the algorithm in the appendix of this chapter, we can approximate the average cycle time and throughput for any WIP level. Figure 6-8 shows that we would just meet demand with a WIP level of 8 jobs corresponding to a throughput of 0.363 jobs/hr.

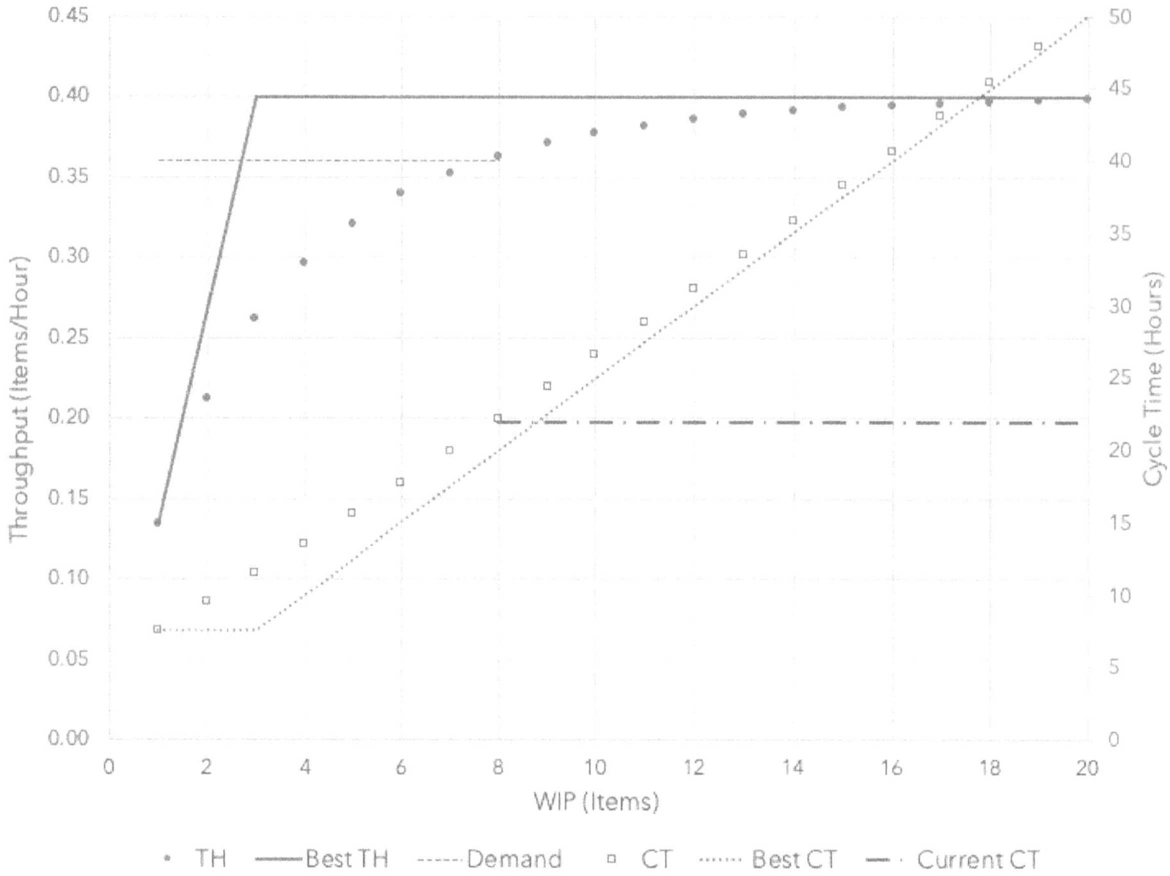

Figure 6-8: Flow Performance Graph

But just meeting demand will not do if there is variability. This is similar to what we saw with the VUT equation, only in this case, the entire CONWIP line is the resource.

$$CT_q = VUT = \frac{c_a^2 + c_e^2}{2} \frac{u}{1-u} t_e$$

As the utilization goes to 1, the queue blows up. Figure 6-9 illustrates what happens in a Monte Carlo simulation when we try to run with just enough WIP and presents the virtual queue and the amount of WIP for a period of 950 days with 16 hours per day. Here the virtual queue reaches a maximum of 101 jobs. Given the bottleneck takes 2.5 hours and we have 8 units in WIP, the 109th job will not finish for 17 days.

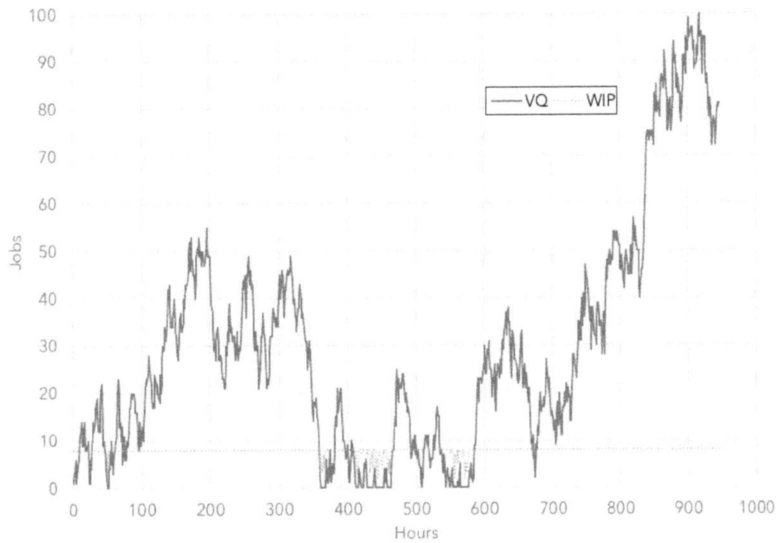

Figure 6-9: WIP and VQ versus Time. CONWIP = 8, no capacity trigger

But we are not using the flexibility of the virtual queue . The main advantage of a virtual queue is to detect problems in time to remedy them. So when the virtual queue becomes too large, we need to add capacity. Consider Figure 6-10 with the same CONWIP level of 8, but whenever the virtual queue goes above 15 jobs, we add capacity at the bottleneck by adding 4 hours per day, increasing its effective rate from 6.4 jobs per day to 8 jobs per day. But we only add this time when the virtual queue is above 15 jobs.

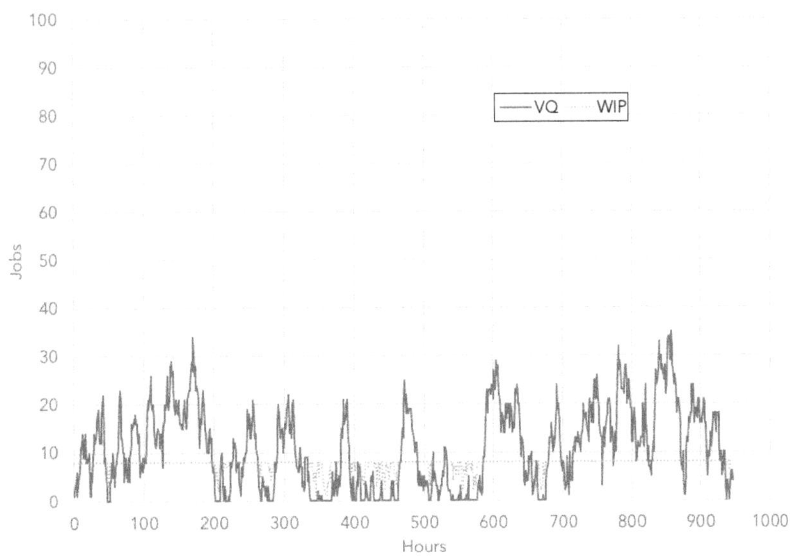

Figure 6-10: WIP and VQ versus Time. CONWIP = 8, capacity trigger =16

Now the maximum virtual queue during these 950 days is 35 (plus 8 for WIP) or 6.7 days. Also, notice that, with this level of variability, the line does not produce the same amount each day. We could force the issue by having so much capacity available that we could always hit the target. But this would be a complete waste of resources. Instead, we produce what we can, monitor our progress, and if our virtual queue starts to become too long, we add more capacity.

In addition to recognizing when we need extra capacity, we also have maximum flexibility for all the jobs in the VQ. We can resequence and even cancel jobs without upsetting the production plan. We can also recognize that we will miss a due date in time to remedy the situation by adding flexible capacity. Of course, if we are running 24/7, then we have no recourse capacity. For this reason, we do not recommend 24/7 schedules.

But we can do better than simply monitoring the number of jobs in the virtual queue. We can quote due dates with a given probability of hitting that due date. Due date quoting provides useful information regarding when to add capacity. As the virtual queue grows, the quoted lead times get longer. When these become longer than our customers will tolerate, we must add capacity to the line to prevent a drop in customer service. The new capacity trigger occurs when the quoted due date becomes too far into the future.

Due date quoting also allows us more flexibility than a simple FIFO allocation of capacity. Instead of giving the next customer the next position in the VQ, we can set aside some emergency positions for those good customers that always come in at the last minute. If that customer comes in, we can accommodate them. If not, we simply move the emergency position further back and allow a number of jobs to finish early.

We now present a method for quoting those due dates.

Due Date Quoting

For make-to-order, we can quote a lead time, ℓ, based on the needed level of service, $\mathcal{s}$, using a formula from *Factory Physics*, section 15.3.2,[ix]

$$\ell = \frac{m}{r_e} + \frac{z_{\mathcal{s}}^2\,\sigma_e^2\left[1 + \sqrt{\frac{4m\,r_e}{z_{\mathcal{s}}^2\,\sigma_e^2} + 1}\right]}{2\,r_e^2}$$

eq 2

where m is the total amount of work in units including the new job, the VQ and the CONWIP line itself, r_e is the effective production rate of the line (the throughput), σ_e^2 is the effective variance of that rate, and $z_{\mathcal{s}}$ is the value of the standard normal variable, Z, such that

$$P\{Z \leq z_{\mathcal{s}}\} = \mathcal{s}$$

The first term, m/r_e, is the average time to perform all of the work currently in the line, including the new job. The second term is the safety lead time that provides the desired service level. The sum of the two is the quoted lead time.

We apply this to a very real problem, using the production line described. Demand has now increased to 0.377 per hour (a good problem to have). However, we are getting close to our constraint rate of 0.4 per hour and need to know what expected lead time should be for a new job. For equivalent comparisons, we will assume the VQ has 10 jobs in all cases. Our effective rate has a mean and variance of 0.377 per hour (see the discussion of demand and throughput in the appendix of Chapter 3). Using the CONWIP algorithm, we find we need a CONWIP level of 11, making $m = 22$ including the new job. The total average cycle time is 22 / 0.377 = 58.4 hours. The safety lead time with 90% confidence requires $z_{0.9}^2 =$ 1.64 and becomes

$$\text{Safety lead time} = \frac{(1.64)(0.377)\left[1+\sqrt{\frac{(4)(22)(0.377)}{(1.64)(0.377)}+1}\right]}{(2)(0.377)^2} = 18.3$$

for a total of 77 hours. Working 16-hour days means we quote 5 days since 77 / 16 = 4.8 days.

Note that because this formula assumes that the bottleneck is always busy, it can result in a quoted lead time that is too short if m becomes very small. So, one should always compare the quoted ℓ against the shortest practical lead time and choose the larger.

We note that r_e and σ_e^2 are the effective production rate and variance. In other words, we have observed our production and estimated these to be the values from experience. In this case, the line is not balanced. With the demand a 0.377 per hour, the utilization of the bottleneck is 94%, but the first station is only 75% utilized, while the last two are utilized even less at only 56%.

The behavior is quite different if we have a balanced line with all four work stations having the same capacity of 0.4 per hour. The result is a system with a bottleneck rate of 0.4 per hour and a raw process time of 10 hours. A balanced line with this level of variability is identical to the marginal case discussed in Chapter 4 with equation 5 computing the throughput for a given WIP level. Using this equation, we solve for the WIP that gives a throughput of 0.377 per hour and find it to be 49 units. This brings the total number of units to be made, m, to 60 (= 49 + 10 + 1). Using eq 2 yields a quoted lead time (LT) of 188 hours (Table).

CASE	AVERAGE LT (HOURS)	SAFETY LT (HOURS)	QUOTED LT (HOURS)
UNBALANCED	58.4	18.4	76.8
BALANCED	159.2	28.6	187.8
% DIFFERENCE	173%	55%	144%

Table 6-2: Comparison of balanced and unbalanced line performance

So, the time required for a balanced system can be much greater than that for an unbalanced line. This is yet another illustration of the OSFR: Where there is variability, there will be a buffer. But the choice of the buffer is up to the manager. The unbalanced line had a significant capacity buffer at the three non-bottleneck stations, whereas the balanced line did not. Consequently, the time buffer is smaller for the unbalanced line than for the balanced line. Table compares the performance of the unbalanced and balanced lines with regard to responsiveness. Both systems have a 90% probability of the cycle time being less than the quoted lead time.

By now the reader is asking, "Wait a minute. I thought a balanced line was always a good thing. We learned all about how to balance the assembly line in school. What is going on?" While it is true that one wants to have the same amount of work at each station on the assembly line, it is also true that the takt time for the line (i.e., the time at each station) will always be longer than the longest average process time. So the line is significantly unbalanced, and this accommodates any variation in process times. Determining the best configuration is guided by operations science and will depend on the relative cost of longer lead times versus extra capacity.

Before we can release parts into the CONWIP line, we need to know what the batch size is. If there are no changeovers (setups), then we could release one unit at a time as we did in the example. But this might not be efficient if we are not using a moving assembly line where all the work stations are next to one another and there is only one routing. If the work stations are not in a straight assembly line or in a cell[x] but significantly far apart, then we want to move more than one unit at a time.

So, we need to determine the optimal process batch (how many parts are processed between changeovers) and an efficient move batch (how many parts we move at one time). These are both discussed in the next section.

Optimal Batch Sizing

We first derive methods to optimize batch sizes in a production facility with significant setup times. To get started, we will review the classic economic production quantity that is similar to the economic order quantity we discussed in the previous chapter but somewhat more complicated because of finite production rates.

The Economic Production Quantity

A search of the Internet and many production and operations management texts will yield a formula for the economic production quantity (EPQ), which seeks to minimize inventory carrying costs plus setup costs. It is remarkably similar to the EOQ we discussed in the last chapter but instead of having inventory appear all at once (like when we receive from a supplier), we see the inventory being produced in the factory as shown in Figure 6-11.

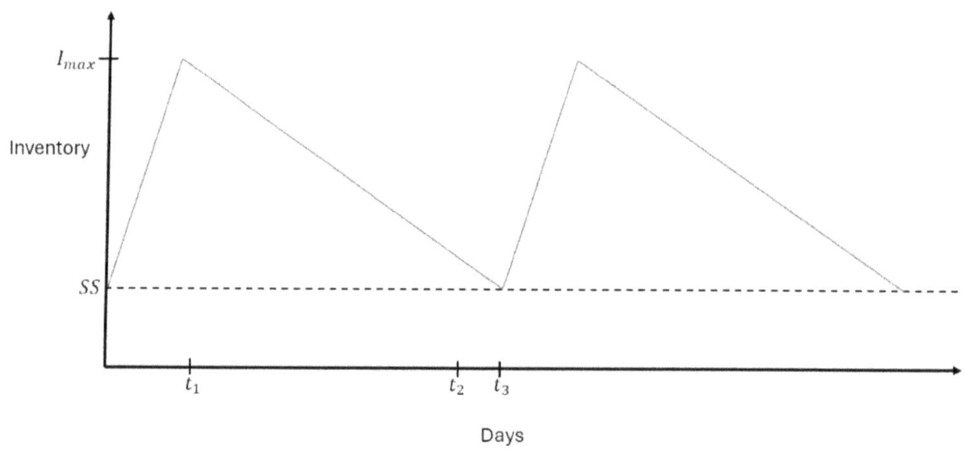

Figure 6-11: Inventory for the economic production quantity

But before we derive the EPQ we must discuss some qualifications. First, like the EOQ, the setup cost should be actual out-of-pocket (OOP) costs that include actual money spent during the setup on, for example, test runs, scrap, and so on. It should not include the cost of labor or machine capacity because that will be considered separately as a constraint. Second, most EPQ formulas include a setup cost but not a setup time. Consequently, the resulting batch sizes may not be capacity feasible. Usually, there are no capacity constraints in the calculation, only inventory carrying costs and setup costs. So, like the EOQ, we may have OOP costs and a shadow price for a capacity constraint. We now derive the classic formula and then add these other features as well (those not interested in the mathematics of this may skip the next few sections and go right to the batch optimization algorithm).

Derivation of the EPQ

In Figure 6-11 we see the trajectory of the on-hand inventory starting at the safety stock (SS) level with a slope of production minus demand, $r - d$. Since we are making Q units, the production ends at $t_1 = Q/r$. After the production ends, there is time that the product is being consumed while the resource is either idle or producing other items. The slope here is $-d$. Little's law gives us $t_3 = Q/d$. We will get back to defining t_2 after concluding the derivation.

Note that the maximum inventory will be

$$I_m = (r - d)t_1 + SS = \frac{(r - d)Q}{r} + SS$$

So, the average inventory will be $(r - d)Q/(2\,r) + SS$. We assume every setup has OOP costs of A, and will see d/Q setups per period. The holding cost (cost to carry one unit for one period) is h, where $h = ic$ with i being the inventory carrying cost ratio for one time period (i.e., the percentage of the cost of the item to carry one unit one period) and c is the unit cost to make the item. So, the total cost per unit time is

$$TC(Q) = h\left(\frac{(r - d)Q}{2r} + SS\right) = A\frac{d}{Q}$$

eq 3

We take the derivative with respect to Q and set equal to zero:

$$\frac{d\,TC}{dQ} = \frac{h}{2}\frac{(r - d)}{r} - A\frac{d}{Q^2} = 0$$

Solving for Q yields the classic EPQ formula.

$$Q^* = \sqrt{\frac{2\,A\,d\,r}{h(r - d)}}$$

eq 4

Note that the safety stock, SS, is not involved as we are assuming it does not depend on the order quantity. This is not exactly true, but the results are close under either set of assumptions, and the analysis is much easier without including the relationship between SS and order quantity.

To incorporate setup times, we recognize that before reaching t_3 (and, also, before the next t_1) there must be a setup for the next run, which takes s periods to complete, so that $s = t_3 - t_2$. Since the time must be $t_3 = Q/d$, we have a constraint on the setup time as it cannot be larger than $t_3 - t_1$ or

$$s < \frac{Q}{d} - \frac{Q}{r}$$

Suppose we have the six items shown in Table 6-3 and want to find the best process batch size.

	d	c	A	s	r
Item	Demand (Items/Hour)	Cost ($)	Out-of-Pocket Cost per Setup ($)	Setup Time (Hours)	Rate (Items/Hour)
100100	2.15	182.00	200.00	8.00	14.00
100200	2.24	147.00	200.00	9.00	12.00
100300	1.64	152.00	200.00	5.00	10.00
100400	1.19	125.00	200.00	10.00	11.00
100500	1.36	138.00	200.00	5.00	15.00
100600	2.12	163.00	200.00	5.00	20.00
	Total = 10.71				

Table 6-3: Batching item data

Consider the item 100100, with a demand of 2.15 per hour and production rate of 14 per hour with a setup of 8 hours. The cost per unit is estimated to be $182, which includes an OOP cost of $200 that is distributed over the number in the batch. The company uses an inventory carrying cost ratio of 25% per year. For a 16-hour day and a 365 day year, the value of our inventory carrying cost is

$$h = \frac{0.25}{\mathrm{y}} \times \frac{\$182/\mathrm{unit}}{365\ \mathrm{d/y} \times 16\ \mathrm{h/d}} = 7.79 \times 10^{-3}\ \$/\mathrm{unit/h}$$

So, the EPQ is

$$Q^* = \sqrt{\frac{2\,A\,d\,r}{h(r-d)}}$$

$$= \sqrt{\frac{2 \times 200 \times 2.15 \times 14}{(7.79 \times 10^{-3})(14 - 2.15)}} \approx 361$$

The utilization of the resource, for this part only, is given by multiplying the demand in batches per hour by the average time to do one batch in hours, namely,

$$u = \frac{d}{Q}\left(s + \frac{Q}{r}\right)$$

$$= \frac{d \times s}{Q} + \frac{d}{r}$$

$$u = \frac{2.15 \times 8}{361} + \frac{2.15}{14} = 0.201$$

But when we use this formula for all six items in the example, the utilization of the resource (which is just the sum of the individual product utilizations) is 101.7%. But this is infeasible, and WIP will continue to grow as long as it is running, until it becomes unmanageable.

Adding a Capacity Constraint

To address this, we add a constraint on the available capacity or, equivalently, on the total utilization, $\sum_i \frac{d_i s_i}{Q_i} + \frac{d_i}{r_i} - u^* \leq 0$, where u^* is the desired utilization. In nonlinear optimization theory, we find that a constrained optimization problem is equivalent to a new unconstrained problem with an added variable (known as the dual price) multiplying the constraint. In this case, our new objective function to minimize is

$$TC(Q) = \sum_j \frac{h_j}{2}\left(\frac{(r_j - d_j)Q_j}{r_j} + SS_j\right) + A_j \frac{d_j}{Q_j} + \lambda\left(\frac{d_j \times s_j}{Q_j} + \frac{d_j}{r_j} - u^*\right)$$

where λ is the dual price. Note that if the constraint is less than zero, the dual price will be zero. This price indicates how much the total cost can be reduced by adding one unit to the constraint. If the constraint is already met (i.e., less than zero), then adding one unit to the constraint does not reduce any cost, so the dual price is zero.

The new expression for each order quantity Q^* is

$$Q^* = \sqrt{\frac{2\,d(A + \lambda\, s)}{i\,c}\frac{r}{r-d}}$$

eq 5

where all of the parameters with the exception of λ and i are unique to the item. Now, we need to find the correct value of λ that gives us our desired utilization. But what should that utilization be?

As we mentioned in Chapter 3, the size of the process batch can have an enormous impact on cycle times, and WIP, and thereby on inventory and customer service. If the batch size is too small, we do too many setups (changeovers) and run out of capacity to make the product. Consequently, the utilization goes up, as does the queue time per the VUT equation. If the batch size is too large, the queue time goes down, but the time to process the batch goes up, so we get a curve like the one shown in Figure 6-12. Thus, there is a sweet spot between 0 and 100% for the utilization. The trick is to find that spot.

The Optimal Utilization

It is fairly easy to find the spot that minimizes cycle time for a machine processing a set of identical (or near-identical) items. For a detailed derivation refer to *Factory Physics*, 3rd Edition, technical note in section 15.3.1.

The example in Figure 6-12 is for a system with a Poisson demand, $c_a^2 = 1$, of 13 per hour with a process time of 3.56 minutes ($t = 0.059$ hours, $r = 16.9$ per hour), a $c_e^2 = 0.5$, and a setup time of 0.5 hours. A good approximation of the optimal batch size is obtained from

$$Q^* \approx \frac{d\,s}{u^* - u_0}$$

eq 6

where $u_0 = d/r = 13/16.9 = 0.771$ and is the utilization without any setups, also called the productive utilization. The optimal utilization for one item and exponential variability is given by $u^* = \sqrt{u_0} = 0.878$. Plugging in the values yields the optimal batch size:

$$= \frac{13 \cdot 0.5}{\sqrt{0.77} - 0.77} = 60.79 \approx 61$$

The formula above guarantees that we will have this utilization if we use the Q^* as computed. With a batch size of 61, we get the minimum cycle time of 25.

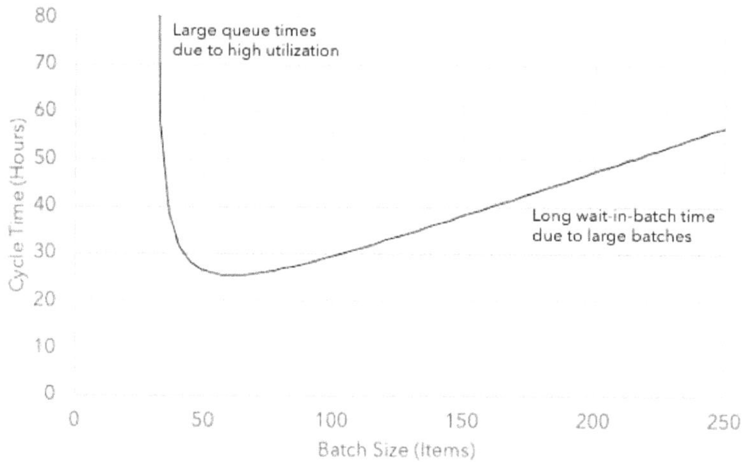

Figure 6-12: Cycle Time versus Batch Size

For multiple products with different demands, setup times, and run rates, the problem becomes much more difficult. First, consider what eq 6 would look like for multiple products. It turns out that if all of the setup times were the same, we would get a similar result, namely,

$$s \sum_j \frac{d_j}{Q_j} = \sqrt{u_0} - u_0$$

So, we develop an approximate solution using the EPQ formula:

$$Q_j^* = \sqrt{\frac{2\, d_j (A_j + \lambda\, s_j)}{i\, c_j} \frac{r_j}{r_j - d_j}}$$

with the qualification that

$$\sum_j \frac{d_j\, s_j}{Q_j} = \sqrt{u_0} - u_0$$

eq 7

Once we find the value of λ that satisfies eq 7, we can compute the individual batch sizes. This procedure is exact when all the setup times are the same and when the coefficients of variation for both demand and production are one. In most other cases, it is a good approximation.

Returning to our example in Table, the productive utilization is

$$u_0 = \sum_{j} d_j/r_j = 0.81$$

So the optimal utilization will be $0.90 = \sqrt{0.81}$ and $\sqrt{u_0} - u_0 = 0.09$ as required. We find that $\lambda =$ 119.8 yields a set of batch sizes that result in this utilization with a total cost is \$16.15 per hour. If we use an optimizer and enforce the same utilization constraint, we get exactly the same results for the optimal batch sizes and also for the total cost.

Adding WIP Carrying Cost

There are times when the cost of having the WIP can be significant and even exceed the inventory carrying cost. Consequently, we need to consider the WIP carrying cost in addition to the inventory carrying and OOP costs. From Little's law, the average WIP for any item will be

$$WIP = d\left(CT_q + s + Q/r\right)$$

where CT_q is the average queueing time for which a conservative value is given by the marginal case with SCVs of demand and process both equal to one, where the VUT equation becomes

$$CT_q = \frac{u}{1-u}\, t_e$$

Then t_e is the expected process time, which is determined by weighting the individual average process times by the demand per batch, d/Q, namely,

$$t_e = \frac{\sum_j \frac{d_j}{Q_j}\left(s_j + \frac{Q_j}{r_j}\right)}{\sum_k d_k/Q_k}$$

Then the total cost, including WIP carrying cost, will be

$$TC = \sum_j i\, c_j\left(\frac{(r_j - d_j)Q_j}{r_j}\right) + \frac{d_j}{Q_j}(A_j + \lambda\, s_j) + i\, c_j\, d_j\left(CT_q + s_j + Q_j/r_j\right)$$

eq 8

We could use a different inventory carrying cost ratio for WIP, but we have probably already defined enough variables for the reader!

The good news is that because eq 7 provides a utilization that minimizes the average cycle time and since Little's law shows that WIP is the product of demand and average cycle time, we should get good results for minimizing the total cost using both eq 5 and eq 7.

Returning to the Table example, we will use the same value of $\lambda = 119.8$, and now, with the WIP carrying cost added, the total cost is \$63.85 per hour. See Table.

Item	Q	Avg CT	Avg WIP	Avg ICC	Avg OOPC	WIP ICC
100100	870	674.8	1,452.3	$ 2.87	$ 0.49	$ 11.31
100200	1057	701.8	1,570.7	$ 2.71	$ 0.42	$ 9.88
100300	694	679.1	1,114.0	$ 1.89	$ 0.47	$ 7.25
100400	837	690.7	824.8	$ 2.00	$ 0.29	$ 4.41
100500	637	652.2	890.1	$ 1.71	$ 0.43	$ 5.26
100600	489	646.6	1,372.9	$ 2.30	$ 0.58	$ 9.58
	Average:	671.8	7,224.8			
	Sum Cost:			$ 13.47	$ 2.68	$ 47.70
	Total Cost:					$ 63.85

Table 6-4: Total cost including WIP carrying cost

Using an optimizer with no constraint on utilization, we get $61.36 per hour. The resulting utilization is 0.911. So, it appears that our optimal utilization of 0.90 is quite close.

But if we want a fast and efficient algorithm, we could use an optimizer to find the λ that gives the lowest cost using the modified EPQ formula. Then, instead of the optimizing over all of the batch sizes, we optimize only one variable, λ. If we do this, we get an optimal $\lambda = 95.8$ with a total cost of $63.38, only 3% above optimal and probably well within the error of our data. Optimizing over one variable can be especially important if we have a large data set that has thousands of items (not just six).

Transfer Batches

We worked at a plant where a planner was opposed to frequent moves of material because, she said, "Moves are a non-value-added activity and waste." Maybe. We also worked at a different plant where many parts were big, machined castings. The carts for transporting WIP in this plant were large, about six feet long by three feet wide by four feet tall, with three shelves so many parts could be loaded on the carts to move between machining centers. A new plant manager showed up and had all the WIP carts cut in half to be three feet by three feet by four feet so fewer parts could be moved at once.

The operations science here concerns the question of transfer batch size. A transfer batch is the number of items completed before moving between processes. A setup might be done for a process batch of 10 items, but each item (transfer batch size = 1) is moved to the next process center once it is finished, rather than waiting for all 10 items to finish and moving all 10 at once.

Small transfer batches can have a significant effect on cycle time, as shown in Figure 6-13. In the figure, the transfer batch size is half the process batch size, and the cycle time reduction effect increases as the process batch size increases. This effect makes sense upon reflection. The larger the process batch size, the longer any one item in the batch would have to wait to be moved if waiting for all items in the batch to be completed. So from a cycle time reduction perspective, the plant manager who cut the WIP carts in half had better intuition than the planner who had the idea that moves were non-value-added and should be avoided.

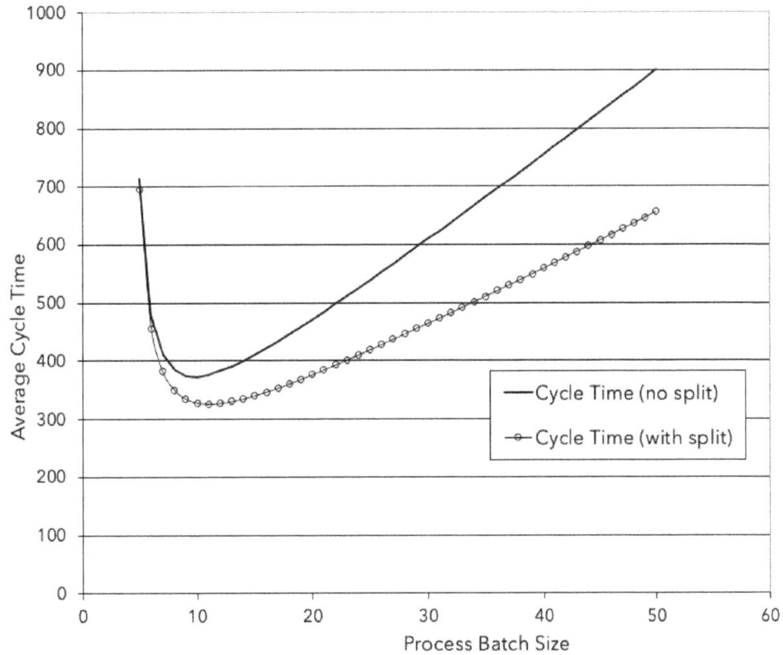

Figure 6-13: Effect of transfer batch size on cycle time

To keep things consistent, one should always make the process batch an integer multiple of the transfer batch. Otherwise, our effective transfer batch will be larger than the optimal. A clever way to ensure this is to make all process batches a power of two, i.e., $Q_p = 2^n$. For example, suppose the process batch is 24. We find n from

$$n = \ln 24 / \ln 2 = 4.58$$

Rounding yields

$$Q_p = 2^5 = 32$$

Let us now see how all of this works with a real firm, which we'll call the Yankee Company.

The Yankee Company

To get an idea of how much improvement can be made we consider the Yankee Company, which sells more than 100,000 different SKUs of carpet. For this exercise, we consider 765 unique items and apply what we have learned about flows and inventories.

While this is a classic manufacturing example, we have additional examples of operations science applications for services and project management in Chapters 15 through 20. The analysis of this chapter applies to those also. Of course, services have fewer options since there are no inventories to optimize.

We use the parameters found in the corporate ERP system and build a detailed model using a forecast of the demand for the next 20 weeks. The model uses the concepts we have covered thus far to compute utilizations, OOP costs, cycle times, throughput, average inventory investment, and fill rates. When we compute the utilizations of all of the process centers, we find that work center T11 is the most utilized at 97%. The item with the highest demand going through T11 has a batch size of 2,000 units. We then compute the replenishment times and, with these, compute the average inventory levels and fill rates. The average cycle time over all items is around 35.5 days with a total of around $11 million in inventory and OOP costs of $1.8 million for the 20-week horizon. The reorder points are adjusted in the model to hit the fill rate targets of 91% and 95% for the two make-to-stock locations.

We now optimize the policies by selecting reorder quantities and reorder points that minimize inventory carrying and OOP costs while hitting the target fill rates. The lot size of the high demand item at T11 increased from 2,000 to 4,400, dropping the utilization from 97% to 95.4%. This reduced the overall cycle time from 35.5 to 14.4 days, while still hitting the fill rate targets. Overall inventory investment went from $16.7 million to $8.1 million. The OOP costs reduced from $310,134 to $280,495. These results are summarized in Table.

Using an inventory carrying cost ratio of 25%, the total of the inventory carrying and OOP costs went from $1,812,046 to $935,472, while hitting the same fill rate targets.

OPTIMIZATION	AVG CT (DAYS)	INV ($ MILLIONS)	OOP ($ MILLIONS)	FR_1 (%)	FR_2 (%)
BEFORE	35.5	16.7	0.310	91	95
AFTER	14.4	8.1	0.281	91	95

Table 6-5: Performance results before and after optimization

Thus, an improvement of more than 48% was achieved with no capital expenditure or extra labor costs. The only things that changed were the parameters in the ERP system!

The company changed its raw material ordering system from one where the planner had to go through hundreds of MRP messages to one with a single spreadsheet showing which items had an inventory position below the reorder point and how much needed to be ordered. This simple application reduced the time to replenish inventory from a day-long job to one that took less than an hour.

The Good, the Bad, and the Ugly—An Operations Science Parable

We now demonstrate how one can apply these concepts in almost any situation. This parable is a combination of many situations we have observed. It also demonstrates the risk of moving forward with improvement efforts without understanding the underlying causes of the existing problems. We begin with the Ugly.

The Ugly

The unit in question has been struggling to maintain profitability with a production line that consists of eight process centers in series producing 10 different products that sell for $650 each. The cycle time is around 32 days even though the total process time is only 6.4 hours. The variability in demand and production is moderate and is typical of random arrivals. The corporate inventory carrying cost ratio is 15%, and there are no backorders. If the inventory is not available at the time of demand, the sale is lost. The cost of this lost sale is estimated at $159 per unit, which is the profit margin plus a $50 charge for loss of good will. Management has always been keen on including both the inventory carrying cost and the lost-sales cost in calculating an adjusted EBITA.

Current performance is barely profitable, with revenues of $5.7 million and an EBITA of $292,000. Moreover, the capacity utilization is only around 80% and inventory turns are at 6.8.

There are, therefore, high hopes for the new manager, whose nick name is Dr. Produktiv. He was hired after his predecessor was sacked for poor performance. The good doctor's first step is to improve productivity to 96% by reducing head count. He also set a new goal of 12 inventory turns. These changes result in a cost reduction of more than 16%. But just as the costs went down, the cycle times skyrocketed, as did the lost-sales cost. Just before Dr. Produktiv was terminated, the lost-sales cost exceeded $1.8 million, and EBITA slid below zero.

The Bad

It was at this point that the COO and her entire staff attended a Lean and Six Sigma conference and learned about the seven or eight forms of waste (or *muda*) as described by Ohno and others. After a value stream mapping workshop, they found that, out of the now 160-day cycle time, less than one day was value added. Dr. Lean was hired and given complete control over operations. Because the operations are very large, it was not possible to convert to a moving assembly line. So instead, Dr. Lean instituted a 2X cycle time policy, which allowed only one unit of WIP in queue and one on each work station. Very soon, cycle times came down, eventually to less than 18 days, an 89% decrease. But turns actually went down to around 5 because more inventory was carried to reduce the lost-sales costs. Unfortunately, to accomplish the low WIP levels, head count had to be increased by around 50% to accommodate the variability, making the unit cost slightly more than the actual selling price. As the unit began to hemorrhage money, Dr. Lean cleaned out his desk and said, "Adieu."

The Good

For Dr. Lean, this was not only a sad event but a rather unjust one as well. Despite losing money, Dr. Lean now had the plant ready to reduce the head count back to its original levels (before Dr. Produktiv). This was eventually suggested by Dr. Ruud, a bright engineer who had studied operations science at a leading university. She pointed out that cycle times were now low enough that some of that improvement could be sacrificed to reduce head count. When the head count was reduced back to the original levels, there was only a slight increase in cycle time (22 days) and no significant loss of turns. At this point, EBITA became around $593,000 per year, more than a 100% increase over the original situation.

Dr. Ruud was soon promoted and went to the sister plant that was languishing in much the same way as this one had. However, instead of "passing through the desert to reach the Promised Land," she decided to optimize the existing plant ERP parameters while the variability was still high and then try to reduce variability. This is generally a good approach since optimizing and implementing new ERP parameters is usually easier than reducing variability in the plant. Also, the use of poor planning parameters can, itself, cause unnecessary variability. She immediately recognized that high inventory turns were not appropriate for such a large lost-sales cost. Applying some operations science principles, she found a sweet spot at 91% utilization with 75 days of cycle time and around one turn of inventory. With no other improvements, the cost of lost sales dropped, and EBITA climbed to $711,000, a 142% improvement. But inventory was high and cycle times long. So, she continued by making changes to reduce variation in the line. Of course, such changes to the line required a second optimization of ERP parameters. This time cycle times dropped to 46 days, and EBITA climbed to almost $1 million at $934,000, a 219% improvement over the performance of the plant when she arrived.

Conclusions

This chapter began with first principles and the OSFR that shows how variability will always be addressed with a combination of the buffers of time, inventory, and capacity. We emphasize the need to start from first principles and illustrate how the earliest implementation of Lean would not have taken decades to develop if one had known the basic principles of OS.

We then described the fundamental elements of an operations planning and execution system that avoids scheduling by creating a set of ground rules that tell when to start new jobs, how much to start, how much inventory to carry, and when more or less capacity is needed. The details of this system included the VQ, a CONWIP production system, due date quoting, and optimal batch sizing. The result is a system that is suitable for production, construction management, and service operations and is more lean than Lean and applicable to a broader range of applications than Lean.

Finally, we showed how millions of dollars can be saved, customer response improved, and inventories slashed by optimizing the basic parameters in the ERP system.

In our next chapter, we discuss how to use operations science with any organization's most important resource—people.

Appendix: CONWIP Algorithm

We present an approximate algorithm for computing average cycle time and throughput for a CONWIP line made up of single-machine stations[xi]. Let t_i be the average time of station i, w be the number of WIP units in the line, and $TH(w)$ and $CT(w)$ be the average throughput and cycle time when there are w units in the line. The algorithm assumes the average queue for a WIP value of w is equal to the average WIP seen when the WIP was $w - 1$. The algorithm is iterative starting with $w = 1$. Inputs include the number of work stations, N; the maximum WIP to be considered, W; and the average process times, t_j. The algorithm is exact for process times with exponential times (e.g., the marginal case). It is also quite accurate for systems with less-variable times and a distinct bottleneck. It is less accurate for balanced lines with less-variable process times.

CONWIP Algorithm

BEGIN

INPUT $N, W, t_j, j = 1, \ldots N$

Set $w = 1, wq_j = 0, \forall j$

DO WHILE $w \leq W$

$CT(w) = 0$

DO FOR j = 1 to N

$wt_j = t_j + wq_j$

$CT(w) = CT(w) + wt_j$

NEXT j

$TH(w) = w/CT(w)$

DO FOR j = 1 to N

$wq_j = TH(w) \times wt_j$

NEXT j

$w \leftarrow w + 1$

LOOP

END

---

[i] As discussed in Chapter 2, scientific "truth" cannot be proven. Here we refer to first principles as truth in the form of generally accepted scientific principles.

[ii] Allysia Finley, "AI's Biggest Threat: Young People Who Can't Think," *Wall Street Journal*, June 23, 2025, https://www.wsj.com/opinion/the-biggest-ai-threat-young-people-who-cant-think-303be1cd.

[iii] Dennis Scimeca, "Industry 4.0 Isn't Quite So Sexy Anymore," *IndustryWeek*, June 6, 2025, https://www.industryweek.com/technology-and-iiot/article/55295201/industry-40-isnt-quite-so-sexy-anymore.

[iv] "Goldilocks Principle," Wikipedia, updated July 2025, https://en.wikipedia.org/wiki/Goldilocks_principle.

[v] Toyota's company website traces the TPS origins back to the 1930s or so. One can debate the origin point, but regardless, the trial-and-error approach took a very long time. See "Origin of the Toyota Production System A production system fine-tuned over generations" in "Toyota Production System," https://global.toyota/en/company/vision-and-philosophy/production-system/.

[vi] Jim Luckman, "Doing Lean Versus Becoming Lean," *The Lean Post, Lean Enterprise Institute,* July 25, 2014, https://www.lean.org/the-lean-post/articles/doing-lean-versus-becoming-lean/.

[vii] Schönberger, Richard J. *World Class Manufacturing: The Lessons of Simplicity Applied.* New York: Free Press, 1986.

[viii] A polynomial solution is one that can find an optimal solution in a time span that is a polynomial function, something like $t_{soln} = cn^k$ where, in this case, n is the number of jobs to be scheduled. A simple scheduling

problem must find the optimal sequence of the available jobs to minimize the total cycle time. This not a schedule but just the sequence. Problems like these have solution times that are more like $t_{soln} = cn!$, which is a factorial function that eventually grows faster than any polynomial.

Suppose we have the fastest computer in the world at the time of this writing, the Frontier supercomputer, which can perform 1.1 quintillion (1.1×10^{15}) calculations per second. If the computation time is $t_{soln} = n^{10}$, it would take 0.0093 seconds to sequence 20 jobs. This would be a near-worst case for a polynomial algorithm as most are more like n^2 or n^3, and we know of no problem that is of the order n^{10}. But for $t_{soln} = n!$, 20 jobs would require 0.611 hours, which is more than 200,000 times longer. For 25 jobs, the n^{10} process would require 0.0867 seconds, and the factorial solution would not be finished until 4.47 centuries later!

So it is quite unlikely that any scheduling algorithm in an ERP system is generating optimal schedules. See section 15.2.4 in *Factory Physics*, third edition, for more detailed discussion on this topic.

[ix] Wallace J. Hopp and Mark L. Spearman, *Factory Physics*, 3rd ed. (Waveland Press, 2008).

[x] Cellular manufacturing is extremely effective when all the different parts require the same resources and take nearly the same time at each station. If this is not the case, we will have idle capacity in the cell, and so its effectiveness will suffer.

[xi] Hopp and Spearman, *Factory Physics,* 10.4.2

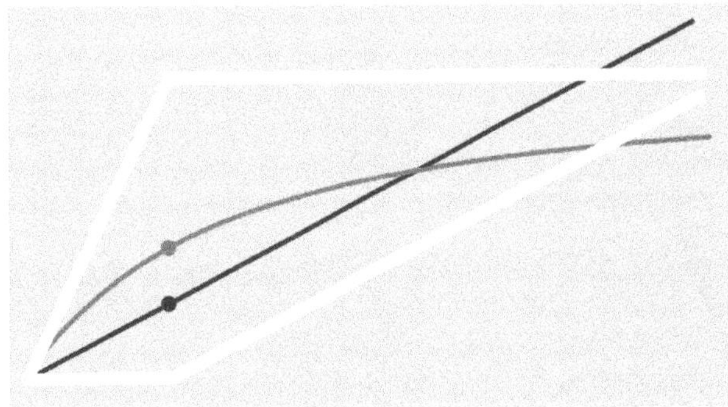

PART THREE

Operations Science Application

We've just covered a tremendous amount of the technical foundation of operations science. In the next six chapters we cover general information required to successfully apply operations science in practice and drive success. The reader will then have a full understanding of operations science which will inform their appreciation of the frontline stories in part 4.

People (Chapter 7)

Management Process (Chapter 8)

Strategy (Chapter 9)

Demand (Chapter 10)

Transformation (Chapter 11)

Managing Variability (Chapter 12)

Optimizing Inventory without Compromising Service (Chapter 13)

7

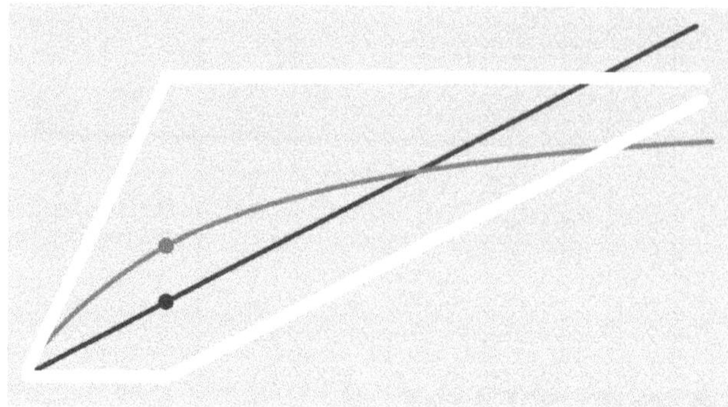

People

Operations science's technical foundation provides powerful concepts to view and manage operations for people at all levels of an organization. One reaction might be that the material is "too theoretical." For extensive discussion on theory versus practice, see Chapter 1 in our earlier book, *Factory Physics for Managers*, and Chapter 2 in this book. To sum it up, we all work by our own theories, and there is nothing so practical as a good theory. The folks at Delta TechOps learned operations science theory and are applying it in practice with great success. Sure, there were engineers in the group and engineers generally love theorizing. However, the largest contingent of the Delta trainees was shop personnel involved in the hands-on efforts to repair and test all parts of the United States' commercial jet airliner fleet that transports 2.9 million people a day in the U. S. alone[i]—with an amazing performance and safety record. The point is that operations science is accessible to anyone.

Learning operations science is like learning to drive a car. To complete a journey the driver needs to understand the basics of braking, acceleration, turning, driving in traffic, filling up the gas tank or charging the batteries, and navigation. If really interested in driving and cars, a person could study car design, internal combustion or battery science, thermodynamics, chemistry and on and on. If desired, one can go way down the operations science technical rabbit hole into queueing theory, statistics, calculus, and, combining all of those, push the boundaries with Dr. Spearman's optimization concepts as described in the previous chapter. A deep dive is not required on the journey to better operations performance, as thousands of people have discovered.

At the Operations Science Institute, we have refined the science and math to a powerfully effective teaching approach for the concepts. The point is not to bowl people over with technical discussion. Effective operations science training helps people understand the concepts in the context of their work while including adult-learning concepts with many participatory examples and group exercises.

Dr. Fred Strathmann, senior vice president of global business for MOBILion Systems and contributor of Chapter 17, describes what operations science has done for him:

> "When I came upon it, it really gave me an appreciation for how to translate [the operations plan] into principles that I could take to administrators, to individuals that held the purse strings to try to explain to them in nonjargon, heavy mathematical laden ways [meaning without jargon or heavy math] what are we talking about here [in operations]. What are we trying to accomplish together and what is going to happen if we don't actually do [the operations plan].[ii]"

Dr. Strathmann, along with many others, has found that the operations science concepts give people throughout an organization a way to understand and explain operations behavior that works to their benefit and aids coordination with others in their organization.

Another aspect of operations science, observed at Delta and in many of the case studies to follow shortly, is that it strengthens intrinsic motivation for people in their jobs. To explore this angle, consider what motivates people. Self-determination theory (SDT) is a concept in psychology that focuses on intrinsic motivation based on what humans find fulfilling rather than extrinsic motivation such as rewards and punishment.

SDT says that people are most motivated when three basic needs are met. As shown in Figure 7-1 these are autonomy, competency, and relatedness.

The need to feel control over one's own actions and decisions

The need to feel effective and capable in one's activities

The need to feel connected to others and experience meaningful relationships

Figure 7-1: Three basic needs for motivation

From the Center for Self-Determination Theory, "Conditions supporting the individual's experience of autonomy, competence, and relatedness are argued to foster the most volitional and high quality forms of motivation and engagement for activities, including enhanced performance, persistence, and creativity. In addition, SDT proposes that the degree to which any of these three psychological needs is unsupported or thwarted within a social context will have a robust detrimental impact on wellness in that setting."[iii]

Knowledge of operations science helps satisfy all three fundamental needs to varying degrees.

Autonomy: Sometimes people don't act with autonomy because they are unsure of the best course of action. Informed with operations science, people make better decisions more quickly.

Competence: Learning operations science is all about strengthening competency. People take great satisfaction in having the science to back up what, many times, their intuition has told them all along.

Relatedness: Admittedly, the connection is less direct, but it still exists. With operations science, a common, scientific framework for understanding operations can be created across an organization. When this gains traction, people can relate to each other more easily through the common framework. They are speaking the same language more often.[iv]

This is well illustrated by the Delta TechOps technician's quote from Chapter 1:

> "We used to think we had 30 problems and 30 solutions because that's about how many people work in the shop, and everyone had an opinion. Now [after operations science training] we have realized we have about four main problems, and we are all focused on solving those problems."

Note the statement, "…we are all focused on…" Using the operations science framework, the group has found an objective approach that allows them to relate to each other in working out the best solutions for the group. Adopting operations science concepts to an organization's processes and terminology provides strong drivers of motivation and focus for its people.

Operations science shapes peoples' intuition to accurately and completely describe behavior of the operations they are managing. As a result, people make better decisions more quickly, which is the best possible use of an organization's most important resource.

Next, we describe a management process as a general approach for applying operations science to successfully manage operations over the long term.

---

[i] "Air Traffic By The Numbers," Federal Aviation Administration"", updated September 9, 2024, https://www.faa.gov/air_traffic/by_the_numbers.

[ii] Frederick Strathmann, "Optimizing Clinical Lab Operations," virtual lecture, November 14, 2024, posted November 15, 2024, by Operations Science Institute, https://opscience.org/operations-science-optimizing-clinical-labs/.

[iii] https://selfdeterminationtheory.org/the-theory/

[iv] In his book "*Drive: The Surprising Truth About What Motivates Us*," Daniel Pink describes the three needs as autonomy, mastery, and purpose. Whether we refer to relatedness or purpose, operations science is less of a direct contributor to the need, but it is a strong enabler of the need.

8

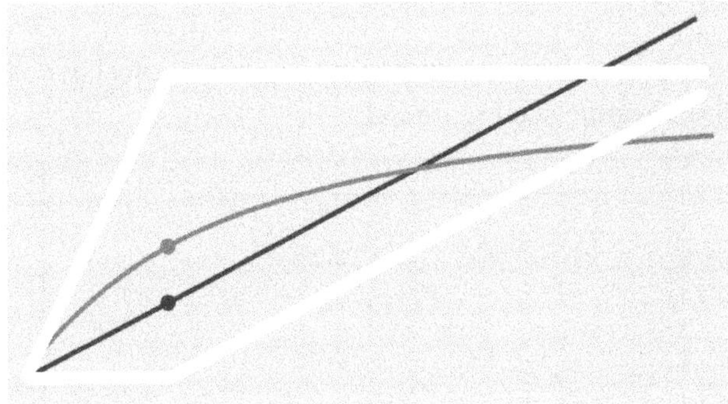

A Management Process and Measures

There are very many organizations that have successful business management systems in place. In such cases, the operations science concepts can be used for auditing and strengthening the existing approach. For those organizations where business management is ad hoc and informal, the operations science framework provides a strong foundation for building a successful, sustainable business management system. By business management system, we mean a structured framework of policies, processes, procedures, and tools that an organization uses to manage and coordinate its activities to achieve its objectives. Our decades of experience have shown us that the management process we will describe here is highly effective for applying operations science to achieve sustainable, best-possible performance. Measures motivate behavior. It is vital to ensure the correct measures are used in managing operations. Too often we have seen organizations institute measures that drive dysfunctional and inefficient behaviors. We will discuss measures and provide recommendations to ensure measures align individual efforts with operations behavior and with business strategies to achieve desired results.

Business Management Systems

From ancient times to the present, organizations have attempted to put structure on their efforts to achieve desired results, see Figure 8-1.

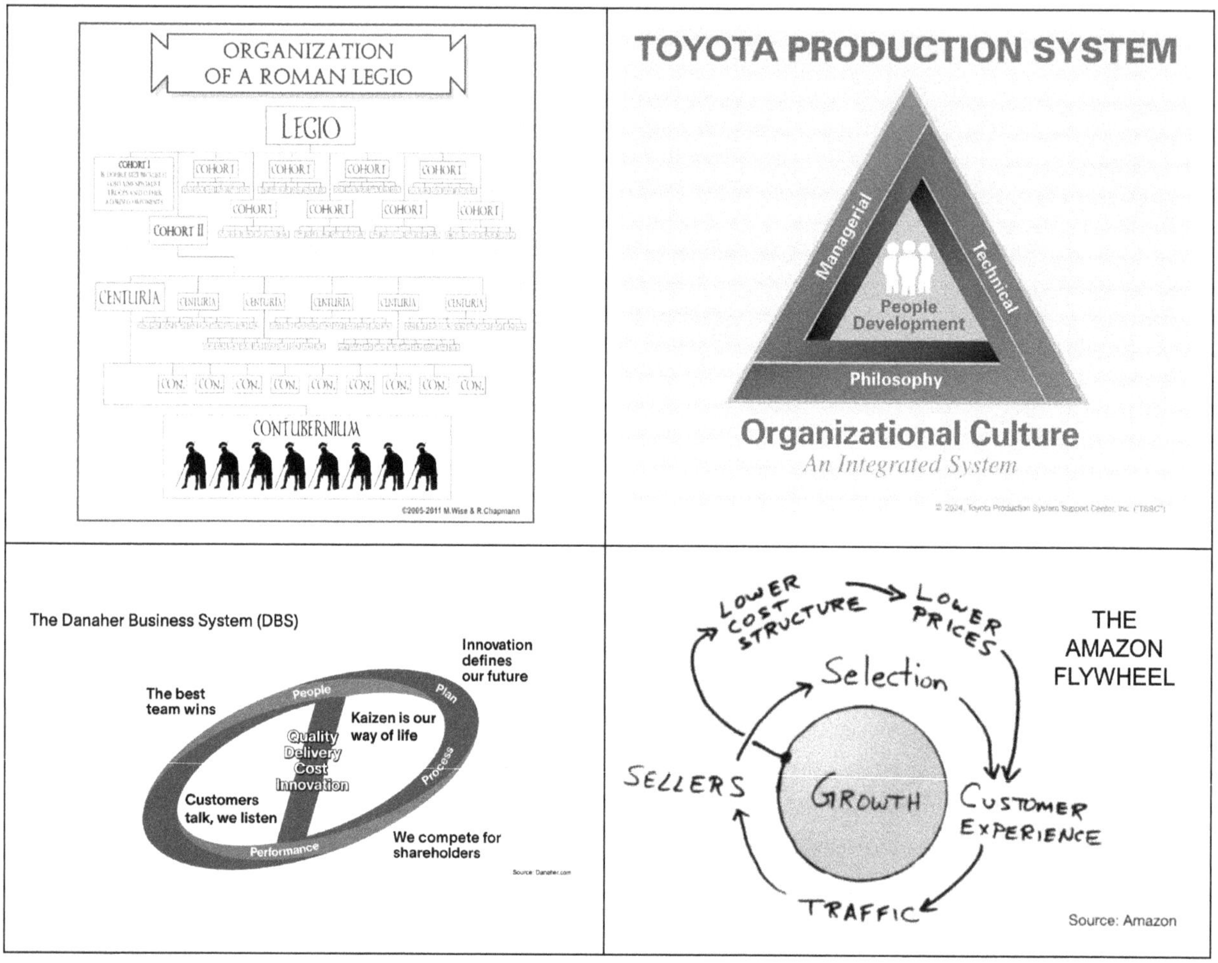

Figure 8-1: Business Management System examples

Every company has a business management system. Some are intentional, many are ad hoc. There are two basic elements of business management systems, what to manage and how to manage, see Figure 8-2. They might be called "first principles" of business management systems.

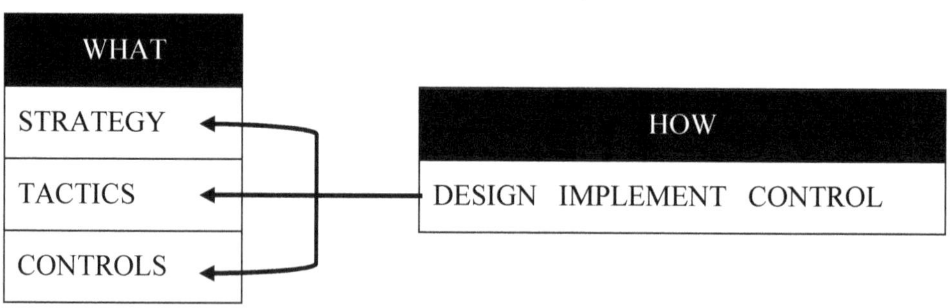

Figure 8-2: Basic elements of business management systems

Following are a few definitions to frame the discussion for your own business management system.

What to Manage

Strategy – plans made to control a company's environment and accomplish its business goals. As we have discussed, operations strategy is the plan for using the fundamental portfolio of variability, inventory, capacity, and time to support business strategy. We will discuss strategy in more detail in Chapter 9.

Tactics – policies or actions implemented to accomplish a task or objective. Examples are capacity resource levels (people and/or machines), safety stock, lead times, overtime levels and, for inventory, reorder points and reorder quantities. Quantified policies, when executed, predictively determine the portfolio of variability, inventory, capacity, and time. In essence, tactics consistent with strategy execute strategy but inconsistent tactics determine a different strategy altogether. Tactics drive outcomes.

Controls (a noun) – in operations science, controls are methods or systems to measure and manage implementation of tactics for achieving desired performance. Examples are inventory position to measure policy adherence to inputs in an MRP/ERP system. The ERP system is itself a type of control for managing and measuring. A CONWIP system of cards is a form of a control. Reporting and tracking Commitment Reliability is a type of a control that can be very helpful in achieving predictable results in project and knowledge work. A Gannt chart is often used as a control, particularly in construction projects, but it is not a very effective control for achieving predictable outcomes. Controls do not have to be IT systems.

How To Manage

Note that in Figure 8-2, all of the "How" are applied to each of the "What." For example, creating a strategy in a large organization can be a complex process. How many months before the end of the current fiscal year does the process for the new fiscal year start? When must it be completed? Who participates in the strategy formulation? What are the inputs, what does the expected output look like? What is the WIP, how is that WIP managed? This is a classic application of design, implement, and control actions to successfully create or update strategy.

Design – analysis, planning, and creation of strategy, tactics, and controls to support an organization's goals. Design is also applied to itself; one should design the design process. Design is applied to implementation, which is a project management process, and to controls.

Implement – put a plan or a decision into effect. This is essentially a project management effort.

Control (a verb) – determine the behavior of a process. Controls are not the same as control.[i] One can do process design well and implement perfectly, but, if poor control is practiced, all the previous effort for design and implement is essentially wasted and desired performance is not achieved. It frequently happens that people use controls like Gantt charts, metrics dashboards, and work schedules as controls for process performance yet do not achieve desired performance.

A project manager might perceive a Gantt chart as a method for establishing control over the project. In reality, Gantt charts are woefully inadequate for managing the day-to-day dynamics of a project and do not provide predictive control over project outcomes. They can be useful for reporting, but that is entirely different than useful control. Gantt charts are deterministic lagging measures and they do not tie to the behavior of the operations they ostensibly control. After the initial and highly useful function of a Gantt chart for documenting project Work Breakdown Structure (WBS), Gantt charts function chiefly as a lagging measure of project performance. Automating the Gantt chart and refreshing it every day does not overcome its severe limitations for control. In any case, daily updates to Gantt charts rarely, if ever, happen on complex projects because it requires tremendous effort. Monitoring and controlling setup efficiency in a manufacturing operation is not directly related to on-time completion and, in fact, usually makes on-time completion worse. Monitoring and controlling overtime in a hospital emergency department does not correlate with better patient care. A key benefit of operations science is helping people understand which tactics, controls, and measures are appropriate for achieving desired cost, throughput, cycle time, and utilization to support their goals. This is accomplished with the elements of great operations plans as described in Table 1 in Chapter 6.

For example, as we discuss in Chapter 14, software vendors often create planning and control systems that are quite capable of achieving the desired process performance. All too often, we see that planners and managers do not understand operations science and drive their system to poor performance levels that are largely self-inflicted.

Once operations science is internalized, the design-implement-control approach becomes a highly useful framework for application. It is simple and powerful. Apply the approach repeatedly for superb results over and over again.

Simply stated, using operations science to develop and deploy operations strategy in a business management system consists of managing the interactions between

- Demand
- Capacity
- Inventory
- Response time
- Variability
- Cost

This process is illustrated in Figure 8-3.

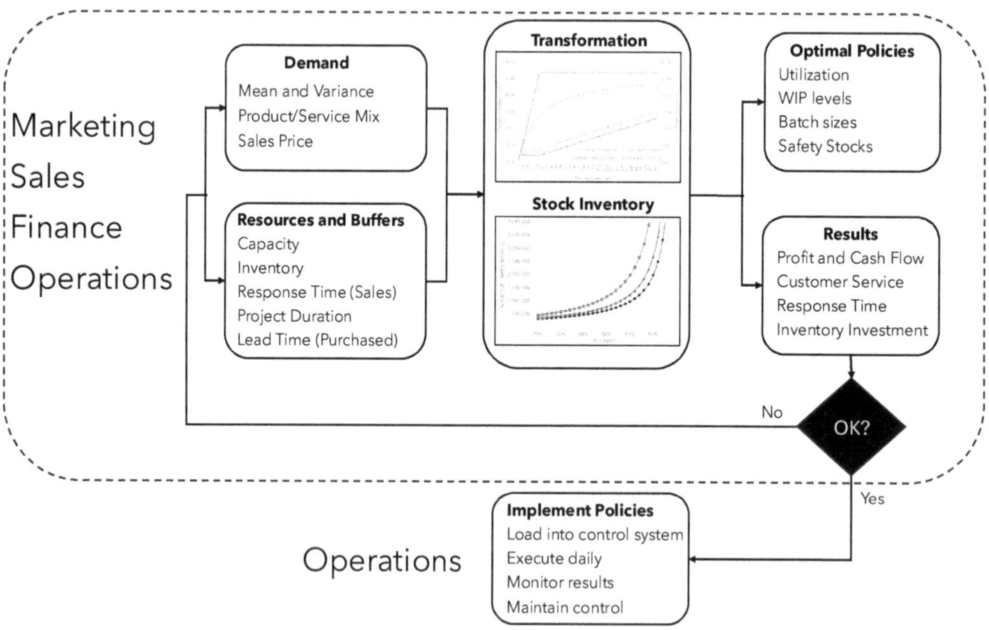

Figure 8-3: Strategy to execution using operations science in a business management system

Business strategy is developed with all the groups listed on the left. Operations contributes an understanding of what it can do with a given level of resources. It is imperative that Operations be able to convey the concepts of operations science to the rest of the organization as a basis for these discussions. If Operations cannot explain the operations science models of behavior to the rest of the organization in a practical, digestible format, then Operations' guidance often gets dismissed and other models take hold, such as the standard cost model of Finance. The standard cost model says that

$$\text{Unit Cost} = \frac{\text{Fixed Cost}}{n} + \text{Variable cost}$$

where n = number of items or services produced.

This means that making more items or providing more services spreads fixed costs across more units which means that unit cost decreases. The inclination from this model is to keep everyone as busy as possible producing as much as possible. We have seen this approach followed even when there is no demand to warrant the extra activity and the result is too much cost, too much inventory, and, at the same time, poor customer service. Note this applies for both physical production, as in traditional manufacturing, and to intellectual work, such as engineering, to services work, such as MRO or diagnostic laboratories, and to project work as in engineering and construction of major capital projects. More on this and a proposed solution when we discuss contribution margin shortly.

In Figure 8-3, moving to the right from the business strategy generates tactical policies that deliver the group's desired service levels, response time, and fill rates. The outcome is a set of predicted results given an expected performance envelope and committed resources. We will discuss the performance envelope in much more detail here and onward, but it is the range of demand and response time for an organization to consider when planning its objectives. If the predicted results are satisfactory for achieving the organization's business strategy, then Operations implements its execution tactics and control to ensure the desired results are achieved.

Managing the Operations Plan

In Chapter 6 we discussed how to create a great operations plan using operations science. Here, we have further elaborated on creating a great operations plan with the Figure 8-3 discussion above. Now we describe a process for managing the operations plan to repeatedly and predictably obtain desired results. The overall operations science management process is illustrated in Figure 8-4.

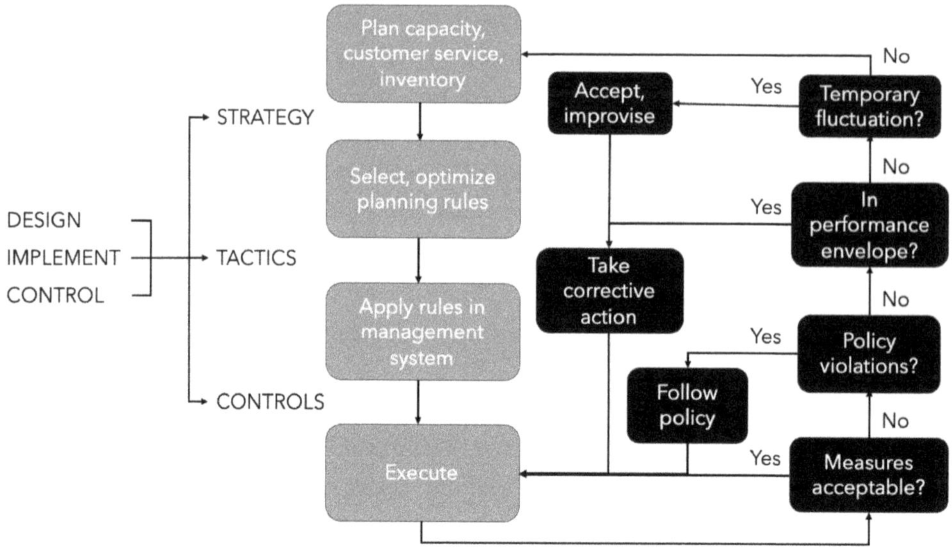

Figure 8-4: Practical operations management process

The general logic of the operations management process of Figure 8-4 is

1. Set expected demand and response time requirements and an acceptable range around those requirements (the performance envelope). For operations strategy, this is determined by the overall business strategy. The conversations described below should be part of an on-going iterative process between marketing, sales, finance, and operations whether for long-term strategy, annual strategy, or ongoing business execution.

   Marketing – "Here are our products and/or services and who we want to sell to."

   Finance – "Here's how much you need to sell and how much is available to spend"

   Sales – "Here's how we want to sell and be supported"

   Operations – "Here's what we can do based on these requirements and resources allocated"

2. Set policies, tactics, controls, and measures to ensure achievement of desired performance. This also feeds into the iterative process of step 1.
3. Execute
4. If measures show good performance, keep executing
5. If measures show bad performance, first check that you are following the tactics and policies as originally planned.
6. If there are policy violations, start following policy
7. If measures are bad and policy violations are not the problem, check if the performance envelope assumptions have been violated, e.g., demand has increased and/or supplier lead times have gotten longer
8. If performance envelope assumptions have not been violated, need to go and see why given policies are not working. This is an operations science-powered investigation

9. If performance envelope assumptions have been violated, determine if the fluctuation is temporary or can be tolerated. If so, improvise actions to deal with the fluctuation. If not, go back and start over on strategy and policies. Replan to the new performance envelope.
10. There will be occasional strategy reviews as a matter of course as strategy may change even if things are running well.

Management Process Outcomes

Let's explore different outcomes in this process. In all cases, the organization's product and service offerings and target markets have been specified. Expected range of demand and required response time, the performance envelope, has been specified for its products and services. Safety has been established as the number one priority. Required quality levels have been specified. Optimal policies have been established to achieve desired outcomes given the expected range of variability in demand and response time. The science of determining many of these policies was described in Chapters 3 through 6.

In other words, the green boxes in Figure 8-4 have been completed and we are looking at the different configurations of outcomes, the black boxes.

Let's describe a few examples of the potential outcomes. To start, if the organization or process is executing and all measures are acceptable, keep up the good work. In Figure 8-4, this means management stays in the bottom loop. Day to day execution is managed, measures are acceptable so keep executing using the tactics and controls in place.

What happens when measures indicate a problem? The decision tree shown is arranged in order from simplest to check and correct to the most extensive measures required for correct, complete replanning.

Policy Violations

If people are accustomed to using policies as controls, such as inventory policies on when to reorder and how much to order, correcting the performance problem is simple. As shown in Figure 8-5, follow the correct policies. Policy violations are rampant when the existing policies are bad policies. There is often a major change management effort to get people to follow good policies since they are often heavily conditioned by work with bad policies so correcting the use of bad policies may not be a simple endeavor.

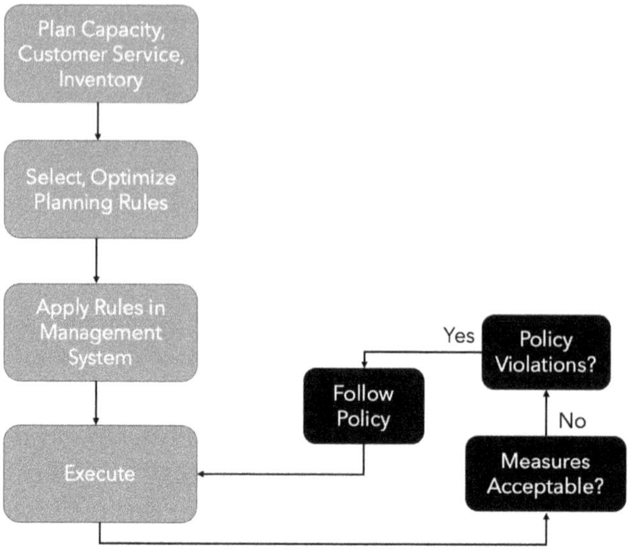

Figure 8-5: Poor performance due to policy violations

Bad means the policies require people to do things that will work against business strategy objectives, e.g., hold too much inventory or not enough inventory or work too much overtime or not enough overtime. When this happens, the policies are ignored and people do whatever they think best which also means they are probably heavily invested in their own spreadsheets for planning and control. Using an IT planning system as an actual control system with good policies often requires that people learn operations science and, often, unlearn bad habits.

A couple of specific examples here include inventory planners not maintaining proper inventory position (see Chapter 5 When to Order discussion) and construction project managers flooding sites with WIP (see Chapter 4 rebar example).

Overcoming the bad habits is a classic change management challenge. One suggestion for overcoming these habits is to find the early adopters who will learn operations science and use optimal policies as designed. Then, support them, and promote their success. The others will then come along unless they like performing poorly, which requires a different solution.

Operations Science Investigations

This outcome of the operations science management process provides interesting insights as it usually leads to discovery of previously unrealized conditions or behavior. As attributed to Isaac Asimov, "The most exciting phrase to hear in science, the one that heralds new discoveries, is not 'Eureka! I found it!' but 'That's funny ...'"[ii] The great thing about the operations science description of operations behavior is that it is both quantitative and scientific. It has a first principles foundation describing operations behavior so when corrective action is required, as illustrated in Figure 8-6, it is often a result of a "That's funny..." moment.

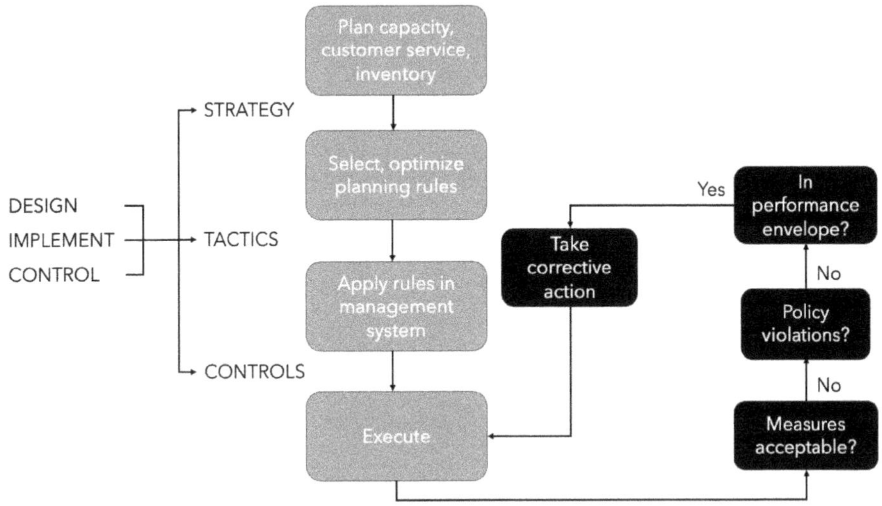

Figure 8-6: Operations science investigation required

Something about the data used to model and create policies or assumptions used to create the analysis are wrong. It could be that completely new scientific discoveries were uncovered—please send an email to info@opscience.org if that happens, we want to know.

For illustration, consider our work at a Medtronic plant. The plant fabricated components for medical devices, assembled them into kits inspected and shipped them as shown in Figure 8-7.

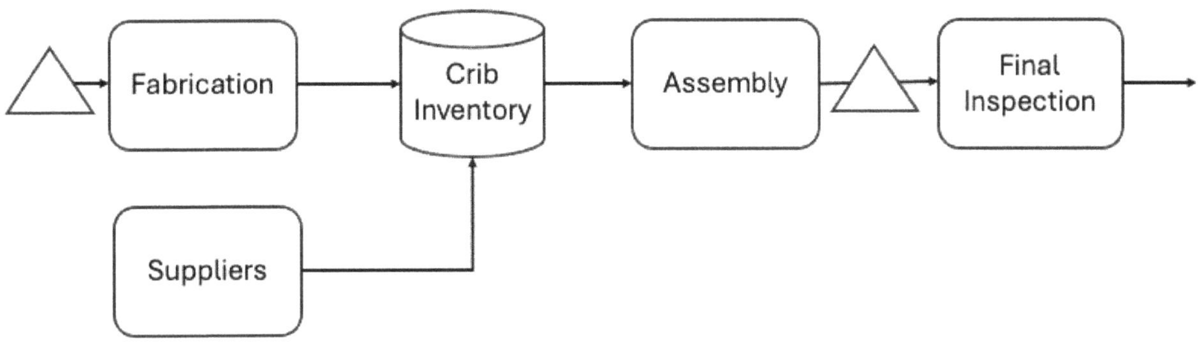

Figure 8-7: Medical device manufacturing

We worked with the site director and operations manager and went through the application of operations science in the management model. We rebalanced inventory from fabrication to the crib so that assembly would have the inventory it needed to meet demand, rather than just making whatever was available from fabrication. Inventory was reduced by millions of dollars through optimization. Planners were trained in operations science concepts, inventory management was put on inventory position control, and the planners began maintaining inventory position appropriately. However, a "funny" thing happened as shown in Figure 8-8.

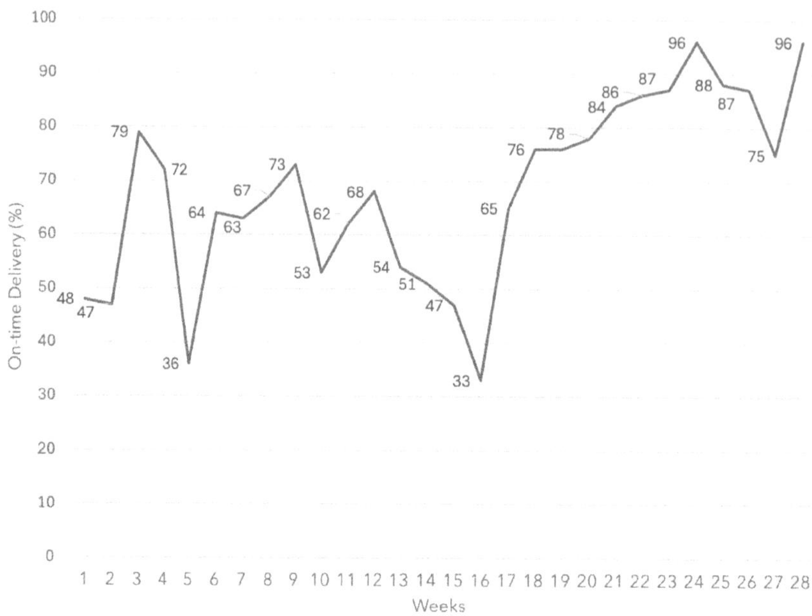

Figure 8-8: On-time delivery performance

The changes to control were implemented around week 12 and our expectation was that on-time delivery was very soon going to the high 90% range. As the chart shows, the exact opposite happened, things got worse. Demand hadn't changed from what was expected. We didn't have any supplier disasters leading to longer than expected replenishment times—we were in the planned performance envelope.

Something was wrong with our model or our assumptions and we conducted an operations science investigation to review the data and assumptions. One assumption, supplied by the site director, was that inspectors in final inspection were skilled and flexible and could cover any requirement that came their way—so, don't worry about inspection capacity. As part of the investigation, we reviewed inspection capacity which showed the site manager's assumption to be incorrect. Inspection had become the constraint.

In this type of outcome, knowledge of the certainty of operations science provides great reassurance. While no one was happy that initial implementation had not achieved desired results, there was no panic. Quick investigation and corrective action—more capacity for inspection—promptly resolved the problem. Ultimately, it was our responsibility for not checking all the assumptions, but lesson learned and problem resolved. The site was able to establish and maintain very high on-time delivery even as demand increased afterwards.

Another takeaway from this example, and one we have repeatedly seen play out, is never assume the people in the process have a complete understanding of the process. People working a process may have complete knowledge about details of work content, e.g., "I have been doing this service or managing this type of project or making this part for 20 years, I have seen it all." Indeed. That experience is extremely valuable. As Jeff Bell likes to say, "The people who do the work, know the work best." However, that knowledge and experience is fundamentally different from understanding how the dynamics of inventory, capacity, time, and variability work to influence organizational outcomes. The operations science approach allows one to see the big picture of operations performance so that deep individual knowledge and experience can be employed most effectively.

When Business Conditions Change

Oftentimes, polices and tactics are set for one set of conditions and those conditions change. As shown in Figure 8-9, there is a decision to make when conditions change.

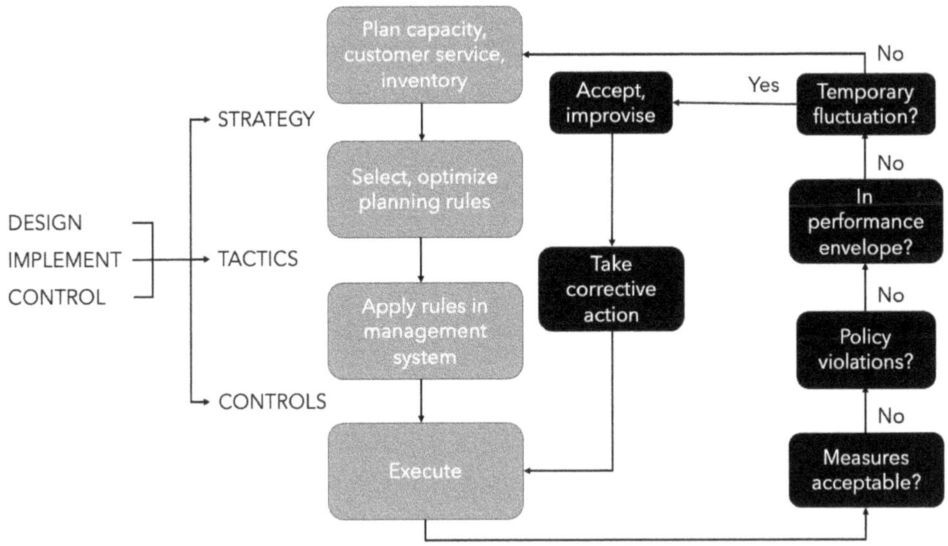

Figure 8-9: When the performance envelope has changed

To reiterate, operations science is about managing variability. Demand will not be deterministic, for example, 30 hours of work per month every month. The demand envelope should specify some expected range around an average, both average and variance—an average of 30 hours per month and a variance of 9. Of course, as operations science shows, more variance requires more buffering to synchronize demand and transformation. "Business conditions change" means either

- demand has exceeded the planned range or,
- demand has reduced to a level which does not justify current resource investment
- replenishment time has reduced or increased to a level which would justify replanning

The replenishment time could be time for suppliers to deliver parts or services. COVID-19 caused a major shift in replenishment time for many organizations. Replenishment time could be the time for an internal process to complete an item or task.

It is not unusual for organizations to restrict their investments in transformation resources, i.e., capacity and inventory, due to financial limitations. This is completely understandable and reasonable. What is also understandable, but not so reasonable is when organizations set resource investment levels to a given performance envelope but expect the organization to respond effectively even when demand and replenishment times are outside of that performance envelope. This behavior is understandable because it is human nature for employees to try and make their organization's plan work even when it is a bad plan. Restricting resources is not in itself bad, not understanding the limitations of performance for given resources is a recipe for stress, waste, and chaos.

It is also human nature for executives who are compensated for achieving performance goals to force the organization into dysfunctional operations decisions to ensure they achieve their goals. Examples of these dysfunctional decisions include

- Continue to produce product to achieve unit cost goals even though there is no demand
- Sell as many commercial airline seats as possible to hit revenue targets
- Dictate arbitrary cost cuts such as reduce inventory 30% or eliminate all overtime

These examples are from actual events; any experienced operations manager can name others.

The question is whether the change is temporary. The decision to make is whether or not to replan or temporarily tolerate less than desirable performance. If a parameter such as demand is being monitored with process behavior charts, there is good quantitative information to answer the question.[iii] Often times the decision to react is made when things simply become unbearable due to customer complaints or excessive cost or unacceptable performance, however that might be defined. In any case, if the fluctuation is deemed to be temporary, the appropriate action is to accept that performance will not be as desired. The organization must then improvise and take corrective action to mitigate the effects of the fluctuation. For example, let customers know ahead of time that demand has spiked, orders are backed up, and lead times are going to be extended to three weeks instead of two weeks for the next three months. The wrong decision is to flog the organization for not responding to changes that it is not capable of responding to. This only feeds variability and stress into the system and increases chaos and waste.

In some cases, companies and entire industries, get tired of going through the cycle of angering their customers and stressing their employees by dysfunctionally responding to fluctuations outside the performance envelope. Or, the government steps in and penalizes dysfunctional behavior. This happened in the airline industry. In 2007 JetBlue had a massive service failure due to a Valentine's Day storm and stranded passengers on planes on the tarmac, some for as long as six hours. JetBlue issued a Passenger Bill of Rights in response and the U.S. Department of Transportation stepped in a few years later.[iv] Airlines have realized that it is better policy to do mass cancellation of flights when severe weather (fluctuation outside their resources' performance envelope) appears likely in the next 24 hours rather than have masses of people come to the airport and have their flights canceled.[v] Customers still miss flights due to weather, but the policy changes prevent passengers from experiencing huge frustration and wasted time waiting at the airport in large crowds of people who are similarly frustrated. See Figure 8-10.

Cancellation effects prior to airline policy changes © Henry Nicholls/Reuters

Cancellation effects after airline policy changes

Figure 8-10: Effects of airline policy choices

For airlines, the nature of weather disruptions is that they usually last less than a day. This has prompted a long-term change in strategy for airlines, mass early cancellations, and is reflected in their day-to-day execution. Covid required the massive medium-term strategy adjustment of grounding fleets. Refer to Chapter One for Delta's experience with the massive adjustment due to Covid.

Of course, if the change in the conditions are expected to persist over the long term, the organization must replan based on the new business conditions. These adjustments to the performance envelope are critical to ensuring long term success.

Operations science enables organizations to use the operations management process described here to sustain predictable, repeatable results and best possible performance over the long term. A vitally important key to the success of the operations management process is selection and use of the appropriate measures.

Selecting and Using the Right Measures

Measures do not just report results, they drive behavior. For an operations plan, measures are quantities used to report performance characteristics of a process, for instance, on-time delivery, capacity utilization, or number of parts or tasks within policy compliance.[vi]

Especially when tied to performance evaluations or financial incentives, poorly designed measures often lead people to take actions that undermine business strategy, create waste, and generate stress across the organization. To avoid this, organizations must be extremely cautious: measures must reinforce, not conflict with, the operations behaviors required for success. A well-designed measurement system grounded in operations science provides predictive control, aligns efforts to strategic objectives throughout an organization, and enables teams to work with focus and confidence. When measures are thoughtfully selected and properly aligned, they reduce chaos, prevent conflicting priorities, and foster coordinated, high-performance execution.

Two Types of Measures: Leading Versus Lagging

Operations science distinguishes between leading and lagging measures. This distinction is critical to avoid a common and costly management mistake: using lagging measures to try to control performance.

Lagging Measures are retrospective. They tell you what happened, after it's already happened. Examples include:

- On-time delivery
- Actual throughput
- Cycle time
- Actual overtime vs. plan
- Quality/rework rate
- Earned Value Analysis for project delivery

A practical example involved a client's reduction in WIP to achieve better cycle time. Cycle time is a classic lagging measure, but it was the measure of interest in this effort. WIP was reduced so cycle time was reduced, goal achieved. However, throughput plummeted. They had unknowingly crossed the "minimum WIP" threshold required to keep the bottleneck fed so were constantly starving the bottleneck and losing throughput as a result. This is the danger of relying solely on lagging measures or misapplying Lean tools without understanding the operations science of process behavior.

Leading Measures are predictive. They reflect current process conditions and allow for proactive control. They focus on causes, not effects. Key examples include:

- Work-in-Process (WIP) – A direct input to both cycle time and throughput via Little's Law.
- Demand Mean and Variance – Critical for capacity and inventory planning.
- Planned Utilization – High planned utilization increases the risk of instability.
- Planned Overtime
- Inventory Accuracy

A pervasive abuse of a leading measure is driving operations to run at or near 100% utilization. With any amount of variability, WIP and cycle times explode at high utilization levels. Think of wait times in emergency rooms when there is a natural disaster, or keeping everyone busy on a project whether the work is needed or not and the chronically late project delivery that results. It is as predictable as the sun rising in the east and setting in the west. We are not suggesting high utilization of resources is a bad goal, it is an understandable attempt to achieve the most return from investment in resources. However, expediting, high levels of WIP and inventory, and poor customer service are also generally the result of overdriving utilization and those are very wasteful in total.

Measures Alignment

Measures alignment is a non-negotiable requirement for any effective management system. Measures should be aligned with process performance, as described by operations science, and aligned with the organization's goals. Misaligned measures cause people to optimize their own metrics at the expense of the organization, classic local optimization. There should also be tiered alignment of measures in that the individuals' measures tie to their managers' measures and so on up the organization and all the measures are tied to achieving the organization's objectives. Successfully constructed measures provide a "line of sight" from individual and team efforts through management levels to the organization's objectives.

Examples of Misalignment

- Measuring developers on lines of code when the goal is functionality
- Incentivizing call center reps to shorten calls when the goal is customer satisfaction
- Driving for 100% utilization when the real goal is short cycle time and low working capital investment in WIP

Principles for Aligned Individual Measures

- The measure should tie to the individual's actual span of control
- The measure should connect to a tactical policy set by the operations strategy
- The measure should support desired operations behavior as described by operations science

Table 1 lists measures which are in the individual's control and tie to process measures required for complete control.

ENGINEERING AND DESIGN	HEALTHCARE SERVICES	MANUFACTURING
Attendance	Attendance	Attendance
Unit throughput	Protocol compliance	Task throughput
Setup time	Documentation accuracy	Rework
Scrap	Error rate	Tasks in process (WIP)
Rework		Commitment reliability
Sequence compliance		
Lot size compliance		

Table 1: Leading measures for individuals

Commitment Reliability is particularly valuable in engineering and project environments where tasks are interdependent and variability is high. People state what they are going to do and when it will be done. That is then measured to see how well tasks and timing are accomplished. A commitment reliability of 100% should not be expected, especially in high-variability work, but 40–50% signals a serious issue. It is also situation dependent. For example, a construction crew on a construction site would measure commitment reliability, before the day's work begins, for the expected tasks to be completed that day. On the other hand, it's more practical for a product design team of engineers to work to a weekly commitment reliability process. A construction crew could be expected a success rate of 90% to 95% while a design engineering team is more reasonably expected to have a 80% to 85% success rate since there is much variability in creating new designs.

Desired Outcomes Measures

Safety is number one and quality is also a desired outcome measure and both influence operations behavior as described by operations science. For example, too much physical WIP lying around can be a safety hazard leading to tripping or forklift collisions, likewise for quality. Metal WIP lying around a long time can rust. Machines on-site too soon for building construction may sit around uninstalled and performance may degrade without adequate maintenance and thermal cycles. However, WIP can be stored and managed safely. Rust can be prevented with appropriate protection from the elements. Machines can be maintained whether they are installed or not. Safety and quality are very important, they are not left off the following list out of disregard. Companies should always report on safety and quality. Following are primary measures that operations sciences governs.

- Cost (Fixed Assets, Working Capital, Direct and Indirect Expenses)
- Profitability
- Cash Flow
- On-time delivery
- Response Time

Policies for optimization to achieve these outcomes include

- Capacity utilization levels
- Overtime levels
- Inventory position (reorder point and reorder quantities or equivalents)
- WIP levels
- Sequence compliance
- Commitment Reliability levels (for project management)
- Availability

When to React to Measures

Measuring throughput or cycle time alone isn't enough. You also need to know if your process is in control. A process behavior chart, see Figure 8-11, helps distinguish signal from noise—and prevents you from making poor decisions based on normal variation.

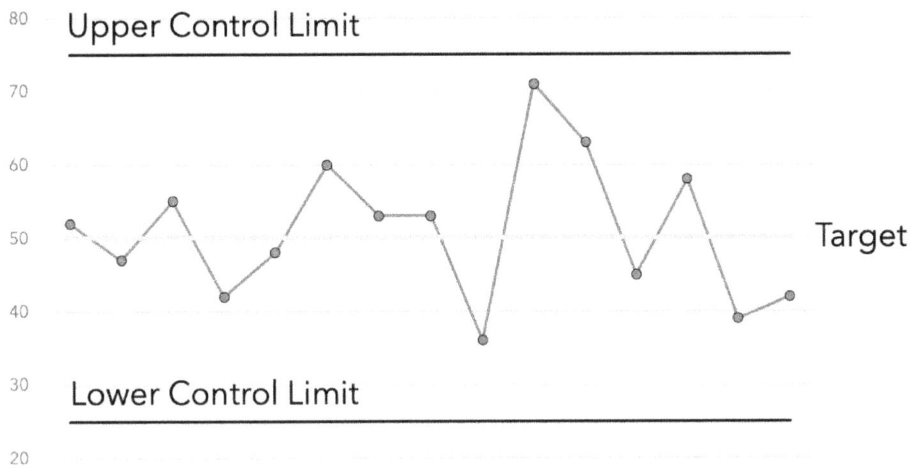

Figure 8-11: Example of a process behavior chart

Key signals of process instability

- A single point outside upper or lower control limits.
- A run of 8 or more points all above or below the average.
- Clustering of 3 out of 4 points near a control limit.

If none of these appear, the process is in control even if it is not hitting a daily target. Without this understanding, organizations celebrate or admonish individuals or teams based on being above or below target, driving unnecessary stress and wasted effort.

One Measure to Rule Cash Flow

Organizations are often highly concerned with profitability. The primary measure to create best possible profitability in operations is marginal contribution at the bottleneck. For more extensive discussion see Financial Performance Driven By The Science Of Operations in Chapter 5 of Factory Physics for Managers, but we summarize here.

If one thinks of the throughput in terms of cash flow rather than units or tasks, then maximum cash flow into an organization is achieved by running the items or tasks that generate the highest cash flow at the bottleneck.

Contribution margin = revenue – variable costs

Profit = contribution margin – fixed costs

For a given set of resources, labor, and expenses as fixed costs, the higher the contribution margin the higher the profit. Operations science shows that the bottleneck resource in any process limits the throughput of that process. So, the most profit to be achieved for a process is driven by the contribution margin at the bottleneck since that limits the amount of cash flow generated by the process.

This is a straightforward concept that has generated fierce debate between operations, finance, and accounting people. Eli Goldratt, the inventor of Theory of Constraints was famous for his statement, "Cost accounting is the number one enemy of productivity."[vii] How does that make all you cost accountants out there feel? Goldratt's reference is to the problem created when the standard cost model described above is used to guide decisions about what to produce. As discussed, the standard cost model can drive counterproductive operations decisions. We propose a straightforward resolution for this debate.

One, stop beating up the cost accountants and finance people for using the standard cost model. That model is recognized by major organizations such as the Securities Exchange Commission and the Financial Accounting Standards Board in the United States and their equivalents elsewhere as a standard financial reporting model and the successful conduct of commerce of much of the world is based on their reporting. These organizations may eventually someday change their reporting requirements, but it seems a quixotical endeavor to run around your organization protesting against use of the standard cost model.

Two, ensure that accounting, finance, and operations people all have a common, accurate understanding of operations science. With an objective, scientific understanding of how to create the best possible cash flow for an organization, all involved can make decisions that are best for the organization. As best possible cash flow is established, finance and accounting can use their standard cost model to report to their heart's content.

As described in Chapter 7 on the effects of operations science in facilitating motivation, then "…we are all focused on…" generating best possible cash flow and profitability will become the operative approach.

Summary Guidelines for Operations Measures

GUIDELINE	DESCRIPTION
Use Leading Measures for Control	Choose measures tied to performance drivers, for example, WIP instead of cycle time
Use Lagging Measures for Validation	Validate performance after the fact. These are not controls
Align Measures Across Levels	Tie strategy to department goals to individual action plans
Ensure Measures are Within Control	Don't hold people accountable for factors they cannot control
Monitor Process Stability	Use process behavior charts to detect real changes, do not react to "noise," natural variability
Use contribution margin at the bottleneck	Operations science shows this generates maximum cash flow. Don't confound its use with reporting requirements of standard cost model. Each has its own purpose.

We have introduced an operations science-based business management system framework and guidance on the appropriate use of measures. This framework provides a proven approach to ensure agile, predictive control is maintained and desired results are achieved in any organization. Operations science concepts and the management process together *will naturally lead organizations to their operations strategy and its sustained, successful execution*. Next, we discuss operations strategy formulation in consideration of operations science

---

[i] For an excellent discussion of this point as applied to Engineering and Construction projects, see https://projectproduction.org/journal/contrasting-project-production-control-with-project-controls/

[ii] https://quoteinvestigator.com/2015/03/02/eureka-funny/

[iii] For a practical, low-tech description of process behavior charts, see Graban, Mark, *Measures of Success: React Less, Lead Better, Improve More, (Constancy, Inc., 2019).*

[iv] The Enhanced Airline Passenger Protections Rule (14 CFR Part 259) was implemented in 2010, requiring airlines to: Limit tarmac delays to three hours for domestic flights and four hours for international flights, provide food, water, and lavatory facilities during extended tarmac delays, and allow passengers to deplane after a certain period of time. https://www.ecfr.gov/current/title-14/chapter-II/subchapter-A/part-259

[v] Zach Wichter, "Airlines Take a Proactive Approach to Potential Weather Woes," *New York Times*, March 14, 2017

[vi] Pound, Edward S.; Bell, Jeffrey H.; Spearman, Mark L. Factory Physics for Managers: How Leaders Improve Performance in a Post-Lean Six Sigma World (p. 199). McGraw Hill LLC. Kindle Edition.

[vii] Goldratt, Eliyahu M., and Jeff Cox. *The Goal: A Process of Ongoing Improvement*. 3rd ed. Great Barrington, MA: North River Press, 2004.

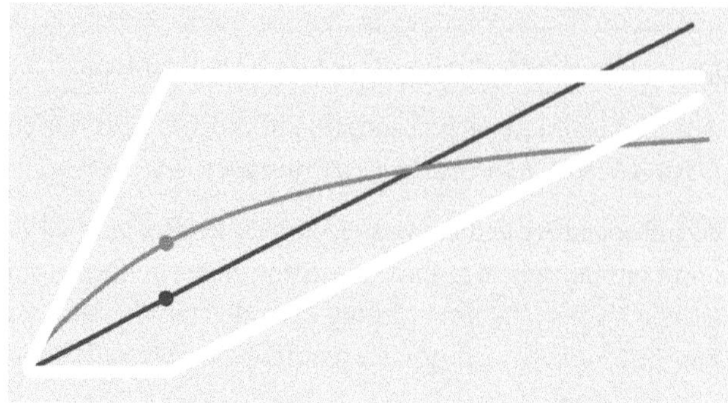

9

Strategy

Countless keystrokes have been devoted to discussion of strategy. Strategy, at its most basic level, is a plan or set of plans to accomplish an organization's objectives. Rather than generating more mountains of megabytes on strategy discussion, we offer the operations science concepts and the management process, which together will naturally lead organizations to their operations strategy and its sustained, successful execution. In this short chapter, we provide context for operations strategy, propose a definition that ties operations strategy quantitatively to its desired outcomes, and offer a case study for illustration.

Business Strategy

Business strategy drives operations strategy. Starting with business strategy and applying operations science through Chapter 8's operations management process will provide operations strategy, tactics, and controls that can be executed to achieve desired business goals.

Executive leadership, marketing, and finance will determine an organization's business strategy, which includes determination of the organization's customers, markets, products, services, and required quality levels, as well as the expected ranges of demand and required range of response time. This is what is meant by an organization's performance envelope. Leadership will also budget the resources, i.e., cost, it believes are required to achieve its goals. Business strategy should include policies and practices for development of its most important resource, its people.

We like the following guide for strategy evaluation. A good strategy should be SMART[i]:

Specific
Measurable
Assignable
Realistic
Time-related

While brief, the business strategy framework of defining a performance envelope, establishing a SMART strategy, and developing people contains the fundamental elements needed to be successful over the long term. As part of the business strategy discussion, there is an iterative discussion with operations about what can be achieved given an operation's available resources and options. So what then is operations strategy?

Operations Strategy

We define operations strategy as the plan to design, implement, and control an organization's portfolio of demand, variability, capacity, inventory, response time, and cost.

Operations strategy is ultimately the plan(s) for managing variability with investments in capacity and inventory to provide an acceptable time buffer. Another term we use for the combination of variability, capacity, inventory, and time is the *operations portfolio*. The term conveys the powerful attitude of risk management, similar to financial portfolios, that operations leaders should add to their toolkit. It takes operations leaders far afield from common brute force approaches such as "load up the process and keep everyone busy" or "we need to be like Toyota and reduce waste." Nothing against Toyota—it is a great company—but the "best practices" approach we discussed in Chapter 2 is a roll of the dice without starting from first principles to evaluate those practices. Future operations leaders managing their operations portfolio using operations science and their own unique experiences will surpass legacy approaches in both performance and clarity. They will select and manage their operations portfolio to support the business strategy and thereby repeatedly and sustainably provide the best desired profitability and service.

Quality and safety are vitally important and never should be ignored. We consider quality and safety key characteristics of the operations portfolio. For example, implementing tighter quality standards may increase costs. If so, it's essential to ensure that customers are willing to pay for higher quality. Quality and safety will be an integral part of operations strategy discussions and can be incorporated into this framework as needed. The framework is adaptable to all organizations.

Next, we provide a case study to illustrate strategic tradeoffs.

Business Strategy Drives Operations Strategy: The Portfolio of Buffers

The branding slogans and financial goals of FedEx reveal a business strategy of delivering packages quickly and getting paid highly for it. Figure 9-1 provides examples of FedEx branding slogans over time.

Figure 9-1: FedEx's marketing slogans promote its business strategy

In terms of the buffer portfolio, FedEx does not want much of a time buffer because customers pay extra for very short time buffers, e.g., overnight delivery. Inventory is not an option for FedEx because package delivery is a service. There is no stock of delivered packages, and FedEx does not know what the end item is until the package is presented to them for delivery. Thus, FedEx needs a significant capacity buffer for its operations to support its branding slogans. The FedEx website lists the following capacity:

- 697 planes
- 600,000+ employees
- 210,000+ motorized vehicles
- the world-famous Memphis Hub and 12 other air express hubs

This is a lot of capacity, but substantial capacity above average demand, a large capacity buffer, is what a short time buffer requires when inventory is not an option and there is variability. FedEx services extensive variability in delivery locations and package types.

FedEx is not the only company in the package-delivery business. Let's look at two of its large competitors and compare business strategies and buffer portfolios to think about operations science from a high-level, strategic view. Table 9-1 compares 2021 business performance (revenue), employees (capacity), and packages delivered (throughput) for FedEx, United Parcel Service (UPS), and the United States Postal Service (USPS).

	TOTAL REVENUE (IN BILLIONS)	EMPLOYEES	VOLUME (PIECES, IN BILLIONS)	PIECES PER EMPLOYEE	OPERATING PROFIT/ (LOSS) IN BILLIONS	REVENUE PER PIECE
FEDEX	$84	600,000	3.1	5,167	$5.9	$27.10
UPS	$97	534,000	6.4	11,985	$12.8	$15.16
USPS	$77	517,000	128.9	249,323	($4.9)	$0.60

Source: Data are from each company's 2021 annual report.

Table 9-1: 2021 performance comparison for package delivery competitors

In the authors' college-level classes and professional development workshops, participants find that the exercise of thinking about buffer strategy is fun and engaging because they get to think like a C-level executive should think. In broad strokes, they are considering where to invest capital and resources (capacity, inventory), with the resulting impact on customer service needed in their industry (time). In some cases, it seems easy to choose: Few students think ambulance services want a big time buffer (long waits), and they recognize that a service can't be inventoried, so a capacity buffer is needed. In other cases, the balance is more subtle: Will customers be willing to pay enough to fund capacity investments and low utilization levels needed for same day shipping? How much is an organization willing to limit its variability in terms of product and service offerings to reduce buffering costs?

The reader can try this exercise for themselves

1. Pick a business (perhaps their own).
2. State the business strategy.

3. Imagine a high-level portfolio of variability, inventory, capacity, and time to support the business strategy.
4. Check to see if their chosen portfolio aligns with what the business is doing and experiencing.

To aid in this pursuit and for evaluation of the organizations described in the later chapters, the Operations Strategy Summary Table (Table 9-2) is used. We have provided a draft of the table information for package delivery businesses cited above. Data is for each individual company's 2021 fiscal year from its annual report.

Operations Strategy Summary Table

	FEDEX	UPS	USPS
DEMAND	Package delivery of all sizes for individual and business customers.	Primarily small-package delivery for individuals and businesses. Supply chain support services such as freight forwarding and truckload brokerage.	Letters and primarily small packages for every individual and business in the United States.
TRANSFORMATION	Transform package location. - 600,000+ employees - 5,500 facilities in 220 countries - 210,000+ vehicles, major hub network - 697 aircraft	Transform package location. Transform information into supply chain management actions. - 534,000 employees - >1,000 facilities in 120 countries - >120,000 motorized vehicles - 595 aircraft owned or leased	Transform package location. - 517,000 employees - 34,451 facilities, all in the US - 232,368 motorized vehicles - 0 aircraft
VARIABILITY	High variability in package size, weight, destination, and demand fluctuations.	Moderate variability in package types and demand. Supply chain service demand.	Extensive variability in delivery locations and mail types.
BUFFERING OPTIONS: CAPACITY	Short term: labor, especially for seasonal spikes, e.g., Christmas package delivery. Longer term: labor, equipment, and facilities.	Short term: labor, especially for seasonal spikes, e.g., Christmas package delivery. Longer term: labor, equipment, and facilities.	Short term: labor, especially for seasonal spikes, e.g., Christmas package delivery. Longer term: labor, equipment, and facilities. Use of other companies' infrastructure, since USPS has no planes.

	FEDEX	UPS	USPS
BUFFERING OPTIONS: INVENTORY	Not an option, as FedEx is a service business.	Not an option, as UPS is a service business.	Not an option, as USPS is a service business.
BUFFERING OPTIONS: TIME	Minimal time buffer: overnight and express delivery as primary service. Can manipulate time buffer to throttle demand. For example, extend required time for promises of delivery by Christmas.	Moderate time buffer: mix of overnight and standard delivery options. Can manipulate time buffer to throttle demand. For example, extend required time for promises of delivery by Christmas.	Longer time buffer: lower-cost services like standard mail with slower delivery. For example, extend required time for promises of delivery by Christmas.
TOOLS AND APPLICATIONS	Advanced logistics, real-time tracking, hub-and-spoke distribution system, automation.	Route optimization, advanced logistics, tracking systems, automation.	National infrastructure, sorting centers, bulk mail handling, automation.

Table 9-2: Business portfolio analysis

Of course, business is about more than supporting a stated strategy for marketing and customer service. In for-profit businesses, it is ultimately about making money. At a high level, it seems FedEx needs more of a capacity buffer than the USPS since each USPS employee (see Table 9-1) accounts for 48 times more pieces delivered than FedEx employees. True but relative comparison of buffers is not enough to drive tactics that execute strategy and predictive control of profits.

The USPS is not close to profitable. The United States government, which established the USPS, will tolerate an operations portfolio that loses money in the course of achieving its mission of providing universal mail service within the country and to military and diplomatic personnel abroad. Can an operations science analysis provide determination if a profitable operations portfolio can be established? Indeed it can, objectively, and it can pinpoint actions required to achieve profitability. That being said, possible actions may have been already considered and rejected to date (e.g., no Saturday delivery), or proposed actions may not be politically palatable.

Operations science helps quantify the size of cost buffers (capacity, inventory) and customer service buffers (wait time) to provide predictive control of profits. As the following news article points out, it is not easy to balance the cost of capacity versus the profit benefit of reduced time for customer service in the face of variable demand. Between 2012 and 2015, both UPS and FedEx struggled during the Christmas season. See "Third Time's Charm for UPS at Christmas, but FedEx Stumbles"[ii] for more details and discussion of the problems and buffer adjustments both companies made.

Following is an non-exhaustive list of buffering measures taken by the companies to ensure on-time delivery during the spike demand of the Christmas season:

- Invest in new facilities, aircraft, and vehicles to increase capacity
- Hire tens of thousands of seasonal workers to handle increased demand
- Implement peak season surcharges to manage demand and encourage customers to ship earlier
- Make supply chain adjustments such as adding more sortation facilities, automate sortation, and increase use of rail transportation
- Invest in technology, such as data analytics, to improve efficiency and accuracy
- Develop contingency plans to address potential disruptions, such as severe weather events

So how has it worked out since then? When considering strategic options, it's good to evaluate the effectiveness of previous efforts. In Figure 9-2, we see on-time delivery results during the peak holiday shopping season by year by company from 2019 through 2024.

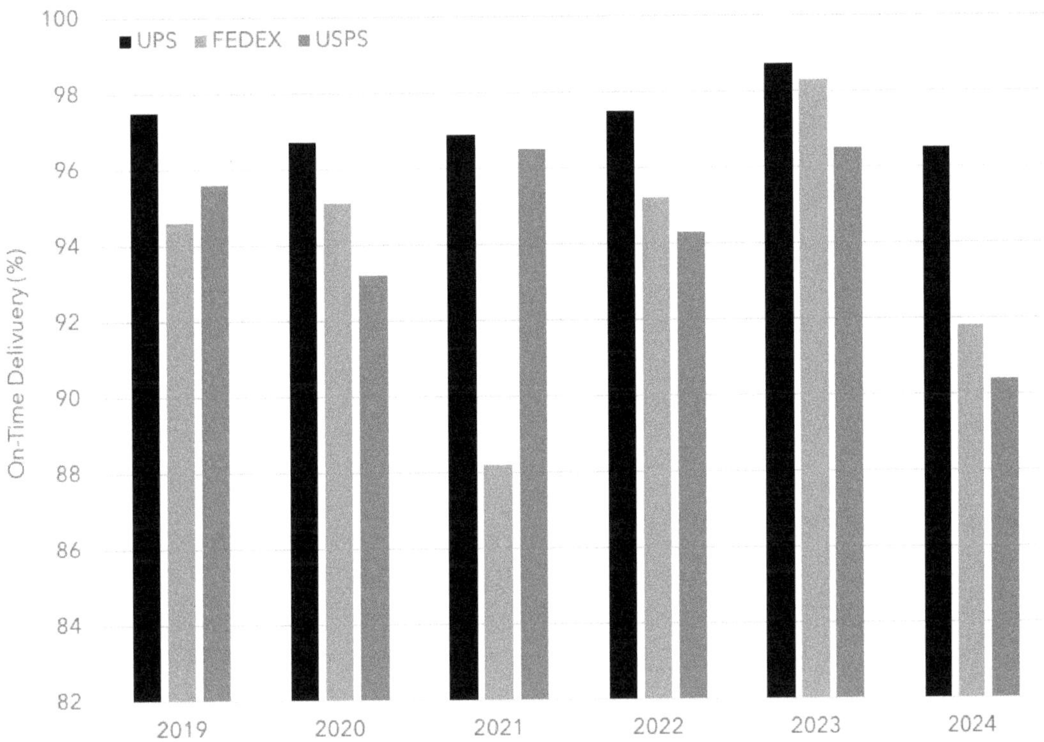

Source: ShipMatrix

Figure 9-2: Percent on-time delivery during peak holiday shopping season, by year

It looks like the portfolio configuration FedEx used during 2021 did not work out so well for them. UPS's operations portfolio appears to perform best for on-time delivery during the peak holiday season.

Figure 9-3 shows the same data by company by year. Initial impressions are that UPS's portfolio works best overall and is most consistent, for the period measured.

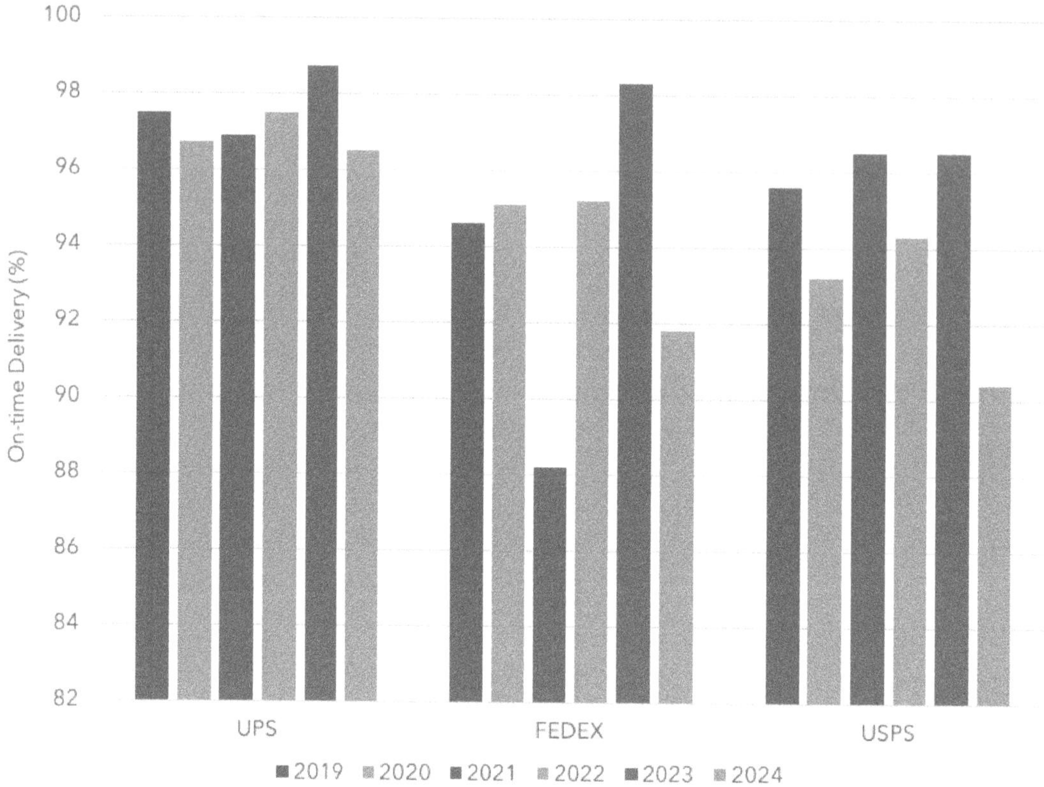

Source: ShipMatrix

Figure 9-3: Percent on-time delivery during peak holiday shopping season, by company

The graphs do not clearly show whether USPS or FedEx ranks second. So, we go to the data as shown in Table 9-3.

	2019	2020	2021	2022	2023	2024	AVERAGE	STD DEV
UPS	97.5%	96.7%	96.9%	97.5%	98.7%	96.5%	97.3%	0.008
FEDEX	94.6%	95.1%	88.2%	95.2%	98.3%	91.8%	93.9%	0.0346
USPS	95.6%	93.2%	96.5%	94.3%	96.5%	90.4%	94.4%	0.0235

Source: ShipMatrix *Table 9-3: On-time delivery data*[iii]

Here the numbers confirm that UPS has the best average on-time delivery *and* smallest standard deviation (least variability). The numbers also show that USPS gets second place, with both a higher average and a smaller standard deviation than FedEx.

An important question is whether the changing results represent statistically significant changes. When considering strategic choices, is the prime objective on-time delivery during the peak holiday shopping season? Profitability is a major objective unless you are backed by the federal government for your losses, but the USPS's operations portfolio does achieve the government's objective of universal mail service in the United States. Which strategic choice is best for a business is not for operations science to say.

The laws of chemistry will not tell you how to clear out skunks from a burrow under your house. You may put mothballs or a lighted stick of dynamite in the burrow. Chemistry will only dictate the results of those two actions. If you are not aware of those consequences and chose to use the dynamite, you will get rid of the skunk burrow, but you will also destroy your house.

Likewise, the laws of operations science will not dictate an organization's business or operations strategy. However, we have seen many organizations do great damage to their efforts due to ignorance of operations science and its consequences when crafting and executing their business and operations strategy. The operations portfolio view offered by operations science provides a powerful, practical framework for successfully and consistently creating, evaluating, and executing operations strategy in support of business strategy.

---

[i] SMART originated with George T. Doran's 1981 article, "There's a S.M.A.R.T. Way to Write Management's Goals and Objectives," published in *Management Review* 70, no. 11 (1981): 35–36.

[ii] Nick Carey, "Third Time's Charm for UPS at Christmas, but FedEx Stumbles," *Reuters,* December 29, 2015, https://www.reuters.com/article/us-ups-fedex-peak/third-times-a-charm-for-ups-at-christmas-but-fedex-stumbles-idUSKBN0UC1KP20151229.

[iii] Data are from ShipMatrix, as cited in Abbas Haleem, "Carriers Struggle with On-Time Performance in 2024 Peak Season," Digital Commerce 360, January 28, 2025,

https://www.digitalcommerce360.com/2025/01/28/carriers-on-time-delivery-performance-2024-peak-season.

10

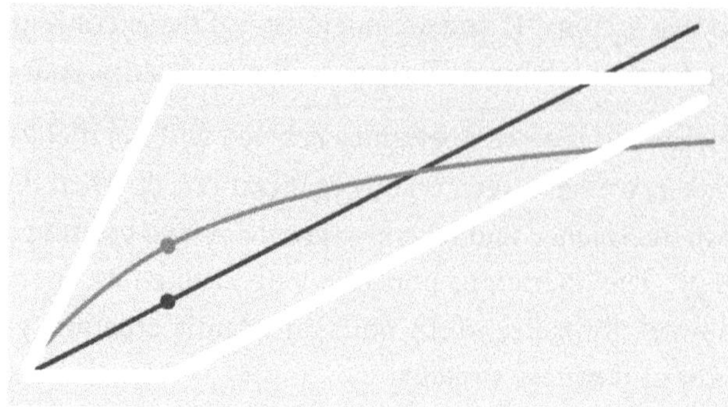

Demand

We have worked with hundreds of organizations applying operations science concepts. When starting work with an organization, after initial introductions and discussions of goals and challenges to provide context, operations science analysis always starts with understanding demand.

To do so, create a process map showing the operations flows which, in turn, defines sources of demand. As illustrated in Chapter 3, Figures 3-1 to 3-4, basic operations structure can be defined at any level of complexity starting with the simplest elements of an operation: transformation, queue, stock, and demand.

Too often, we see organizations move straight to discussion of continuous improvement initiatives for transformation, queues, and stocks without taking the time to understand the demand profile of the business. Another indicator of the absence of demand profile understanding is that, at the market level, the sales group offers all products with the same lead time to customers, regardless of product characteristics such as complexity, work content, and demand level. Ultimately, understanding the profile of demand and its variability is critical for making powerful decisions to improve competitiveness. The decisions drive needed investments in both transformation capacity and stock inventory, which delivers desired response times to beat the competition.

For example, in Chapter 15 we discuss a flowmeter business that has two distinct segments with different demand profiles. Flowmeters are devices used to measure the flow rate or quantity of a liquid, gas, or steam moving through a pipe or conduit. They are vital in many industrial and commercial businesses. The water meter on your house or apartment is a type of flowmeter to track amount of water used. In our Chapter 15 case study, the company grouped demand for flowmeters into two segments—new plant installation projects and emergency repair projects. In the example of a new plant installation project, there was typically significant visibility to demand and an ability to schedule with fixed lead times. That said, it was critical in these cases that on-time delivery be high so projects would stay on schedule. In emergency repair projects, short lead times were essential because plants could be shut down if a meter failed and needed to be replaced. Without this understanding of demand by segment, an organization might set a fixed lead time for all segments. Advertising a general fixed lead time of three weeks is nonsensical when the emergency response segment needs a one-day response time and the project segment is okay with a three-month lead time. Finally, although there were potentially thousands of different configurations for this engineered-to-order product, a Pareto (80/20) analysis of demand showed that much of the demand was driven by a subset of products that made a stock inventory viable. Without this understanding of the Pareto of demand, an organization might assume inventory was not a buffer option in an engineered-to-order product.

Chapter 16 focuses on a machining business that has three distinct market segments with different demand profiles. The prototype segment consists of truly engineered-to-order fabricated components with highly variable cycle times since the requested parts are being made for the first time. Customers typically want these prototypes as fast as possible to drive short product development time-to-market for new products. This requires a significant capacity buffer because an inventory stock is not possible and time is of the essence. The demand profile also includes an ongoing production segment that consists of repeat products where an inventory stock buffer is possible. Finally, there is a sporadic production segment where the same parts are only required every few years. In this case, there is typically some visibility to demand, so a time buffer is an option; however, inventory is less desired because it does not turn over quickly, and product version changes might occur during the time between orders which could obsolete existing inventory. Understanding the overall mix of these demand profiles is critical to developing operations strategy.

In summary, the concept that business strategy should drive operations strategy starts with understanding demand as driven by business strategy execution in the market. We provide considerations for employing operations science to sharpen demand analysis as a powerful competitive weapon.

Quantifying Demand

Demand should be quantified as a mean and variance for purposes of planning. Deterministic planning of demand that forecasts demand period by period over the planning horizon is an exercise in futility (see Table 10-1).

WEEK	1	2	3	4	5	6	7	8	9	10	11	12
DEMAND	5	20	35	28	16	23	45	47	38	23	16	7

Table 10-1: Deterministic demand forecasting

Firstly, a person with the ability to accurately forecast this way would be rich because they would also be able to forecast the stock market. Artificial intelligence will not, cannot, fix this. Secondly, this type of forecast requires a huge amount of labor to complete for any more than a few items or services. Artificial intelligence could assist in this area, but it would simply automate something that should not be done at all. Therefore, deterministic forecasting is the worst of both worlds. It is highly inaccurate, and it requires extensive time and effort.

Helpful hint: Calculate means and variances using periods of equal length. Using months is troublesome because some months have four weeks, and some have five weeks. Our recommendation is to use days, weeks, quarters, and so on because they have the same time content from one period to the next.

For fast, effective demand designation, specify a mean and variance of demand. For the data in Table 10-1, average demand is 25.3 and variance of demand is 189.1.

So instead of doing all the work to argue about and specify the numbers for each week, argue about and specify mean and variance. It is beyond the scope of this book but choosing the right distribution to fit the mean and variance can be a challenge. That's where you want expert statistics help.

Another consideration is planning horizon and seasonality. If your actual demand pattern year in and year out looks something like the data in Figure 10-11, it makes no sense to plan on the average and variance of demand for the entire year. A final consideration for demand planning is managing large, one-time orders. We worked with a company that took large orders, equal to about 20% of yearly demand, from big box retailers. There is nothing wrong with that; it is good revenue. Until it is not. The sales team then allowed the customer to push out its order whenever it wanted to. This is extremely disruptive to working capital and customer service management because there were long lead time items that had to be ordered for the one-time order and simply managing the reschedules for materials was an onerous task. Sales might dismiss operations' complaints about the schedule volatility with the retort, "Operations is not customer focused." However, a smart executive will realize that sales is not financially focused because of the significant expense and cash tied up with managing the material. Also, this self-induced variability adversely affects service to other customers, which belies sales' claim of customer focus. These competing tradeoffs must be managed, or chaos will ensue along with degradation to customer service and financial performance.

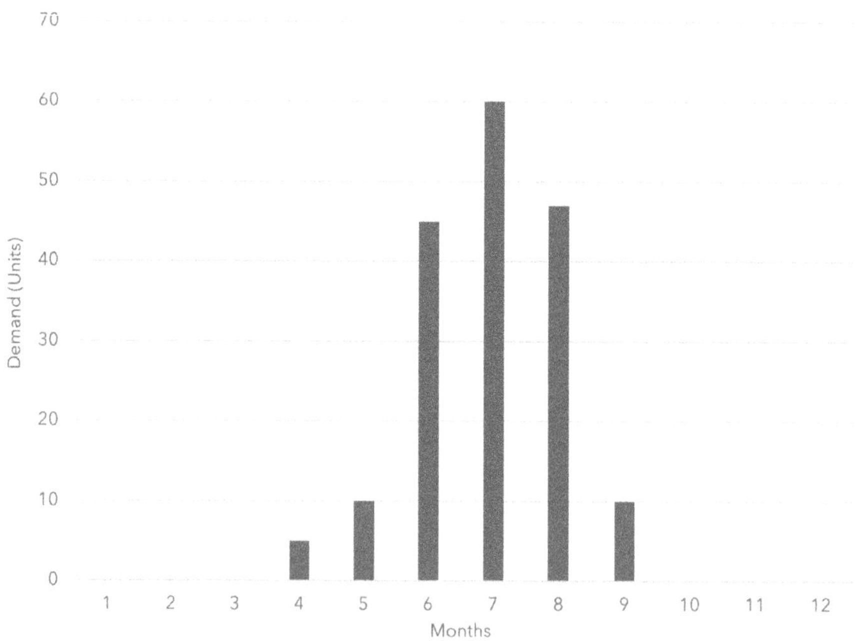

Figure 10-1: Seasonal demand

Cumulative Distribution of Demand

Illinois Tool Works (ITW) has become a Fortune 200 company, in part, through focusing on the 80/20 rule. This means focusing on the relatively small percent (20%) of customers, or products, or processes that provide a large (80%) percentage of value for the company, thereby focusing on the cumulative distribution of value being pursued. A cumulative distribution of demand provides a nice visual when analyzing demand (see Figure 10-2).

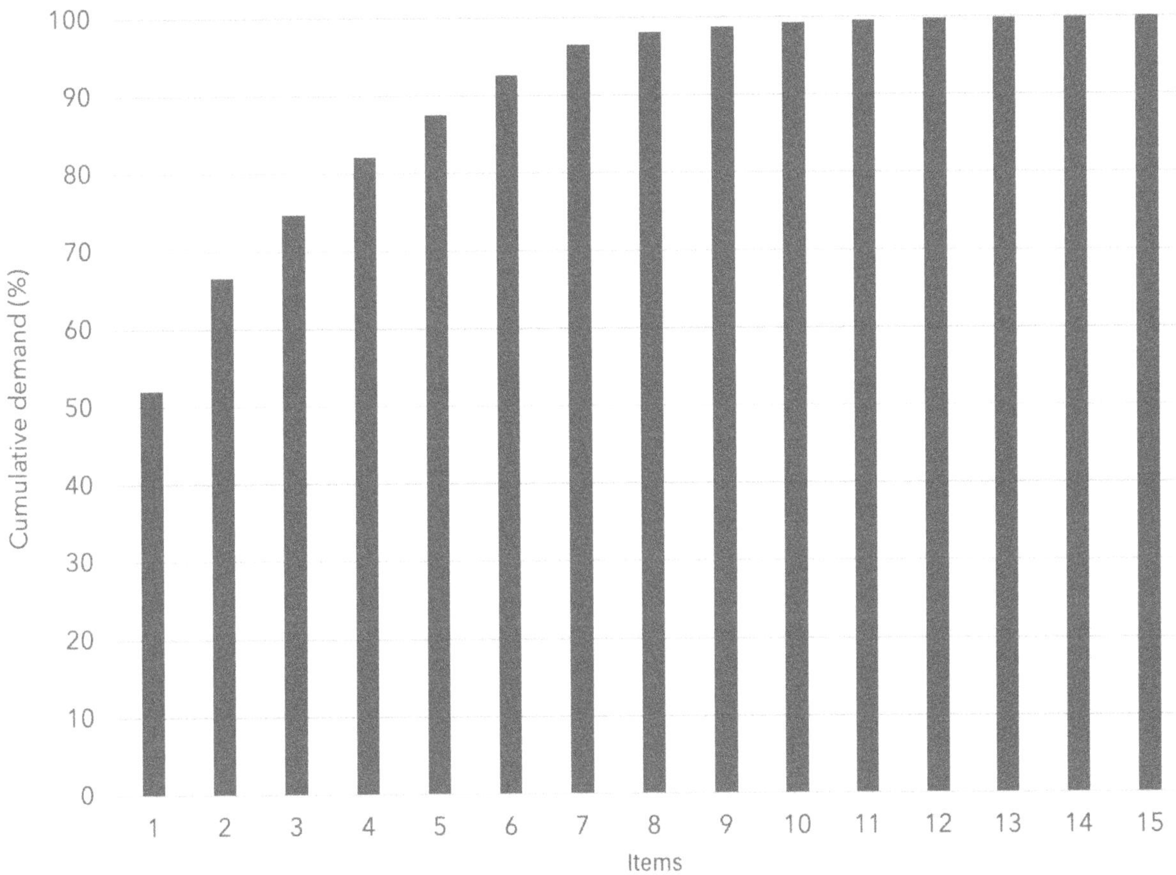

Figure 10-2: Cumulative distribution of demand

To create a cumulative distribution, use the following steps (see Table 10-2 for sample data):

1. Sort items or services or tasks by weekly demand, in descending order. Get the total.
2. Divide each item's weekly demand by the total.
3. Add each succeeding percent of demand to get the cumulative distribution. For example, 66.59 = 14.56 + 52.03.
4. Plot the cumulative distribution.

	WEEKLY DEMAND	PERCENT OF DEMAND	CUMULATIVE DISTRIBUTION
Item 6	431	52.03%	52.03%
Item 14	121	14.56%	66.59%
Item 15	67	8.09%	74.68%
Item 10	61	7.42%	82.09%
Item 4	45	5.41%	87.50%
Item 3	43	5.15%	92.65%
Item 1	32	3.87%	96.52%
Item 2	13	1.54%	98.06%
Item 5	6	0.68%	98.74%
Item 11	4	0.51%	99.24%
Item 12	2	0.29%	99.54%
Item 8	1	0.12%	99.66%
Item 13	1	0.12%	99.78%
Item 7	1	0.11%	99.89%
Item 9	1	0.11%	100.00%
TOTAL	829		

Table 10-2: Example for cumulative distribution calculations

In the example shown, more than 80% of demand occurs with only four products or services. Naturally, these four would be the place to start analysis for establishing best possible performance. The beauty of operations science is that the analysis does not stop there. We have seen Lean kaizen events where similar analysis is done, flow lines are established for the top four products, and typical results include reduction in cycle times and inventory. However, when questions come up about what to do for the other 11 items, guidance is vague. This is where operations science shines. It provides a comprehensive approach to understanding the tradeoffs between variability and buffering so a complete solution for all items is achieved. That doesn't mean the Lean efforts are discarded or dismissed, on the contrary, they are strengthened. Operations science strengthens and completes any continuous effort so that a comprehensive solution can be reached for best possible performance.

Forecasting

This section includes some general laws and guidelines on forecasting.

If the time to respond to demand by making a product is longer than the expected response time for providing the product, forecasting is required. In this situation, if one doesn't forecast, customers will always be unhappy because product cannot be provided within the expected response time. This tends to take care of the demand problem because customers go elsewhere. Services are required to forecast since, by definition, they can't respond to demand until the demand appears. As mentioned many times, services cannot produce stock inventory so they are not forecasting that; they must forecast capacity. Physical production processes must forecast capacity and inventory requirements.

There are some basic laws of forecasting:

1. The forecast is always wrong
2. The forecast will always change
3. The further out in time one tries to forecast, the less accurate the forecast
4. The more complex levels of detail one tries to forecast, the less accurate the forecast

Unfortunately, forecasting must be done at the level of the entity that consumes time on transformation resources for forecasting to be useful in operations planning and control. As Law 4 describes, this requirement induces inaccuracy in forecasting. For example, Ford Motor Company financial planning takes market research and the organization's financial goals into account to forecast the number of F-Series truck to be sold in a year. Operations must manufacture an individual truck with its accompanying options. If marketing and finance forecast 800,000 F-Series trucks are to be sold in a year, operations must work with those numbers and decide what the breakdown will be among F-150, 250, 350, and 450 models and their associated option packages.

Reporting high forecast accuracy at the F-Series level is of utility because it helps convince Wall Street and the public that company management has its act together. However, that level of forecast accuracy reporting is of little use to the people planning and running operations.

Finally, using mean absolute deviation (MAD) or mean absolute percent error (MAPE) to measure forecast accuracy may have some utility, but those measures will not provide predictive control when forecasting inventory levels. As we will show, Mean squared error (MSE) is required as forecast error measure to ensure quantitative and predictive control of inventory and service level.

Improving the Utility of Forecasting

For operations planning and control, forecasting must be provided at the entity level that consumes time on transformation resources. In a straightforward example, if a factory makes three different types of candy on different lines, it is not particularly useful for the factory to be provided a revenue forecast to hit without any guidance on which candies to make. At the same time, the laws of forecasting are in effect.

So, the need for understanding how a resource's time will be used creates a dilemma where a planner is caught between the effort required to forecast in detail and law 4 above. Following are some contrived examples for illustration.

CANDY FACTORY FORECAST 1 (Useful for operations planning and control)

PRODUCTION LINE	MONTHLY DEMAND (IN NUMBER OF BARS)	MONTHLY STANDARD DEVIATION (IN NUMBER OF BARS)
SUGAR DADDY BARS	20,000	3,000
SUGAR MOMMA BARS	40,000	5,000
SUGAR BABY BARS	80,000	8,000

CANDY FACTORY FORECAST 2 (Not so useful for operations planning and control)

$360,000 per month

A finance or sales person may think the financial forecast is just fine for their purposes, and it probably is. The problem is that the forecast provides very little information for the transformation part of the business to plan effectively. This is the classic argument:

Sales says to Operations, "You never make what I want to sell."

Operations says to Sales, "I'd make what you want, but you can't tell me what you want; you are always changing it."

Here we come back to the idea of a performance envelope. The performance envelope consists of an average and range (variance) of demand. If demand is within that range, and operations misses delivery, that's the fault of operations. If demand is outside the agreed envelope, operations is not to blame. The decision needs to be made to either replan for higher demand or accept the hits to customer service caused by the unanticipated demand.

Of course, this is the logical, rational operations approach, but people are not always logical and rational, and operations is no exception. For example, when a sales person is rewarded for total sales, they sometimes keep accepting orders way above capacity because it is in their financial interest. They are acting rationally, at least in the short term, because of the incentives they have been given. Over the long term, continued poor performance by operations loaded above capacity will drive customers away and hurt the sales person's financial interest.

While we are here, the same goes for front-end engineering and design (FEED) engineering capacity (measured in hours) on a large construction project.

ENGINEERING HOURS FORECAST 1 (Useful for project control)

TASKS	MONTHLY DEMAND (IN HOURS)	MONTHLY STANDARD DEVIATION (IN HOURS)
DESIGN AND SIMULATION WORK	2,000	300
PROTOTYPE BUILD COORDINATION	4,000	500
FIRST-RUN STUDIES	8,000	800

ENGINEERING FORECAST 2 (Not so useful for project control)

$720,000 per month

The same tradeoff between forecast usefulness and forecast effort and accuracy goes for staffing nurses at a hospital.

NURSING HOURS FORECAST 1 (Useful for planning nurse staffing)

SPECIALTY	MONTHLY DEMAND (IN HOURS)	MONTHLY STANDARD DEVIATION (IN HOURS)
POST-ANESTHESIA CARE UNIT	2,000	300
EMERGENCY DEPARTMENT	4,000	500
INTENSIVE CARE UNIT	8,000	800

NURSING FORECAST 2 (Not so useful for planning nurse staffing)

$540,000 per month

Finally, the good news is that the forecasts don't have to be accurate to achieve predictable results. In each of the cases above, the focus is on forecasting transformation capacity required to meet demand.

Quantifying Forecast Error

There are a few considerations to ensure that the use of forecast error supports scientific control of operations.

First, forecast error should be measured as MSE.

$$= \frac{1}{n}\sum_{i=1}^{n}\left(D_i - \widehat{D}_i\right)^2$$

where

n = number of data points

D_i = actual demand

$\widehat{D}_i$ = forecast demand

The reason for use of MSE as opposed to other common errors such as MAD or MAPE is that MSE is a variance measure. As discussed in Chapter 5, the standard deviation of replenishment time demand

$$\sigma_{RTD} = \sqrt{\ell\, \sigma_D^2 + d^2\sigma_\ell^2}$$

is a key parameter in determining inventory policy. A component of σ_{RTD} is σ_D^2 which is variance of demand.

If one is trying to predict the amount of inventory needed in the future, it is a reasonable assumption that the past forecast error is a good proxy for the variance in demand for future requirements. If the format of MAD, MAPE, or others are used, they do not fit the statistical description of variance needed to do a statistical analysis of inventory requirements to establish optimal inventory policies.

Second, forecast error should be measured at the replenishment time of a project or service. This should be self-evident. There is little utility in measuring forecast error for demand at one week or six months if replenishment time in responding to demand is three months. However, it is highly common for forecasting packages to provide forecast errors along an entire planning horizon. Software providers may do this to demonstrate the vast computational power of their software rather than because such detail is of any use. For the unsuspecting user, such displays of analytical complexity may be perceived as evidence of the power of the software when, in fact, it merely demonstrates the ability to perform calculations whether they are useful or not.

When Forecast Error Makes Things Worse

If forecast error is greater than the historical variance of demand, then using forecast error increases variability—making things worse. In this case, the use of forecast error for future demand variance will lead to, for example, more projected on-hand inventory requirements than are actually needed.

As introduced in Chapter 5, a simpler approach is to use the historical variance of demand and the assumption that past variance-to-mean ratio is equal to future variance to mean:

$$\frac{v_h}{m_h} = \frac{v_f}{m_f}$$

where

v_h is historical variance of demand,

m_h is mean historical demand,

v_f is future variance of demand, and

m_f is future mean demand.

In this calculation, the two historical quantities are known. The future mean demand is specified as a forecast, so the future variance of demand can be calculated, which allows calculation of optimal inventory policies.

Conclusion

With a firm understanding of demand, an organization can focus on the investments needed to meet it. Quantifying demand and its variability enable straightforward and effective analysis of operations strategy options for transformation, stock inventory, and response time, as covered in subsequent chapters.

11

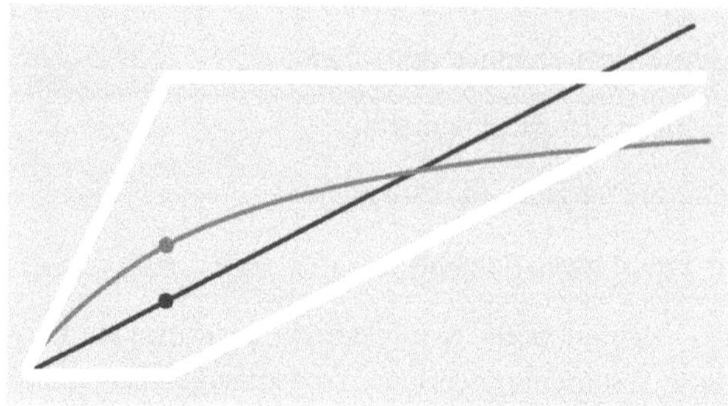

Transformation

Transformation is the use of a set of resources to convert one or more attributes of a given entity to satisfy a given demand, that is, to produce desired throughput. The work of transformation occurs in flows as defined in Chapter 4.

Parameters for understanding and controlling transformation to produce desired throughput are

1. capacity
2. work in process
3. variability

Cycle time is also a primary concern for people who manage flows. Recall from Chapter 3 that cycle time is the average time for an item to traverse a flow, from the time it is released until it is completed. Cycle time is the sum of all process times, queue times, batch formation times, down times, changeover times, and so on.

A simple but profound result of operations dynamics is that cycle time is a lagging indicator. One must have capacity resources and WIP in the flow to produce throughput, and only when the transformation is complete does one know cycle time. We have seen organizations focus on improving cycle time by measuring and monitoring cycle time; this is like driving by looking in the rearview mirror. Certainly, one must measure cycle time to know if and how it is changing, but controlling and predicting cycle time requires calculating and controlling capacity and WIP for expected levels of variability. If variability can be controlled, that is another lever for managing cycle time.

To aid operations managers, in this chapter we elaborate on the technical science of chapters 4 and 6 to provide less technical, but still scientific, insight on transformation. A comprehensive list of the elements of cycle time is provided. Queue time and batch time are usually the two largest drivers of cycle time. We will summarize those in this chapter. An open queuing network provides a special case in practice illustrating the combined effects of variability, utilization, and process time when determining queue time.

WIP control is a powerful method for managing cycle time and throughput. We will build on the earlier discussion. Practice and definition of WIP control to date has led to much obfuscation of the science, particularly in the descriptions and practices that have emerged from the Lean movement.

In what seems to us a simple and painfully overlooked solution, we will provide resolution, once and for all, to the great push-versus-pull debate and, in the same fell swoop, eliminate the seemingly endless confusion around definition and use of the term *pull* in operations management.

Finally, we cannot discuss operations science without discussing its connection to business finance and accounting. There is typically somewhat of a contentious flavor to the relationship between operations and finance. Some of this is natural creative tension, but too often the groups are in a tug-of-war that drives dysfunctional, wasteful behavior. Many propose that accounting and finance need to change and do "Lean accounting." This is an unnecessary and quixotic quest. We describe a solution that has been proven to reduce conflicts between operations and finance, unite efforts, and accelerate success.

Capacity and Utilization

Capacity is the bedrock buffer. Without capacity, one can't deliver goods and services to meet demand. Proper capacity calculation is vital to establishing control over operations, yet we have seen many people confused about the calculation. People in service organizations and project managers in any organization seem especially confused about capacity management and calculation. It's understandable, given the virtual nature of the work. However, an engineer performing finite element analysis or a nurse caring for patients is an expensive capacity resource that is governed by the laws of operations science.

We were working with a hospital emergency room to decrease patient flow time, and one of the nurses said, "We are treating *people* here; it's not the same as building widgets." Our thoughts on that line of reasoning? Quality control at many hospitals is not nearly as good as it could be. "Scrap" is buried, and "rework" means more revenue for the hospital. Medical personnel are wonderful in their dedication to helping others, but focusing solely on patient care without understanding the best way to manage and control the delivery process ultimately undermines patient care.

Project managers also get tied up by a lack of understanding and control of capacity. The desire to keep everyone busy all the time as a proxy for efficiency is especially damaging to project delivery cost and schedule performance.

Managers may think the worst thing they can see is to walk out on the production floor, or the hospital floor, or the project site, or the office floor, and see people with no "productive" work to do. However, if work is scheduled for, say, 80% utilization to account for variability, a manager should expect to see resources with no productive work around 20% of the time. That also does not mean workers are sitting around having bon-bons and espressos. They can be doing many other things such as working on process improvement, acquiring new skills, and attending meetings. This is not poor management or a cause for frustration; it is good control.

If capacity utilization is not properly calculated and controlled, all bets are off for maintaining control and achieving best possible profitability. Proper capacity calculation yields three critical operations insights:

1. Bottleneck identification, it is not always the slowest or most expensive resource

2. Determination of utilization levels to understand expected cycle time
3. Quantification of improvement opportunities through scenario analysis

We covered the calculation of utilization in extensive detail in Chapter 4, to review:

$$Utilization = \frac{time\ used}{time\ available} \text{ or } \frac{rate\ of\ arrivals}{effective\ rate} \text{ or } \frac{demand}{capacity}$$

Utilization is dimensionless and always between 0 and 1 or 0% and 100%. The difference between the actual utilization and 100% is the capacity buffer. When resource utilization is planned at 80%, this means there is a 20% capacity buffer. Let's look at a simple example for vital managerial guidance.

Would you and 100% of the people responsible for running your operations be able to calculate the capacity of the process Figure 11-1? Hint: It is not 50 per day. 50 per day is the capacity of each individual process center.

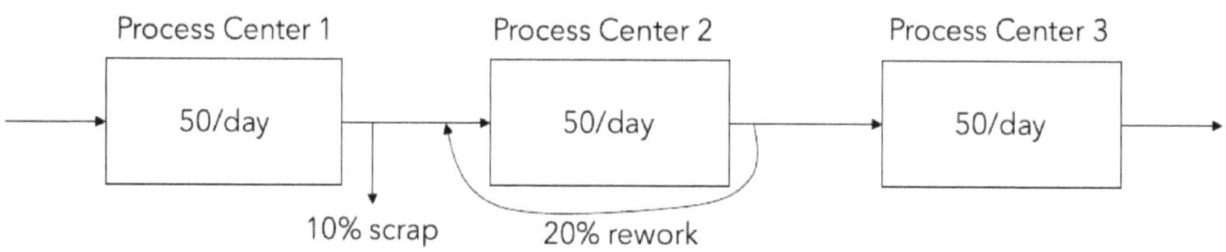

Figure 11-1: A simple utilization calculation example

We have found that, often, people are confused about the calculations for this simple process. Imagine what that means for managers running complex real-world processes or projects. Put the following exercise to your people and see how well they do with it.

In Table 11-1, we use 40 per day as the demand and focus only on utilization and capacity and examine data for process centers (PC) 1 to 3. You can use the VUT equation as described in Chapter 4 to calculate expected cycle time.

	RELEASES	PC1	PC2	PC3
STARTS	= 40/0.9 = 44.4	44.4	40	40
THROUGHPUT		40	40	40
UTILIZATION		= 44.4/50 = 0.89	= (40 × 1.2)/50 = 0.96	= 40/50 = 0.8

Table 11-1: Simple example of capacity calculations

Since there is 10% scrap, we must start 40/0.9 = 44.4 to get 40 completed. For PC1, we are assuming all work is done and then the part is inspected and scrapped if defective.

Next ask what the bottleneck is. The answer is process center 2, which has the highest utilization at 96%. Finally, ask what the capacity of the process is. The answer is the throughput of the process divided by the bottleneck utilization so, 40/0.96 = 41.7/day. One hundred percent of your people managing your production, services, or projects should understand this calculation exercise. If they do not, you are probably wasting time and money.

A common mistake is focusing solely on value-added time or the most expensive equipment when analyzing capacity. The operations science approach provides a more complete picture by considering all factors affecting resource utilization. The operations science fundamental relationship shows that striving for highest possible utilization levels (very small capacity buffers) is usually not a cost effective option.

Cycle Time

Cycle time is a critical performance parameter for most organizations. "When will I get it?" for products or "When will it be done?" for services are common questions for any customer.

For physical products, if a buffer of stock inventory is available, the answer usually is, "You will get it in the *x* days it takes for shipping and delivery." Stock inventory is usually of great interest to organizations, due to its role in satisfying customer demand and its financial implications on the balance sheet.

As discussed in Chapter 5, cycle time is a primary driver of stock inventory. So even if your organization relies on stock inventory for quick customer response, understanding how cycle time (or replenishment time) affects stock inventory is a key driver of success.

In looking at cycle time, we will also provide clarity on cycle-time drivers and classifications that seem to have been lost or confused in the popular best-practice efforts of the last few decades—such as Lean, Six Sigma, Agile, Theory of Constraints, and their ilk. First, we list the elements of cycle time.

Cycle Time Elements

Understanding the elements of cycle time provides great insight into process improvement opportunities. Since process cycle time is a lagging indicator, it is a *symptom* of process performance. Cycle time is not a *cause* of process performance for good or bad.

The elements of cycle time are known; it is not a black box. One doesn't need to conduct designed experiments to "discover" what drives cycle time. Table 11-2 lists the elements of cycle time; definitions are in the glossary. Queue time and batch time commonly comprise the largest portion of cycle time. We will review queue time calculations here and demonstrate batch time calculations.

CYCLE TIME ELEMENT	ABBREVIATION
BATCH TIME	BT
DOWN TIME	DT
MOVE TIME	MT
PROCESS TIME	PT
PLANNED TIME BUFFER	PTB
QUEUE TIME	QT
SHIFT DIFFERENTIAL TIME	SDT
SETUP TIME	ST
WAIT-TO-MATCH TIME	WTMT

Table 11-2: Elements of cycle time

One element that is an exception to cycle time as a symptom of process performance is planned time buffer. This is a scheduling tactic commonly used in construction where waits are deliberately placed in a schedule as a management policy decision.

Queue Time

A major component of cycle time in any process is wait time, or queue time. We have already covered the VUT equation extensively in Chapter 4 as an approximation of queue time but list it here again for convenience:

$$CT_q = \frac{c_a^2 + c_e^2}{2} \times \frac{u}{1-u} \times t_e$$

where

c_a^2 = squared coefficient of variation of time between arrivals

c_e^2 = squared coefficient of variation of effective process time

u = utilization

t_e= effective process time of a process batch

The following observations concern the VUT equation:

1. It explicitly applies for open queuing systems where WIP is introduced to a process (the arrival rate) regardless of the amount of work already in the process. This behavior is often labeled as a "push" system. However, the dynamic interaction between variability, utilization, and process time occurs in any system, so the VUT equation is a great model for informing your intuition.

2. It is for a single resource process center. As described in Chapter 4, there is an approximation for multiple parallel resources in a process center, with m = the number of resources. Note that if m = 1, the equation reduces to the same VUT as above:

$$CT_q = \frac{c_a^2 + c_e^2}{2} \times \frac{u^{\left(\sqrt{2(m+1)}\right)-1}}{m(1-u)} \times t_e$$

3. The VUT approximation is not accurate for squared coefficient of variation (SCV) values greater than 3. In any case, it is a powerful illustration of operations behavior dynamics whether or not the accuracy of the mathematical result is compromised by high SCVs.

4. The VUT shows the queuing effects of variability, utilization, and process times are multiplicative. If utilization goes from 90% to 95%, the utilization factor goes from 9 to 19. Combining high variability with high utilization is a recipe for very high cycle times.

5. t_e is effective process time for a process batch, not for one item. Doubling process batch size obviously doubles batch processing time, but it also doubles batch queue time.

Look at the impacts of variability graphically when considering the value-added construct. If we look at the cycle time over the entire range of utilization, we get the graph in Figure 11-2.

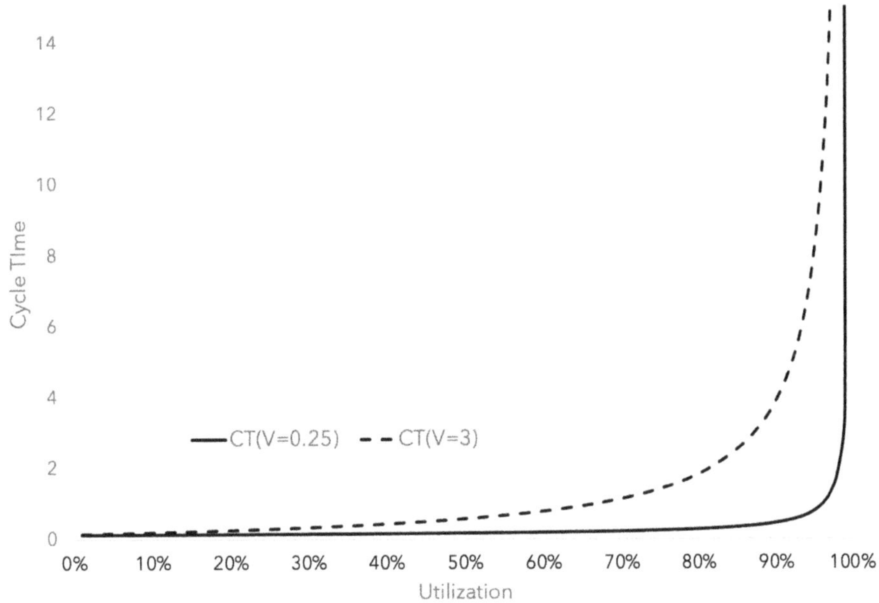

Figure 11-2: Cycle time with queueing

The graph shows how high variability blows up cycle time much sooner than low variability. This is a very important insight from operations science. The V values are the V in VUT,

$$V = \frac{c_a^2 + c_e^2}{2}$$

We have tested intuition about this relationship with literally thousands of people, who have consistently demonstrated great confusion as was illustrated in Chapter 2, Figure 2-2.

Next, we return to the value-added concept, as shown in Figure 11-3. Note that utilization only goes to 80% on the x-axis and that the y-axis scale is much smaller than for Figure 11-2.

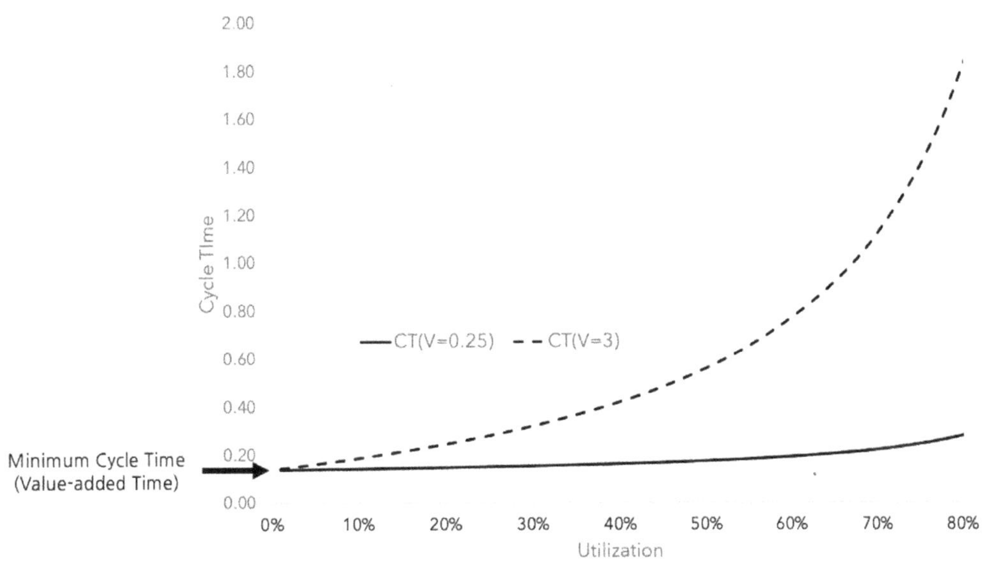

Figure 11-3: Process time, aka value-added time, and cycle time

Value-added time is the minimum cycle time, the process time. This is commonly called the touch time and we have seen only touch time entered when calculating time used for determining utilization—this is a serious mistake. See the utilization calculations above and the Chapter 4 utilization calculations to understand why. A few more observations from Figure 11-3:

1. All the work on capacity calculations only gets one to the starting line in good operations management. Even so, it is vitally important to calculate capacity correctly so you are at the correct starting line. Imagine showing up for the 100 meter dash only to find out you are entered in the 1,500 meter race—no gold medal for you, maybe a trip to the hospital instead.
2. However, the capacity calculation says nothing about the effects of variability, which is where many management efforts falter.
3. As shown in Figure 11-3, value-added time *is* achievable if there is little utilization of equipment and very little variability. If, in pursuit of value-added time as process cycle time, your company can be profitable by making only one product that is ordered in nearly the same amount every day and you can also afford low utilization of capacity, go for it.
4. Note that even at low utilization levels, variability effects are significant. In Figure 11-3's high-variability scenario, cycle times are twice the low-variability value, at about 30% utilization.
5. Since cycle time is visible WIP, consider the curves an indicator of WIP levels in a process. At high utilization levels, WIP "explosions" are the result, and WIP is everywhere. Preventing WIP explosions is why WIP control is important and a main topic of the next section.

Batch Time

Before we get to WIP control, let's look at the other major contributor to cycle time—batch time. Batching is a perfectly logical tactic to use, and we covered batch size optimization extensively in Chapter 6. The Chapter 6 discussion was heavily technical. Here we step back and look at the effects of batching from a conceptual level.

Batch time is the additional cycle time caused by batching items as they complete a process. It is different than queue time. Batch time originates with the batch size selected and results in items waiting to form a batch. For example, if the batch size is seven, the first item must wait until the remaining six items arrive, the second item waits for the remaining five, and so on. When a batch arrives at a sequential processing center, the last item must wait for the preceding six items to complete before it is processed. If the process center is a batch processing center, such as heat treat, many items can be loaded at once so the wait to process time is zero assuming the entire batch can be processed at once. There is no algorithm for calculating batch time as it depends on the configuration of process times and set up times at process centers in a routing.

To illustrate the effects of batching, we start with the simplest case, a batch size of one. The simulation process consists of three process centers, each with a setup time and a process time. The black shape represents a setup. See Figure 11-4.

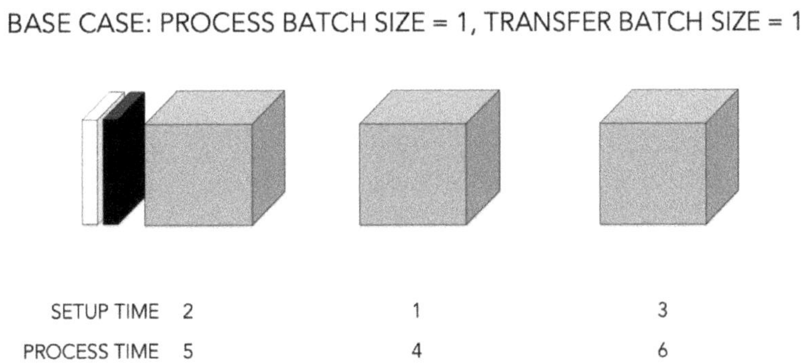

Figure 11-4: Batch time simulation, base case (t=0)

After seven hours, the item is through the first process center and ready for setup on the second process center (Figure 11-5).

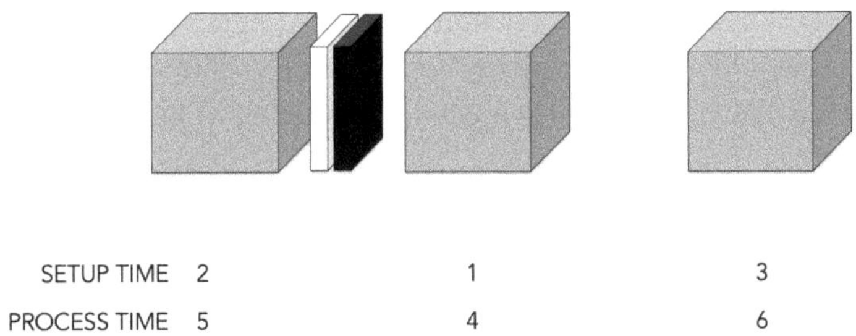

Figure 11-5: Batch time simulation, base case (t=7)

After 12 hours, the item is through the second process center and ready for set up on the third process center (Figure 11-6).

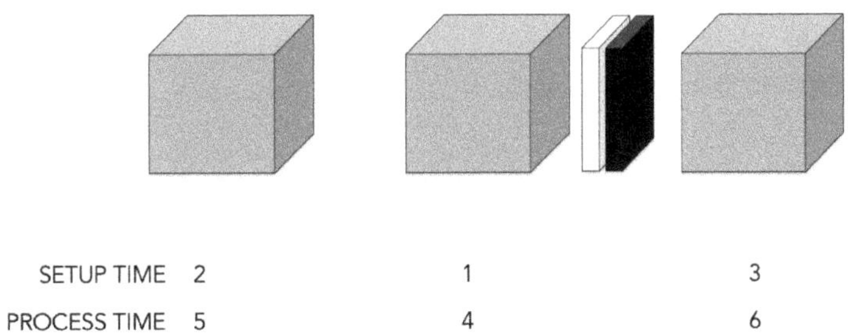

Figure 11-6: Batch time simulation, base case (t=12)

After 21 hours, the item is finished (Figure 11-7). Total cycle time is 21 hours; batch time is zero.

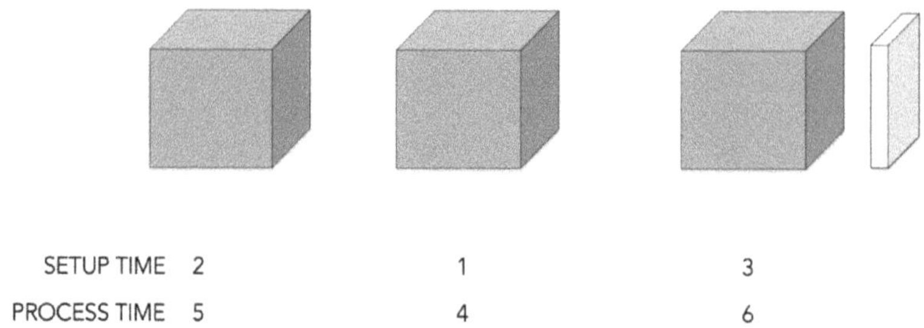

Figure 11-7: Batch time simulation, base case (t=21)

This simple illustration shows the effects of one-piece flow and why, in operations with setups, that one-piece flow implementation can meet with strong resistance. Certainly, it would be great to get rid of setups or make them nearly instantaneous, but often that is not possible. Formula 1 racing teams think the tradeoff in cost and resources is a good investment to radically reduce pit stops, which are essentially setups, to install new tires and fuel up. A company manufacturing railroad tank cars and using a 2,000-ton hydraulic press to fabricate tank car endcaps is much more tolerant of batch sizes greater than one.

Next we illustrate what happens when we increase the process batch size to seven. The transfer batch size remains one. (Figure 11-8).

SCENARIO 1: PROCESS BATCH SIZE = 7, TRANSFER BATCH SIZE = 1

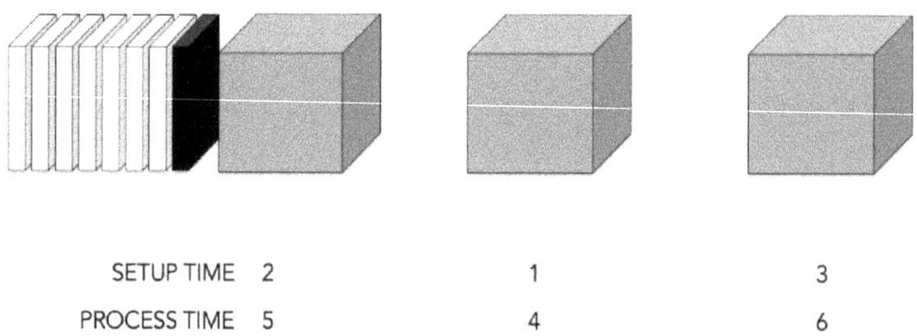

Figure 11-8: Batch time simulation, scenario one (t=0)

We will not show all states of this simulation. You can see a quick video of the simulation at https://opscience.org/batch-time-drives-cycle-time/.

Figure 11-9 shows what the process looks like at 21 hours. The first item has finished at the same time (21 hours) as in the base case, but there are other items at various locations in the process.

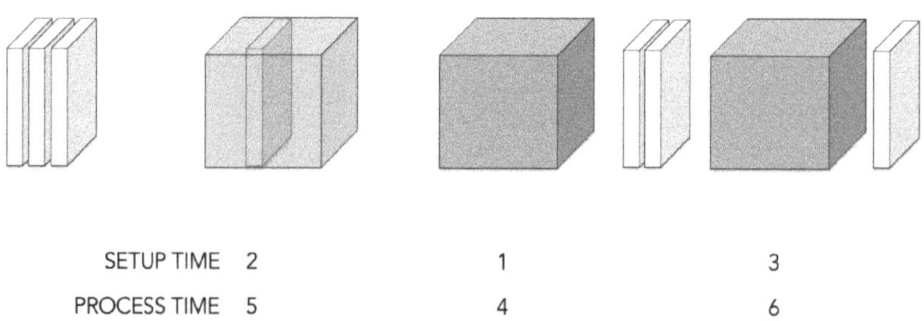

Figure 11-9: Batch time simulation, scenario one (t=21)

After 33 hours, Figure 11-10, three items are complete and all items have started.

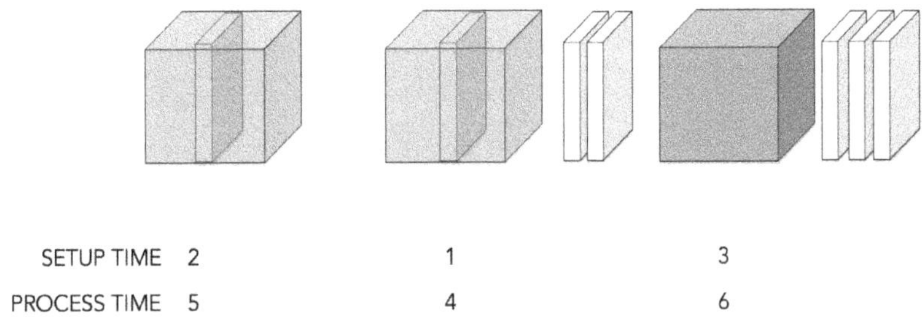

Figure 11-10: Batch time simulation, scenario 1 (t=33)

All items are complete after 57 hours (Figure 11-11).

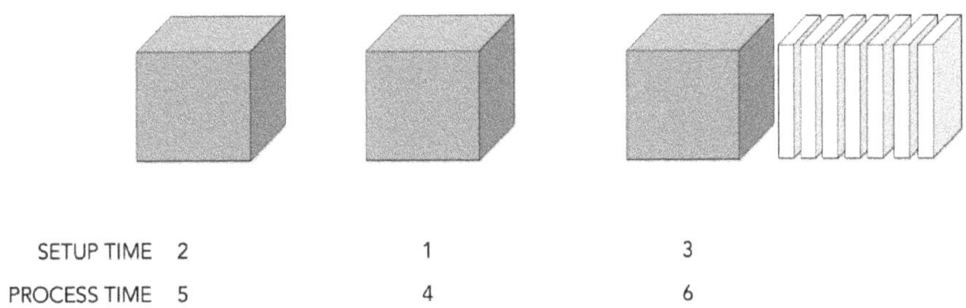

Figure 11-11: Batch time simulation, scenario 1 (t=57)

Table 11-2 displays the data for the individual items in Scenario 1.

	PROCESS TIME (HOURS)	CYCLE TIME (HOURS)	BATCH TIME (HOURS)
Item 1	21	21	0
Item 2	15	27	12
Item 3	15	33	18
Item 4	15	39	24
Item 5	15	45	30
Item 6	15	51	36
Item 7	15	57	42
Average	16	39	23

Table 11-2: Scenario one data

Note that item 1 gets all the setup time as part of its process time. After that, each item's process time is the same. As would be expected, the last item in the batch has the longest cycle time. The difference between an item's cycle time and its process time is its batch time. The cycle times would be longer if the transfer batch sizes were greater than one since there would be even more wait time for accumulating and waiting in transfer batches.

One more observation is that batch time is a direct result of the specific configuration of process times and setup times. In Scenario 2, Figure 11-12, we swap the first and third process centers in our simulation and get different results as shown in Figure 11-3, even though we have not changed the process times or setup times.

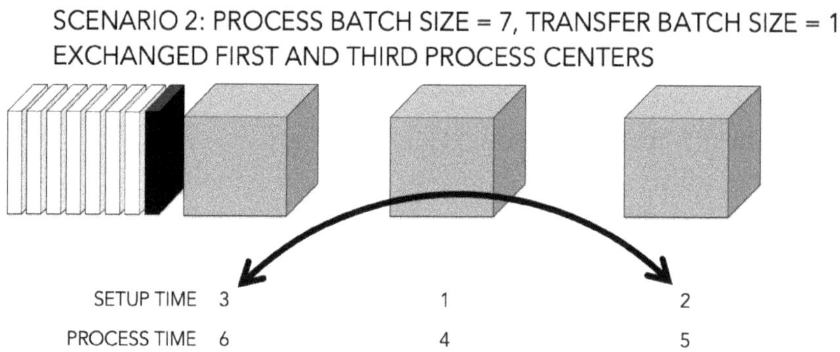

Figure 11-12: Batch time simulation, scenario two, swap PC1 and PC3

		SCENARIO 1	SCENARIO 2	SCENARIO 1	SCENARIO 2
	PROCESS TIME	CYCLE TIME	CYCLE TIME	BATCH TIME	BATCH TIME
ITEM 1	21	21	21	0	0
ITEM 2	15	27	26	12	11
ITEM 3	15	33	31	18	16
ITEM 4	15	39	36	24	21
ITEM 5	15	45	42	30	27
ITEM 6	15	51	48	36	33
ITEM 7	15	57	54	42	39
AVERAGE	16	39	37	23	21

Table 11-3: Effect of rearranging sequence on batch time

While batch time is a distinct phenomenon, its effects are not considered if using the EOQ since the optimization is for inventory carrying cost versus setup costs. However, we discussed batch time optimization for minimizing cycle times in Chapter 6. People often do not use complex batch size optimization calculations when setting batch sizes, they go with rules of thumb or simple estimates. A classic mistake is trying to minimize setup time by running large batches.

Setting batch sizes influences WIP levels in a process. In the design, implement, control approach, it is vitally important to design and implement effective WIP control. One may have determined an optimal batch size, but if there are too many optimal batches in process at any one time, the results will still be longer-than-optimal cycle times.

WIP Control

WIP control is a massively important lever for performance control. We will first add detail to the concepts introduced in Chapter 4. Then we will provide practical observations and calculation examples for the reader to use in their processes and projects.

In any process, if WIP arrives at a rate faster than the process' bottleneck rate, work is going to back up, and cycle times will explode depending on variability and the difference between the arrival rate and the bottleneck rate. When traveling through downtown Atlanta, Georgia, on Interstate 75 at 2 a.m. on a Wednesday morning, speeds of 70 or 80 miles an hour are common. It's a different story at 5 p.m. on a Friday afternoon: Average speeds of 20 miles per hour or less are common. Let's indulge in some commuter fantasy. Imagine that we limit Friday afternoon highway traffic in Atlanta to the maximum number of cars that would facilitate traffic flow at 60 miles per hour. It would probably never happen, but it is possible. For the Atlanta highway driver, this would mean shorter cycle times, i.e., higher speeds. Certainly there would also be very many unhappy drivers who were prevented from highway travel. The other parameter to consider is how many cars would be able to get through downtown during rush hour, that is, the throughput. The point is that limiting the amount of WIP in a system to the system's capability to process WIP eliminates much chaos and waste and decreases cycle times. This cycle time reduction must be balanced against throughput requirements. Operations managers and project managers ignore these fundamental behaviors, to their lasting pain. It is also common for project and operations managers to understand that limiting WIP reduces cycle time. However, they still get in trouble because they can't quantify the trade-offs between WIP, cycle time, and throughput, so they fly blind and their results are unpredictable. Depending on good luck is not a recommended strategy for success.

CONWIP control is of particular interest because it is more flexible, easier to manage, and more efficient than other WIP control protocols. Figure 11-13 illustrates the basic CONWIP control approach of controlling WIP in a flow. We discussed the interaction between CONWIP control and the virtual queue in detail in Chapter 6. Here we drop back to the basic CONWIP mechanics to simplify for additional insight.

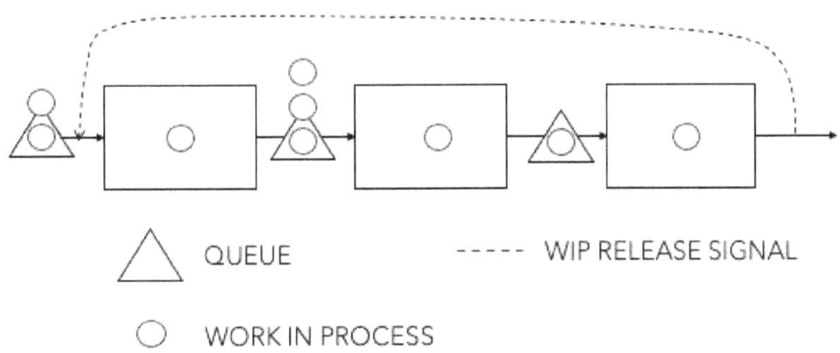

Figure 11-13: CONWIP control

CONWIP is an abbreviation of CONstant WIP. In CONWIP,[i] control of WIP is not at a process-center level. WIP is controlled for a process flow. This provides advantages over the other methods. First, it is simpler to control. Release WIP into the flow when the flow's CONWIP level falls below the targeted level. There is no need to track cards or WIP levels at each process center as is done in classic kanban. A second powerful advantage is that WIP will naturally accumulate at the bottleneck so there is no need to continually recalibrate where the bottleneck is if it changes due to product mix or other reasons. For example, in the simple three-step process in Figure 11-13, if PC3 is the bottleneck but PC2 breaks down and becomes the bottleneck, WIP will begin to accumulate in front of PC2, not PC3. This same effect happens as the bottleneck changes with product mix changes. Another benefit of CONWIP control is that it enables the use of flow performance curves.

Flow Performance Curves

The development of operations science led to the groundbreaking definition of flow performance curves, as discussed in Chapter 4 and illustrated again in Figure 11-14.

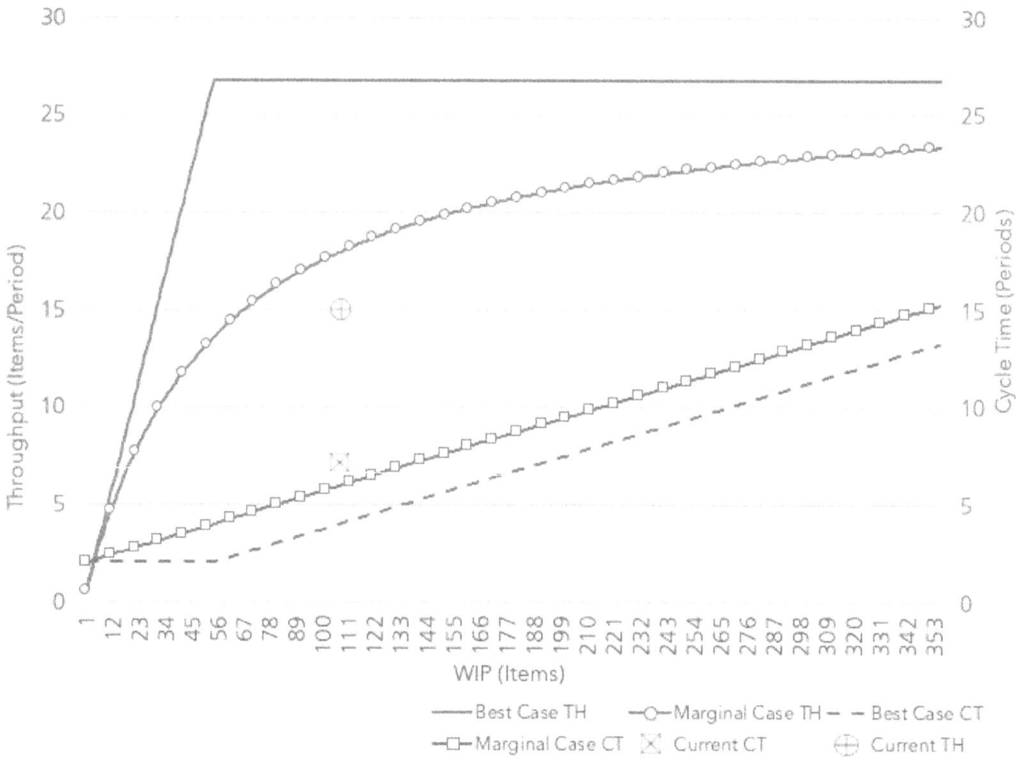

Figure 11-14: Flow performance curves

Flow performance curves are a powerful analytical tool that are simple to create and apply. They are a graphical representation of Little's law and provide valuable insight into process behavior.

To provide a short review, in Figure 11-14, the left *y*-axis is for throughput. The right *y*-axis is for cycle time. The *x*-axis is WIP. So, we see the parameters of Little's law.

$$Cycle\ Time = \frac{WIP}{Throughput}$$

The uppermost and lowermost lines are best case performance. There might be variability in characteristics, e.g., for physical production, one item is square and the next item is round; or, for project management, one task is to check a drawing, and the next task is to write a specification, but for best case performance, there is zero variability in process times and in time between arrivals.[ii] As the reader will recall from the VUT discussion in Chapters 3 and 4, this means the best case has zero queuing.

The following are some initial observations from the flow performance curves:

1. Only two pieces of data are required to calculate flow performance curves, raw process time and bottleneck rate, information easily obtainable in any organization. See Chapter 4 for the performance curve equations and more detailed calculation examples. With two more pieces of data, average throughput and average WIP, a process's performance can be compared to the curves. This provides a simple, objective benchmark of performance.

2. All WIP reductions are not equal. In Figure 11-14, when WIP is reduced by 100 from 353 to 253, there is minimal effect on marginal case throughput, a reduction of about 1 item/period. However, if WIP is reduced by 100 from 144 to 44, the throughput reduction is 7 times greater. In both cases the cycle time reduction is about 4 periods.

   This explains why many one-piece flow implementations are abandoned. Cycle times are very short with low WIP levels but at the cost of much wasted capacity and a huge reduction in throughput. We were on a plant tour where the continuous improvement group's guide, proudly pointing out their one-piece flow line, made the statement, "It has worked out well. We have really reduced cycle time. We are having some problems with customer service levels, but we're working on that." No surprise.

3. Increasing capacity at the constraint moves the horizontal portion of the best case throughput curve up and rotates the diagonal portion of the best case cycle time curve down.

4. Reducing variability moves any non-best case throughput performance curve up and to the left and any non-best case cycle time performance curve down and to the right.

5. The amount of separation of a process' performance curves from the best case curves indicates the levels of variability in the process.

6. In practice, a manager may not be able to control the amount of WIP coming their way. The technicians in the metal spray department at Delta TechOps, for example, have no control over the demand for their services, so an initial response to the flow performance curves was, "That's nice, but WIP comes to us whether we want it or not." Indeed. They can't control when WIP shows up, but they are in complete control of how much WIP they introduce to the department to be worked on. Simply introducing WIP into the workflow because it arrives at the department is a recipe for disaster. With the flow performance curves, the appropriate amount of WIP in the department at any time can be known. Additionally, using operations science, the manager can use Little's law and observed shop throughput to let anyone with WIP waiting in the queue know what their expected completion date is as Chapter 6's Due Date Quoting section describes.

In Chapter 4 we defined the equations for the best case and marginal case curves. With bottleneck rate and raw process time, one can create the curves for most any process. Then with the data of average WIP and average throughput, one can plot where their process performs compared to the curves as absolute benchmarks. This is not benchmarking against someone else's process by collecting information from another company—how would one know if the "best practice" company is performing as well as is possible. The flow performance curves provide is a simple, quick analysis that can be very powerful or, sometimes, it points to the need for further analysis.

There is one more attribute that can be added to the flow performance graph, the estimated actual performance curves. It's one thing to have the point data for current WIP and throughput, it is quite another to decide how much to adjust WIP from current performance to change performance. Following is an approach for adding a curve of estimated actual performance of throughput and cycle time based on current performance.

Estimated Actual Flow Performance Curve

Recall that the marginal case performance curve assumes exponential processing times, i.e., mean = standard deviation. Until now, we have shown that one can plot a process performance point relative to the marginal case curve by plotting the process' average throughput at its average WIP level. If the point is above the marginal case curve, the inference is that process time variability is less than exponential. If below the curve, process time variability is greater than exponential. However, the question often arises as to what the WIP level should be for given current performance variability levels because, if throughput is not being achieved due to low WIP levels, it is generally a quick and straightforward matter to increase WIP to increase throughput.

To aid in that analysis, here is a method for estimation of the actual flow performance curve for a process based on its current performance data.

1. Using the current data for bottleneck rate, average WIP, and average throughput, solve for the critical WIP using the marginal case throughput equation (Chapter 4, eq 4)

$$TH_{MC} = \frac{w \times BNR}{CW + w - 1}$$

2. Since critical WIP = (raw process time)(bottleneck rate), and bottleneck rate is known, solve for the estimated raw process time. This is the raw process time the process would require to perform at its current level. If this number is much greater than the actual raw process time measured for the process, it may point to an opportunity to reduce variability and improve performance. A caution, throughput may be below marginal case throughput simply because demand is low. In this case, the relative values of estimated and actual raw process time are not pointing to the presence of excessive variability.
3. Then use the calculated raw process time in the marginal case throughput equation to plot the estimated flow performance curves for the process' performance. This curve will pass through the average throughput point in step 1.

The same can be done with cycle time using the calculated raw process time and the marginal case cycle time equation. See Figure 11-15.

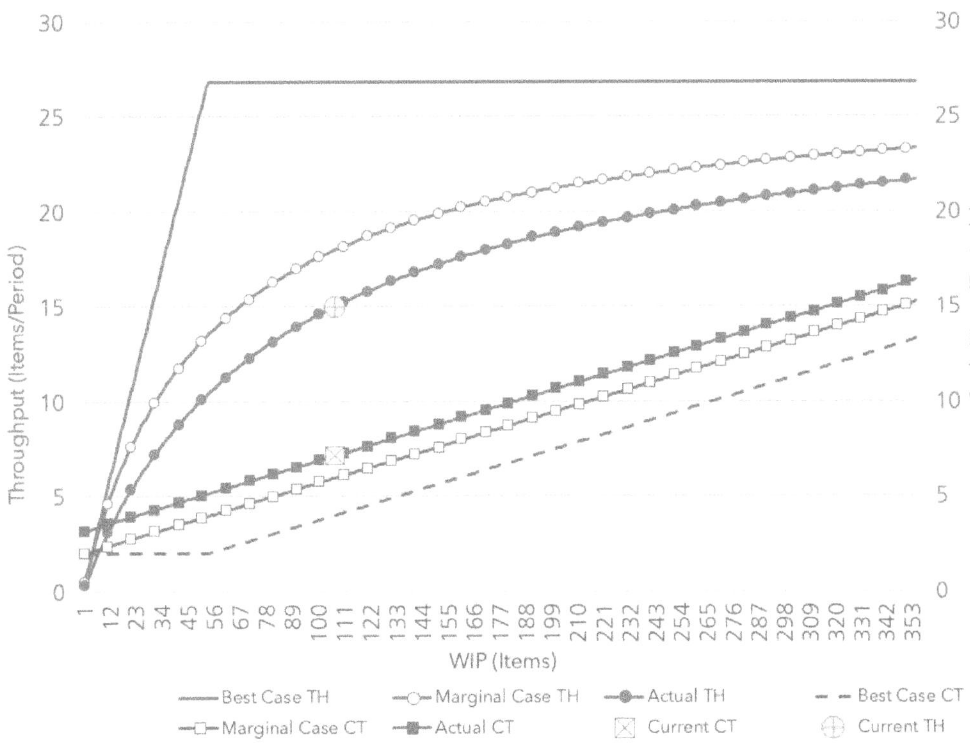

Figure 11-15: Flow performance graph with estimated actual performance curves

For example, in Chapter 14, we showed a case where the raw process time was estimated at two *days*. Current throughput was 48 units/month. Required throughput was 63 units/month. However, using the analysis above, the estimated raw process time was one *month* (22 days in a month). That is poor performance and calls for an operations science investigation to determine causes and corrective actions.

End Endless Discussions on Push versus Pull and the Definition of Pull

One of the distractions that dilutes and obfuscates the application of operations science is the use of imprecise terms and concepts.

The discussion of "pull" as a form of WIP control is highly ingrained in the contemporary industrial and project management zeitgeist. We don't have the data behind the chart in Figure 11-16, but we'd bet dollars to donuts that the increase in the use of the term "*pull*" was driven by the Lean movement as it emerged in the late 1980s.

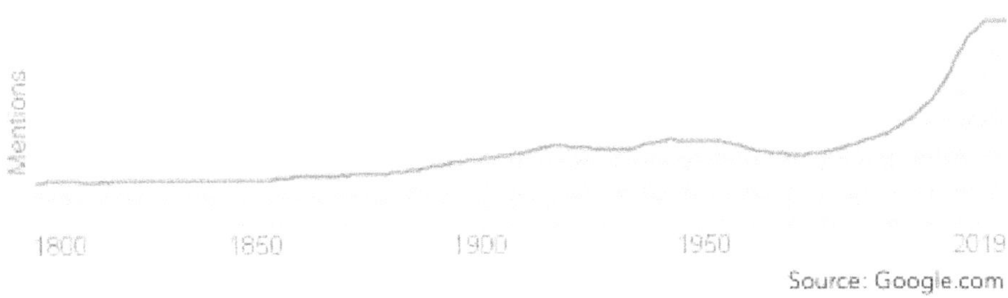

Figure 11-16: Use over time for the word "pull"

The amount of confusion around the term is immense. One would think that, given its widespread use, there would be general agreement and a widely accepted, standard definition. A quick survey of the web for definitions of *pull* reveals a huge variety of interpretations. We are here to put that confusion to rest.

Pull is one of the "best-practice" terms that we referenced in Chapter 2. When it first emerged as a term in operations management, there was strong debate about the advantages of "push" systems, generally associated with MRP/ERP IT systems for planning and control, and "pull" systems, for which the classic configuration was Toyota's kanban system. A push system was generally associated with an MRP-generated production schedule, in which work was released into the process according to the schedule. A pull system, like kanban, controls work release based on some sort of signal from the process that indicates it is ready for work, e.g., a kanban card.[iii]

The general attitude was "push bad" and "pull good," because, among other things, pull systems ostensibly synchronized to demand, while push systems were thought to be concerned with a schedule disconnected from demand. In any case, the argument seems to have been long ago won in pull's favor. "We have a great new initiative going to install a push system" is a statement you now hear from literally no one.

A much greater source of time-wasting discussion is the debate about what a pull system is and what *pull* means. *Pull* is an imprecise term, but it sounds cool. Some common exhortations for pull advocates are "No one upstream should produce until someone downstream requests production,"[iv] "Pull to demand!" (who wouldn't want to do that?), and "One-piece flow is the goal!" We propose a simple but effective solution to the confusion and thereby, an elimination of the time wasted debating the highly imprecise and subjective term.

Now that you have made your way here through the core operations science technical discussions, the relationships between WIP, cycle time, capacity, and variability should be familiar. In light of this new understanding, the solution to resolving the confusion about pull and, if it is still of interest, the debate about push versus pull is easy to see: *Stop using the terms.*

Certainly there will be pundits and people who want to talk about pull because that is their mental model. Let them. If a discussion of pull breaks out, excuse yourself for a cup of coffee or some other diversion until the discussion is over. Then there can be a reasonable, quantifiable, scientifically based discussion about how to achieve the best-possible system performance.

With an understanding of operations science, it is obvious that the amount of WIP in a process is a design parameter for process performance. There are other design parameters for process performance such as capacity utilization, setup time, batch size, and so on. Whether one calls the WIP level in a process and WIP control *pull*, *push*, *kick*, *shout*, or *jump* does not matter.

The flow performance curves, Figure 11-15, perfectly illustrate this. Generally speaking, as WIP levels increase, throughput increases to a point and cycle time increases indefinitely. The secret to controlling performance is not in using some vague, subjective term such as *pull*. It is in understanding the science behind the operations being managed and how to design, implement, and control the best performance given a specific set of resources and goals.

It has been our experience that the more people become familiar with operations science, the less they use nonspecific terms like *pull*.

Operations Science in the Financial Statements

We conclude our overview of the operations science of transformation with a vital discussion of the financial implications of operations science. We always encourage those we coach in operations or project management to involve their financial counterparts in a joint effort to learn and apply operations science.

It is quite common for the relationship between operations and finance to be adversarial. The operations folks may see the finance people as stingy watchdogs always on guard against spending money to make operations more productive or setting unrealistic goals, e.g., budget variances that focus on spending less money but getting higher output. Finance may see their operations counterparts as always wanting to spend money on the latest technology or to hire more people to make the job easier. There is generally an ongoing contentiousness. To illustrate, we provide a couple of quotes from famous operations leaders:

Eli Goldratt, inventor of the theory of constraints, titled an article he wrote, "Cost Accounting: The Number One Enemy of Productivity"[v] and Taiichi Ohno, inventor of the Toyota Production System is frequently quoted as saying, "Costs do not exist to be calculated; costs exist to be reduced," though the original publication context is not documented.

After resolving the confusion around pull just now, we are going to double down by proposing a solution to eliminating the friction between finance (we include accounting in this group for simplicity in description) and operations. This solution is not quite as simple as the solution to the confusion around pull, but it has massively larger potential upside because it will coordinate and focus operations and finance's efforts to major benefit of an organization's goals.

One root of the problem has to do with the different models the two groups use to understand operations (see Figure 11-17). Finance uses a standard cost model, and operations, hopefully, uses an operations science model to make operations decisions.

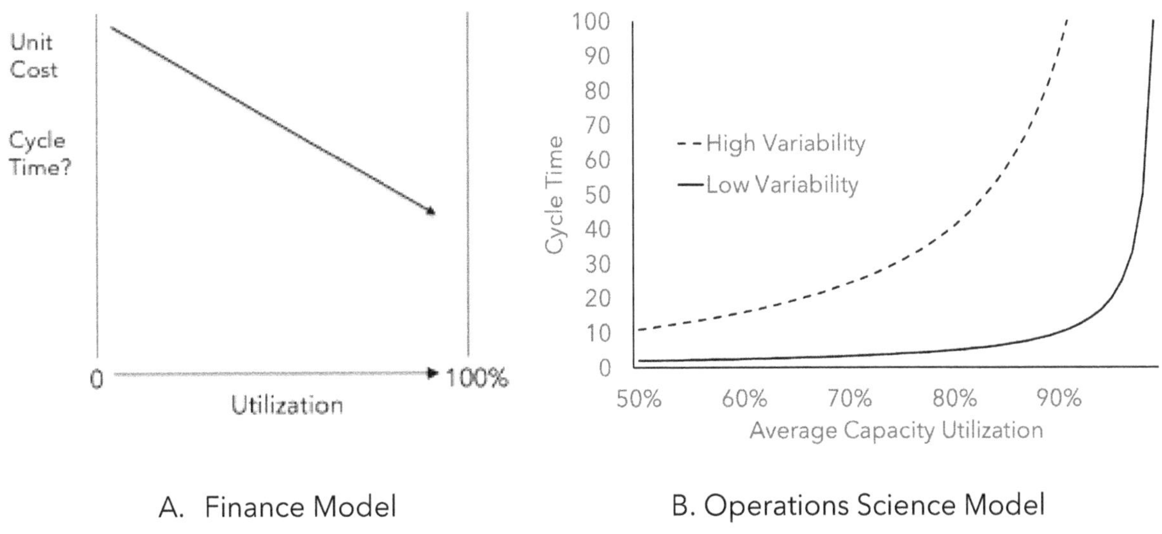

Figure 11-17: Different models for understanding operations performance behavior

In the world of finance and accounting, the standard cost model is used to reconcile revenues with expenses. Why is that? Because powerful standards organizations such as the US-based Securities and

Exchange Commission (SEC) and the Financial Accounting Standards Board (FASB) require use of this model or similar to ensure the integrity of financial reporting standards to maintain confidence in the conduct of commerce.

Those operations people who are familiar with operations science use their understanding of the relationship between utilization and cycle time to ensure best-possible cycle times, which increases customer service and optimizes inventory levels.

Note that the two models convey different but overlapping information. In the standard cost model (Figure 11-17A), higher utilization for production of more goods spreads fixed costs over more units, which means lower unit costs. The standard cost model says nothing about cycle time. In the operations science description of real-world behavior (Figure 11-17B), higher utilization leads to longer cycles times, which drives up inventory requirements and cost, leaving customers usually dissatisfied with longer response times.

The two models have their own legitimate purposes. However, if operations personnel do not explain the relationship between utilization and cycle time, and other operations science relationships, Finance personnel are going to default to use of their standard cost model. In general in business, the people who have the money are the ones who write the rules. So, in the absence of a mutual understanding of operations science, operations personnel will be held to the standard cost model of operations and suffer accordingly.

If the objective for operations is to generate the most cash flow possible for a business, the focus should be on contribution margin at the bottleneck. To get the most profit out of a transformation flow, one should produce the products or services that are the most profitable at the bottleneck. As an example of transformation flow, consider the following exercise: Suppose that stacks of $20, $50 and $100 bills are laid on a table. Each stack contains only one type of bill, and all stacks are the same height (Figure 11-18). The table contains multiple stacks of currency, each stack a different denomination. A person is told to take as much money as possible off the table within two seconds. However, the stacks of bills must be picked up one at a time and placed in a bag, using only one hand. The person can't just sweep money off the table. The money's location is transformed from the table to a bag.

Figure 11-18: A cash flow maximization example

Naturally, the participant is going to pick up the stacks of $100 bills first until they are all gone. If time allows, the participant will then pick up whatever the next-highest denomination stack of money is lying on the table until time's up. Why? Because this provides the most cash to the participant, given the constraints of the operation. This is exactly the same idea as maximizing cash flow at the constraint in production.

contribution margin = revenue – variable costs

profit = contribution margin – fixed costs

Variable costs depend on the time frame considered. For a 20-year time frame, whole production facilities are variable costs. If a manager is planning for only the next month or the next six months and is not planning on major numbers of new hires or layoffs or new equipment acquisitions, capacity and all the indirect costs associated with it are fixed. Variable costs are those out-of-pockets costs tied directly to the production of the product or service. For instance, raw materials are often the primary variable cost. Or if a pharmaceutical company tests batch quality with expensive reagents, the cost of the reagent is a variable cost because it depends on the number of batches made. Contribution margin ties directly to the income statement in terms of revenue and the variable costs found in the cost of goods sold. Ultimately, maximizing the contribution margin at the constraint increases profit, as expressed in net income. The contribution margin at the constraint directly affects the return on assets on the balance sheet by increasing the return on the asset of capacity. As is the original intent of driving performance with contribution margin at the constraint, more cash is generated from operations, less cash is tied up in investment in capacity, and the results also show up in the cash-flow statement.

In any case, maximizing cash flow requires understanding how to determine the constraint and how to manage the buffers to ensure lowest-possible fixed cost.

Operations science provides a structured framework for understanding how day-to-day operational decisions affect financial outcomes. It focuses on measurable relationships among flow, inventory, and variability in processes and offers a framework that unites the efforts of finance and operations managers to make evidence-based decisions about resource allocation, scheduling, and, if applicable, stock inventory levels. These decisions directly influence key financial metrics such as cost of goods sold, working capital, and cash flow. By quantifying trade-offs—such as the required resources for utilization levels to maintain acceptable cycle times—*operations science equips leaders to optimize operations in a way that balances efficiency and financial performance.*

At the heart of this connection is the ability of operations science to translate physical process behavior into financial language. For instance, excess inventory is not just a storage issue—it ties up capital, increases carrying costs, and can lead to write-downs , all of which hit the balance sheet and income statement. Conversely, improving throughput without increasing input costs enhances margin and revenue potential. Operations science provides the equations and logic that map these process dynamics to financial results, enabling operations managers to speak the same language as CFOs and make a credible case for process investments.

Finally, operations science helps organizations forecast the financial impact of operational changes with greater accuracy. It supports scenario analysis and risk assessment grounded in system behavior, not just intuition or static spreadsheets. This makes it especially valuable in environments with high variability or constrained capacity, where small decisions can have outsized financial consequences. By aligning operations strategy with financial objectives, operations science acts as a bridge between the workplace and the boardroom, making it a critical discipline for any organization that wants to sustain a high level of performance.

There will always be creative tension between finance and operations because cash is not infinite and is usually precious for organizations. The solution to greatly reduce unnecessary contention between the two groups, unite their efforts, and accelerate success is for both to understand the science describing operations performance and work together to apply the science as works best for their organization.

As mentioned in Chapter 1, after attending 'operations science training, the finance department at Delta TechOps assigned one of their personnel full time with the SQCI group to combine operations science and finance efforts for improving performance.

Dr. Fred Strathmann is senior vice president of global business at MOBILion Systems and author of this book's Chapter 17, which describes the application of operations science in the service business of diagnostic labs. As Strathmann describes, "

> [Operations science] levels the playing field in my mind to the point where now '[operations and finance are] both using facts and...' numbers and mathematics and science to understand... what we need to get to so we can have that healthy back and forth and find out how we're actually going to do right by patients, which is really the most important aspect."[vi]

---

[i] First described by Mark L. Spearman, David L. Woodruff, and Wallace J. Hopp in "CONWIP: A Pull Alternative to Kanban," *International Journal of Production Research*, 28, no. 5 (1990) 879–94.

[ii] For more description of the flow performance curves, see Chapter 7 in *Factory Physics* and Chapter 3 in *Factory Physics for Managers*. Where they are called production flow graphs. We have renamed it the flow performance curves because they apply generally, not only to production which is typically associated with traditional manufacturing. For description and a simulation illustrating the dynamics, take the online operations science course Basic Flow Dynamics at learn.opscience.org.

[iii] There is a massive amount of literature on the topic. For a detailed discussion, see Chapters 3, 4, and 5 in Wallace J. Hopp and Mark L. Spearman, *Factory Physics*, 3rd ed. (Waveland Press, 2011).

[iv] James P. Womack and Daniel T. Jones, *Lean Thinking: Banish Waste and Create Wealth in Your Corporation* (Simon & Schuster, 1996).

[v] Eli Goldratt, Cost Accounting: The Number One Enemy of Productivity, Reprinted from the Proceedings of the 1983 APICS International Conference, https://www.scienceofbusiness.com/wp-content/uploads/2018/10/Cost-Accounting-The-number-one-enemy-of-Productivity.pdf accessed August 2025.

[vi] Frederick Strathmann, "Optimizing Clinical Operations," Operations Science Institute, November 15, 2024, https://opscience.org/operations-science-optimizing-clinical-labs/. For an audio clip of quote, see https://bit.ly/3OkhiGx.

12

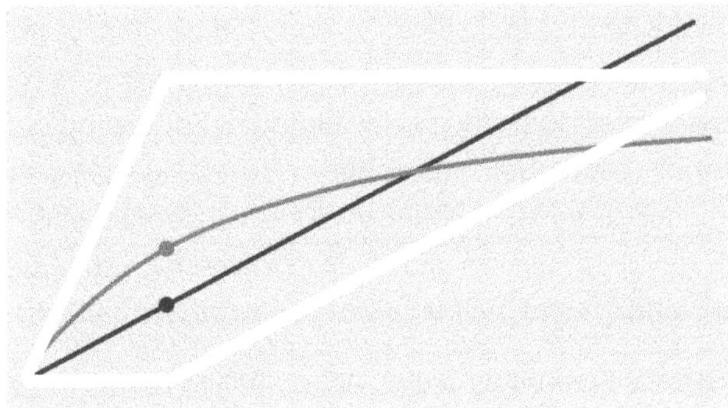

Managing Variability

Variability Buffering Options

In our experience working with people and their operations, common protests to the goal of establishing predictability in performance are "There's too much variability; nothing is the same from day to day" or "Your example works because it is so simple; there's too much complexity in my operations." These are understandable and part of human nature. Operations science validates the persistent challenges managers face in managing variability and provides powerful, practical guidance for addressing those challenges.

Typically, people at all levels of the organization have managed those challenges with sophisticated IT systems to track all the detail. Then the overall control decisions (e.g., no more overtime, cut inventory by *x*%) are made by gut instinct, or edict. Results become unpredictable because people are *responding* to variability rather than *managing* variability.

As an analogy, consider an event planner trying to organize an outdoor party by looking out the window every day to see what the weather is like. If it is rainy and cold two weeks before the event, the planner orders a tent set up, heaters, and umbrellas for the guests to use as needed. That is not how event planners work: They look at the weather models available and account for possibilities such as rain if it looks likely on the day of the event. They don't eliminate variability; weather forecast accuracy is in the 50% range two weeks out, which makes for significant variability when planning for event-day weather. They manage variability within a structured framework that will account for the variability.

If you think this is a flawed analogy, check how often your planning system regenerates. We have been to companies who take pride that their MRP system regenerates twice a day. There is no way planners and suppliers can respond that frequently to changes. For those thinking, "No, but AI will be able to," sorry; that won't happen either. AI can't change the laws of nature. One can't make changes to operations faster than the operations can respond and expect to get desired results of the change. By operations here, we mean the entire supply chain, production process, construction project, or service operations. In control theory, this is called feedback overrun, and common results are system instability and, in physical systems, excessive energy consumption. For example, feedback from a thermostat that goes too rapidly to a heating, ventilation, and cooling (HVAC) system will result in excessive on-off cycling, wasted energy, and unnecessary wear and tear on system components. For planning systems and planners, wasted energy from feedback overrun shows up as excessive expediting; frequent war-room meetings; poor predictability; and high levels of chaos, stress, and cost.

In Chapter 8, we described a management approach that enables predictable management of variability. An organization must pick the level of variability it is willing to handle, set policies to account for that variability, and then use controls and measures that will ensure predictable results. It may sound strange that part of developing organizational strategy is to pick a level of variability. Many management gurus agree with former General Electric CEO Jack Welch, author of *Winning*, who wrote, "variation is evil,"[i] Thus, shouldn't a manager's objective be zero variability? This, as discussed in Chapter 3, is impossible, as there is always some level of variability. A primary law of operations science is that variability requires buffering to manage performance predictively. A corollary to this is that the amount of buffering required is a function of the amount of variability. More variability requires more buffering. However, following the "variability is evil" thinking unnecessarily limits options for improved performance. An often-overlooked idea is that *increasing* variability and buffering costs can increase profitability! If through increased variability, revenue increases more than the accompanying buffering costs increase, profitability increases. Operations science and the management approach discussed in Chapter 8 focus on understanding and managing variability with buffers to meet demand as best suits your objectives. You may choose more or less profitability depending on the trade-offs.

Throughout your process of designing, implementing, and controlling your business strategy and supporting operations strategy, operations science is used to ensure strategy, policies, and controls do not violate natural law. That being said, there is no remedy for leaders who, due to either ambition, ignorance, or organizational politics, choose a path for the organization that exceeds the organization's ability to execute. One can try to educate the leader. Failing that, in the final measure, at least the person knows that the leader is endangering the organization's health and can react accordingly. We know of cases where managers have seen the train wreck coming due to a leader's decisions and have made a job change as a result. If changing jobs is not a preferred option, at least one can avoid the danger of hitching their wagon to an apparently rising star only to find that star turn into a flaming meteor that burns to a crisp against atmospheric friction. Forewarned is forearmed. Let's return to what can be done in the friendly confines of natural law and the operations science framework: What are options for managing variability?

At the most basic level (one resource at 100% utilization), variability is like an incompressible fluid. As described by the operations science fundamental relationship, and shown in Figure 12-1, this can be represented by the area of a rectangle with the capacity buffer being one dimension of the rectangle and the time-inventory buffer being the other.

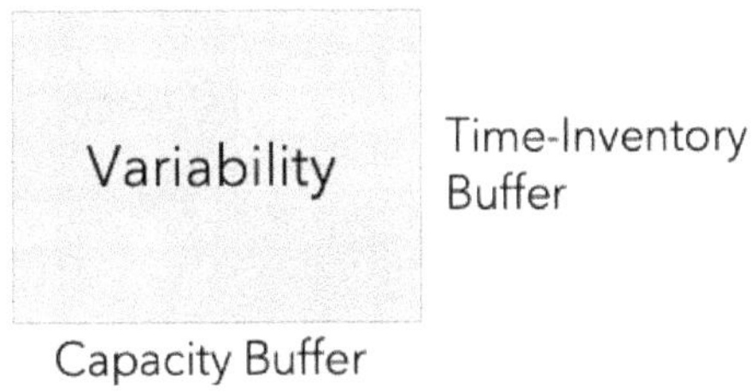

Figure 12-1: An illustration of the operations science fundamental relationship

There are many different combinations of the two dimensions that will address the variability. For a rectangle, length × width = area, and so capacity buffer × time-inventory buffer = variability. As an example, we arbitrarily set the area of the rectangle to 12. The buffer combinations in Table 12-1 are all possible.

VARIABILITY	CAPACITY BUFFER	TIME-INVENTORY BUFFER
12	1	12
12	12	1
12	3	4
12	4	3
12	2	6
12	6	2
12	4.8	2.5
12	2.5	4.8

Table 12-1: Some buffer combinations for variability equal to 12

This representation provides a simple illustration of the many different combinations of buffers available.

In the simplest case of a single resource, operations science shows that the optimal total cost occurs when the capacity buffer cost equals the cost of the time-inventory buffer. The world is much more complex than the simplest case. There are many capacity configurations that occur.

In practical terms, having multiple resources in parallel greatly decreases the effect of variability on queue time. Your wait time at a supermarket (see Figure 12-2) will be shorter on average if you go to the single queue, multi-server (SQMS) group of self-checkout machines rather than trying to pick out the shortest line for a lone cashier, or single queue, single server (SQSS).

Figure 12-2: Capacity configurations at the grocery store

Shortly, we provide a general list of options for managing variability. First, let's look at a method to determine the best opportunities to pursue for improving performance.

The VU Analysis

A problem in many situations is determining where to focus first. It's great to tell everyone to reduce waste in their area. However, with limited resources (time and money), it is more efficient to first determine where the best opportunity lies to achieve the highest return on your investment. The VU analysis is a simple and powerful approach for helping prioritize improvement opportunities.

The VU analysis works with some simple algebra, using the VUT equation and then practical considerations for getting data about performance.

First, we start with the equation for cycle time:

$$Total\ Cycle\ Time = Process\ Time + Queue\ Time$$

$$CT = t_e + VUT$$

since the capital *T* in the *VUT* is the same as effective process time, we get

$$CT = t_e + VUt_e$$

A little manipulation yields

$$VU = \frac{CT}{t_e} - 1$$

This is where practical considerations about data come in. It is commonly difficult to get data about variability. It may be less difficult to get utilization data. In any case, it should be straightforward to get data for cycle time and effective process time. This means it is straightforward to calculate VU.

Once that is done, the first place to focus on for improvement opportunities is in the areas with the highest VU numbers. Of course, the VU number may be high because the U factor is high and there is not substantial opportunity to change that. If that's the case, move on to the next highest VU number.

We used this with a leading semiconductor manufacturer to quickly identify cycle time improvement opportunities. Semiconductor manufacturing is fiendishly complicated with wafers requiring 1,000 or more steps for completing the microchip manufacturing process. This is not a problem with the VU analysis. It led us to substantial, low-cost improvement measures in an unforeseen area.

General Options for Managing Variability

The following are ideas for managing variability. These expand to many different possible solutions based on each organization's unique business model and goals. Solutions for managing variability are only limited by the laws of operations science and your creativity in application.

1. **Stock inventory in anticipation of demand:** Stock inventory is a very common option for buffering variability. However, as Uwe Göhring will describe in Chapter 14, companies invest millions of dollars in expensive inventory planning and optimization software yet waste incredible amounts of money and create great chaos in operations and supply chains through ignorance of operations science and of the mechanics of their software. Services cannot be stocked as inventory. For example, an emergency department cannot stock healed arms for when someone walks in with a broken arm. However, there may be supplies of stock inventory, like syringes, drugs, and bandages, to support the services.

2. **Throttle demand:** This is not a good option for emergency departments but might be for some other organizations. Our experience with hospitals is that they will serve whoever shows up; it's part of their mission. Telling someone who arrives at the emergency department, "Sorry; we are very busy. Come back tomorrow," does not further the hospital's mission. One option for managing emergency department demand is diverting that demand. The patient is not turned away but, depending on their condition, may be directed to a nearby facility that is not overloaded.

   An organization like the Chicago Marathon can limit the number of participants to explicitly throttle demand for the number of resources needed to manage the race.

   Another often-overlooked way to throttle demand is to raise prices. Uber does this with surge pricing. When demand for rides exceeds the available supply of drivers in a specific area, Uber's algorithm kicks in and increases prices. Surge pricing does two things to manage utilization: It attracts more drivers because they can make more money during surge pricing, and it dampens demand because some riders don't want to pay the more expensive fares. However, this can cause negative customer perceptions, as Uber has found out.[ii] Uber's surge pricing creates a flex increase in capacity because it attracts more drivers. Capacity management is a major lever for improving service performance.

3. **Add capacity:** The blunt force approach is to buy machines and hire people. However, there are other options to consider before taking that step, depending on the timing and duration of capacity-constrained situations. Flex capacity is a great alternative. Either have people on standby at home to

call in if needed or cross train workers so that people from less busy areas can help in capacity constrained areas. See the Chapter 20 case on St. Mary's hospital for an insightful study of flex capacity in services (hospital nursing staff). Another way to create more capacity is to reduce the underlying process times—a great continuous improvement opportunity. Things such as eliminating process steps or automating them can free up capacity.

If equipment is needed to address capacity constraints, consider buying used or inexpensive equipment. As long as the equipment can hold quality standards, it can be a valuable resource for relieving capacity constraints. Remember the behavior of cycle time (Figure 3-6, Chapter 3). At high levels of utilization, even a small addition of capacity can provide dramatic benefits. We worked with a client that recycled exotic metals like titanium and Inconel. Demand had put them over 100% utilization at their facility in upstate New York. They also had an operation in South Carolina that did not do the same processing, but they were considering investing $8 million in a new facility to expand capability and capacity there. At an executive training session, we covered these concepts and asked why they didn't set up a small makeshift processing operation in South Carolina to offload the New York facility. For whatever reason, that had not occurred to them, and they did just that. It wasn't a high-tech solution but doing so relieved the worst of their capacity problems and, about one year later, the bottom dropped out of the market. They never started on an $8 million facility and eventually scrapped those plans altogether.

4. **Reduce variability:** This option is often the most difficult to manage. Variability is inherent to processes. As the VUT equation shows, queueing is affected by variability in arrival or demand and process time. Process-time variability is often difficult to reduce. The tools of Six Sigma can be very helpful here. Arrival variability is different. Often external demand variability cannot be controlled. As mentioned above, pricing is one way to manage external demand, though managing average demand is not the same as managing demand variability.

   It is easy to create simple VUT models a user could use to model different levels of variability. For example, reduce SCVs from 1 to 0.25 to see the accompanying effect and then decide if the variability reduction efforts are possible and if they are worth the effort.

   In general, variability is often self-inflicted. Planning and control IT systems are notorious variability generators. From the dawn of the computer age and MRP nervousness[iii] to today's ability to regenerate schedules much more quickly than the process being scheduled can react, leaders are enamored with the idea that more data and more data faster means better control. This is a basic control systems problem writ large, yet it often evades notice. Executives tout their ability to update their planning systems daily yet ignore the unintended consequence of generating many more planning messages than their operations, planning staff, and supply chain can manage.

   Another source of self-inflicted variability is batching. As we have discussed, batching is a perfectly logical operations tactic, but organizations often go overboard with large batches in the belief it will make them more efficient.

5. **Let people wait longer:** Time is the default buffer. If no capacity or inventory are added, people will wait. This is often an unintentional consequence of executive decrees. Declarations like "No more overtime" or "Cut inventory by 30%" without a basis in science tends to lead companies into the whack-a-mole loop:

a. Executives direct the organization to work on an initiative to improve financial performance, for example, "Cut inventory by 30%!"

b. Inventory is reduced too much, and customer service suffers.

c. Executives now focus on customer service, increase inventory, and work more overtime.

d. Costs are too high, so focus reverts to reducing inventory and cutting costs, such as overtime.

e. Go back to a.

One option for dealing with long wait times is to dynamically publish wait times. Due date quoting, as detailed in Chapter 6 is a powerful lever for dynamic lead times. If people must wait longer due to high utilization levels, this tactic at least manages their expectations and, as with surge pricing, can reduce or redistribute demand: People self-select out of a process when they don't want to wait.

Sometimes waiting longer can be associated with high quality, as James Womack found out with his custom bicycle shop venture.[iv] Shortest possible cycle time is not always the goal. Sometimes the customer does not care if a product gets delivered in five days or seven days: It may sit on a shelf in inventory once they receive it.

When operations science is not well understood and applied, variability is not effectively managed. This leaves an organization at the mercy of time buffers rather than in control of them.

We have provided a list of variability management options in the context of operations science. We are sure you can come up with others. As mentioned, solutions are limited only by your creativity and the laws of operations science.

---

[i] Welch, Jack; Welch, Suzy. Winning: The Ultimate Business How-To Book, p. 252 (Chapter 15). Kindle Edition.

[ii] Utpal M. Dholakia, "Everyone Hates Uber's Surge Pricing—Here's How to Fix It," *Harvard Business Review,* December 21, 2015, https://hbr.org/2015/12/everyone-hates-ubers-surge-pricing-heres-how-to-fix-it.

[iii] Hopp and Spearman, *Factory Physics*, 136.

[iv] James Womack, "Beach Reading," Lean Enterprise Institute, August 12, 2003, https://www.lean.org/the-lean-post/articles/beach-reading/.

13

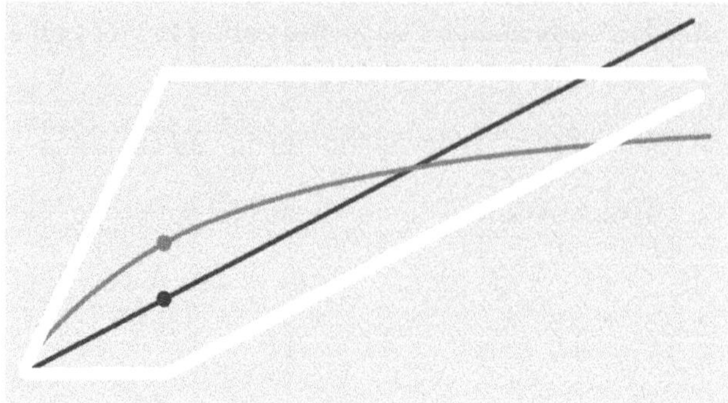

Optimizing Inventory Without Compromising Service

Introduction

This chapter builds on Chapter 5, Stock Inventory, to show practical application of operations science for inventory optimization. First, we go back to the Bubba's Lumberyard example to show how a pragmatic, quick-hit analysis of reorder points can provide great value without requiring highly technical analysis. This doesn't replace the more comprehensive optimization shown in Chapter 5. The more technical approach can provide greater additional value than a quick-hit analysis, but the trade-off is in complexity and time. We are not suggesting the reader forego the comprehensive analysis, only that there is a practical, quick-hit approach that is easy to use. Next, we analyze a more complex case study using the management approach presented in Chapter 8. After describing the comprehensive optimization, we will also show how a pragmatic, quick-hit application of the science of reorder points and safety stocks offers the proverbial low-hanging fruit.

Bubba Revisited

Recall the original Chapter 5 example of Bubba's Lumber, shown here in Table 13-1.

Item	Cost ($/item)	Demand (items/yr)	Lead Time (days)	Lead Time Demand (items)	Reorder Quantity (items)	Reorder Point (items)	Order Frequency (order freq)	Fill Rate	Back Order Level (items)		Inventory ($)
4 x 8, 3/4" Ash	100	100	60	16.4	25	17	4.0	95%	0.13	$	1,319
2 x 12, 10' Pine	250	500	30	41.1	125	83	4.0	100%	0.00	$	26,101
2 x 4, Stud bndl	20	50	100	13.7	13	8	4.0	54%	1.83	$	48
4 x 8, 1/4" util	10	100	15	4.1	25	17	4.0	100%	0.00	$	254
1 x 4, 10' Pine	100	1000	30	82.2	250	167	4.0	100%	0.00	$	20,981
		1750					20.0	98.4%	1.96	$	48,703

Table 13-1: Bubba Lumber redux

Note that a quick check of reorder points compared to average replenishment time demand (RTD) reveals two items (2 × 12, 10′ Pine and 1 x 4, 10′ Pine) that have reorder points much greater than average RTD. Those two items also have a 100% fill rate and are also the most expensive items. A planner at Bubba's Lumber, armed with the knowledge of operations science, realizes this is a mismatch and explains why Bubba never runs out of these items. It's good to never run out of items but too much inventory on-hand means too much cash tied up in inventory, which is a bad thing. To cover variability, one needs to order a little before the inventory position reaches average replenishment time demand. Recall from Chapter 5 that the amount added to average RTD to account for variability is safety stock

$$r = d\ell + s$$

where r is the reorder point, d is demand, ℓ is average replenishment time for the item, and s is safety stock. RTD = $d\ell$. So, the items referenced above will have 42 units and 85 units of safety stock.

2 x 12, 10′ Pine safety stock = 83 – 41 = 42

1 x 4, 10′ Pine safety stock = 167 – 82 = 85

How much safety stock should the planner choose to determine the reorder point? For detailed optimization, follow the approach in Chapter 5, but a simple, effective approach is for the planner to make a conservative guess based on her experience with the supplier. In the example of these two items, let's say the planner knows the supplier is very dependable and always delivers on time and that demand for this item is steady without much variation week to week. The planner could add a week's worth of demand as safety stock. This would drop the reorder point for 2 x 12, 10′ Pine from 83 to 51 and drop the reorder point for 1 x 4, 10′ Pine from 167 to 101. This results in a 30% reduction in inventory ($48,703 to $34,074) with essentially no drop in overall fill rate (98.4% to 98.3%), as shown in Table 13-2, where the only change is the reorder point for the two highlighted items:

Item	Cost ($/item)	Demand (items/yr)	Lead Time (days)	Lead Time Demand (items)	Reorder Quantity (items)	Reorder Point (items)	Order Frequency (order freq)	Fill Rate	Back Order Level (items)		Inventory ($)
4 x 8, 3/4" Ash	100	100	60	16.4	25	17	4.0	95%	0.13	$	1,319
2 x 12, 10' Pine	250	500	30	41.1	125	51	4.0	100%	0.00	$	18,030
2 x 4, Stud bndl	20	50	100	13.7	13	8	4.0	54%	1.83	$	48
4 x 8, 1/4" util	10	100	15	4.1	25	17	4.0	100%	0.00	$	254
1 x 4, 10' Pine	100	1000	30	82.2	250	101	4.0	100%	0.00	$	14,423
		1750					20.0	98.3%	1.97	$	34,074

Table 13-2: Bubba Lumber simple policy adjustment results

It is true that the optimization in Chapter 5 dropped the inventory all the way to $18,708, so making the investment to do a comprehensive optimization of all reorder points and quantities has great value. However, this example shows how use of operations science–based intuition can bring significant initial benefits. Using operations science–based intuition to do a quick check of inventory policies by exporting reorder points from an ERP system and comparing them to average replenishment time demand is a pragmatic way to yield significant benefits. Imagine if you are in a different company where the inventory values are in millions instead of tens of thousands, and a check like this of all items in your ERP system showed an opportunity to reduce inventory from $48 million to $34 million with no loss of service!

SpatialPrint Case Study

The company in the next example, which we'll call SpatialPrint, is a leader in additive manufacturing. It faced a critical challenge common to most product companies: reducing inventory levels while maintaining a very high level of customer service. SpatialPrint's existing inventory policy required holding three months of stock for each item to ensure a 100% fill rate. The result of the policy was excessive inventory costs, which tied up capital and warehouse space. This is the classic inventory discussion among finance, production, and sales in all companies with stock inventory. In general, finance wants to spend as little as possible on inventory, Sales wants so much inventory that inventory is always available for every product, and production wants to do whatever is easiest and most efficient.

There are competing objectives here, and they don't exist in a vacuum. A major consideration is the natural concerns people have about their job security and career advancement. This chapter provides another example of the power of operations science to help companies advance performance while navigating the personal dynamics associated with jobs and careers.

In Chapter 5, there was a detailed technical discussion about calculating and optimizing inventory policies. Here we don't go deep into calculations.

Back to SpatialPrint: There was reason for the three-month on-hand inventory policy even though the number of months was arbitrary. Why not six months or two months, and why one number for all parts?

SpatialPrint is the only supplier of the material (consumables) used in its additive manufacturing machines (3D printers). The additive manufacturing machines are its primary product offering. Since this makes the company a sole-source supplier of consumables, its customers have contractual stipulations that SpatialPrint maintain 100% availability of material. Without that, customers will develop material sources elsewhere. And, more importantly, lack of service on consumables may open the door for customers to consider other sources for additive manufacturing machines. The challenge in this case was to implement a strategy that would optimize consumables inventory without jeopardizing service reliability.

Context

Providing a goal without a viable execution strategy is a WeWork approach to business. WeWork drove to expand globally, with a goal to open in every major city, and ended up 2023 valued at $361 million, down from $47 billion just four years earlier.[i]

SpatialPrint suspected that it was carrying more inventory than it needed. Using operations science to provide the finding "There is too much inventory" was not very helpful. SpatialPrint was prepared to carry excess inventory to protect its sole-source status. Going further, operations science enabled

SpatialPrint to see that theoretically it could reduce inventory by 50%. That sounds great, right? Well, not so fast. Imagine you are the planner, manager, or VP responsible for inventory and are told that you could cut 50% of your inventory right now and still maintain your service commitments. What would you say? For people who are concerned about their jobs, careers, and the politics within an organization, the initial response is typically not "Damn the torpedoes; let's do it!" The response is more like "That is interesting: Prove it." It's a completely natural response in risky situations. We will describe execution of the concepts in practice to realize the predicted gains while addressing the concerns of managers and employees who will be the ones to realize the benefits of success or suffer the consequences of failure.

Key Measures

The process for producing consumable inventory is fairly simple. See Figure 13-1.

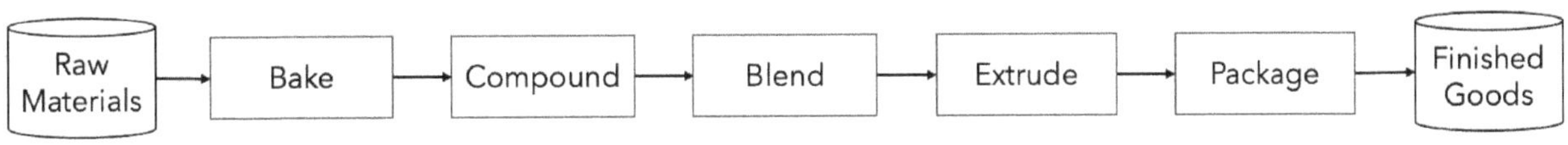

Figure 13-1: SpatialPrint's consumables production process

There is an additional level of complexity not illustrated in Figure 13-1, multiple parallel resources exist at blend, extrude, and package. As explained in Chapter 4, this parallel capacity is of great benefit in reducing total cycle time. Total cycle time, replenishment time for inventory in this case, is a primary driver for inventory requirements. The average replenishment time through the process was 8.5 days, with a standard deviation of 7 days.

We started with an inventory of 71 stock-keeping units (SKUs). The total value of the inventory was $1.22 million. An initial analysis of the inventory policies and practices related to the SKUs produced the performance curves in Figure 13-2.

Figure 13-2: Inventory performance curves for a group of 71 SKUs

Some important observations on SpatialPrint's policy and results reflecting the four key measures of inventory performance described in Chapter 5:

- Average on-hand inventory
- Fill rate
- Back order time (when back ordered)
- Replenishment frequency

1. In Figure 13-2, the actual inventory level (red circle) is much higher than would be predicted (yellow circle) given the stated policy (3 months on hand at all times). There are couple of common reasons that could happen:
   a. There could have been a one-time build for a large campaign, and the data collected included this outlier inventory. To prevent this in analysis, the best inventory levels to use for comparison are average daily inventory on hand over a few months or whatever is the appropriate time horizon.

   b. Planners are not following policy. This occurs in the vast majority of cases and is the first check when desired performance is not being achieved. This is the policy violation described in the management approach of Chapter 8. That being said, it's hard to blame the planners (unless they created the policy) because the policy is a random creation and does not correspond to the expected performance envelope as described by operations science.

2. The actual fill rate as reported is 99.9%. The predicted fill rate is 99.89%. These are both extremely high, which is great for customer service. Only one in every 100 demands occurs without inventory on the shelf to satisfy the demand. That seems to be good enough for SpatialPrint because that is the fill rate they have been maintaining, and their question is not "How can we increase fill rate?" but "How do we reduce inventory and maintain a very high fill rate?"

   For anyone considering inventory performance, there is an important characteristic of fill rate to remember. If fill rate is not 100%, there will be occasions when demand appears and there is no inventory in stock. This is called a stockout. If fill rate is set to 95%, and everyone goes into panic and recovery mode when 1 in 20 orders results in a stockout, they are missing the point. Setting 95% fill rate means that, *by design*, there will be 1 stockout in every 20 orders. Sales might think this an untenable situation, but the picture is incomplete if not looking at both fill rate and back-order time.

3. One can achieve 100% customer satisfaction without having 100% fill rate. This is where back order time when back ordered (BTB) should be considered when determining required fill rates.

   A quick technical note—the average back order calculations in Chapter 5 are for average back orders over all periods, even periods where back orders are zero. What managers and sales people are really interested in is BTB. That calculation is

$$BTB = \frac{\bar{B}}{d(1 - FR)}$$

Where

$\bar{B}$ = average backorders (see Chapter 5)

d = rate of demand

FR = fill rate

If setting fill rate to 95% saves large amounts of inventory investment, and BTB is one day or less, or even a couple of days, does the customer care? They might. They might not. That is a decision unique to every organization, but, at least with application of operations science, planners and managers can make an informed decision. That is operations strategy at its best.

4. The Figure 13-2 analysis shows three curves, each corresponding to number of orders (20, 10, and 40 production batches, in this case) per period, or replenishment frequency. The period is one week. On average, SpatialPrint is releasing 20 batches per week to production as calculated from the data, using the calculations detailed in Chapter 5. This replenishment frequency is over 71 SKUs, so obviously SpatialPrint is not releasing a batch of each SKU to production each week.
Replenishment frequency, as described by operations science, contrasts with the Every Product Every Interval (EPE) policy for production, which is part of Lean Manufacturing lexicon[ii] and promoted by many. EPE is the idea of leveling production, or *heijunka*. The idea is to reduce variability in production to better synchronize supply with demand, but it has a deep flaw as an optimal policy. The flaw is the assumption that production can be perfectly synchronized with demand. If there is little demand for a product, why would one make the product on any interval? If there is demand for a product, and it is highly variable, a fixed interval of production will be ineffective. Lean manufacturing, and thereby Lean, originated with the Toyota production system. For Toyota and most major auto manufacturers, the final assembly line speed is set at a takt time, the interval between completed cars rolling off the line. As previously mentioned, takt time is just the inverse of the rate at which cars roll off the line. For example, an 88-second takt time and an eight-hour day, with no breaks, mean that 327 cars a day roll off the line in a day. We guarantee that actual demand for those cars is not 327 cars every day. What Toyota and other automakers using takt time have done is to isolate variability from transformation (the assembly line), which facilitates higher utilization of capacity. The operations science approach is to set up policies and a control approach that naturally responds to variability within a chosen performance envelope at minimal cost.

Doing fewer orders per week means releasing larger batches to production. This is not an EPE approach. The constant is the average number of batches per week. The mix of batches will change to respond to changes in demand. As described in Chapter 6, true batch size optimization is complex. As we will describe, the optimal policies meant that SpatialPrint produced batches to account for demand and demand variability, it was not trying to synchronize exactly to average demand. SpatialPrint's capacity was aligned to what was needed, making it more efficient, rather than trying to achieve some arbitrary EPE production interval. Figure 13-2 provides a nice illustration of fundamental operations behavior: Sometimes there is no need for smaller batch sizes. All three curves are basically on top of each other. In this case, fewer batches per week work well. For 10 orders per week (green curve), the batch size would be twice as large as the current release of 20 orders per week (red curve). This also means there are half as many setups to switch from one product to another. This type of behavior is highly common at high fill rates, which is what organizations usually want anyway, so that's a happy coincidence.

Policy Analysis

Let's dig a little deeper to see the results before and after policy execution. Figure 13-3 shows the results of the initial practices and a policy of three months of inventory on hand. The SKU data are illustrated by cost in descending order from left to right.

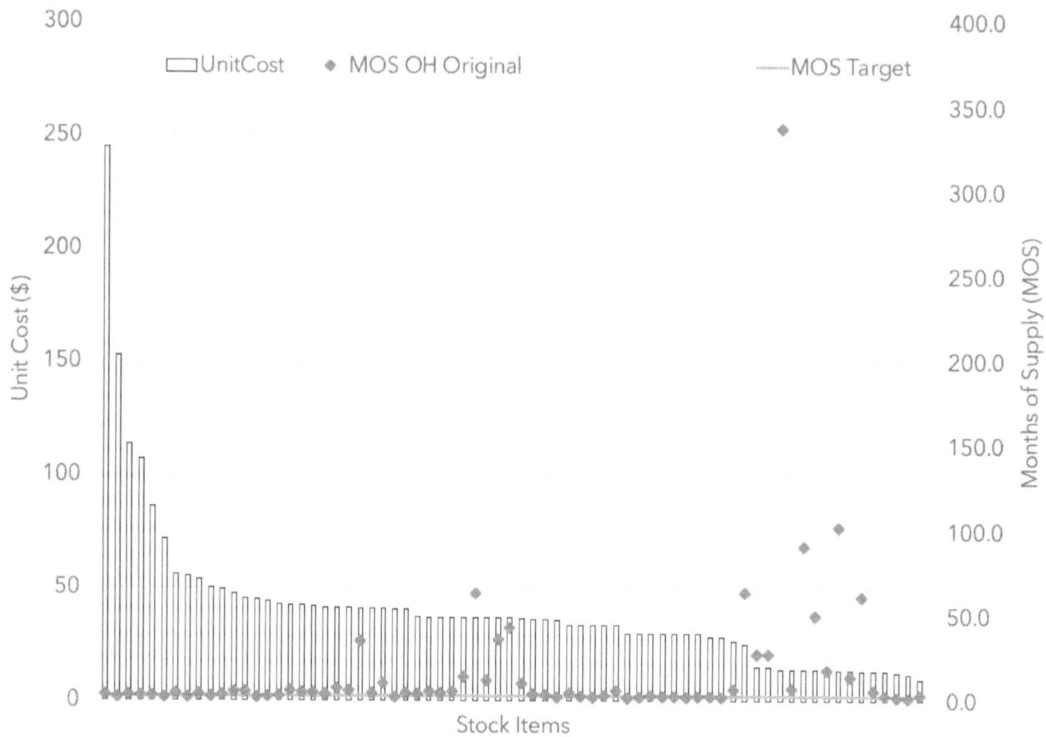

Figure 13-3: Results from initial policies

As we saw in Chapter 5's Bubba Lumber example, general rules for minimizing inventory cost while maintaining fill rate for a given replenishment frequency are

1. keep as little on hand as possible for expensive items, using frequent replenishments and
2. use infrequent replenishment for inexpensive items that will result in more on hand.

Looking at Figure 13-3, this seems to be what the SpatialPrint planners were doing, with the exception of five or six items. The red diamonds show each SKU's months of supply (MOS) value of inventory on hand. The expensive SKUs seem to have low MOS close to the three-month MOS target (blue line). It turns out this is a matter of scale.

Figure 13-4 shows a different picture. When looking at the 22 most expensive SKUs, about a third of them have MOS values of five months or greater and the minimum MOS value is two months.

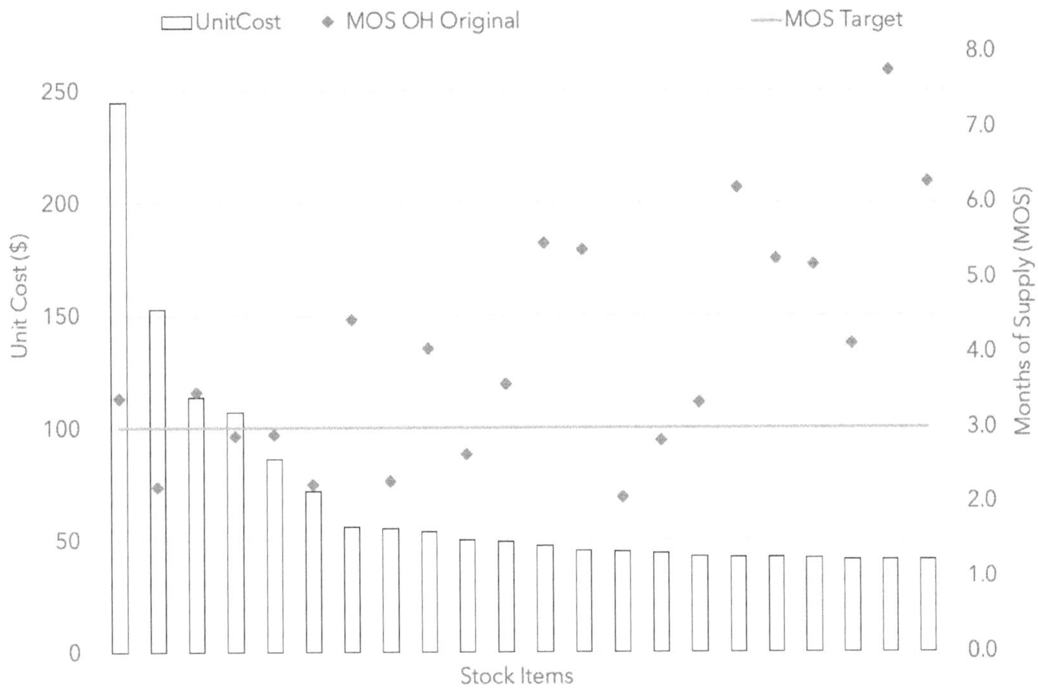

Figure 13-4: Initial policy results for 22 most expensive SKUs

We're going to skip ahead to the final state for policy comparison. To elaborate, SpatialPrint was originally using an 8.5-day replenishment time for determining optimal inventory policy. With the implementation of the initial optimal policies (more on this shortly) and an operations science-based inventory management approach, the reduced load on production enabled SpatialPrint to achieve a 5-day replenishment time, which was used to determine the final state inventory policy.

Figure 13-5 shows initial and optimal MOS results for all 71 SKUs. Optimal policy resulted in a reduction to under $600,000 of inventory from the original $1.22 million. The black line shows the optimal MOS for each part. We realize the data are not continuous, so a line plot is not technically correct, but it does provide good contrast so it's easier to interpret. The high MOS of some of the SKUs' initial policy throws off the scale but, in general, the optimal policies produced lower MOS than the initial policy and practices.

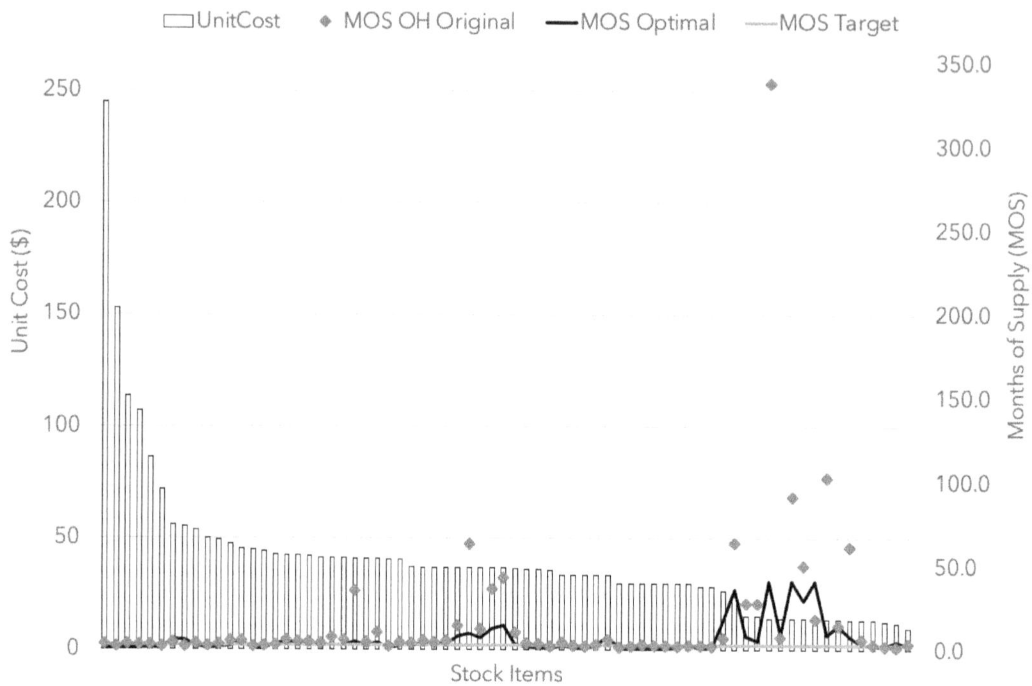

Figure 13-5: Comparison of original and optimal MOS for all 71 SKUs

Figure 13-6 shows initial and optimal MOS for the top 22 most expensive parts. Note in Figure 13-6 that the top six most expensive SKUs have much lower MOS optimal targets than the initial policy target of three months.

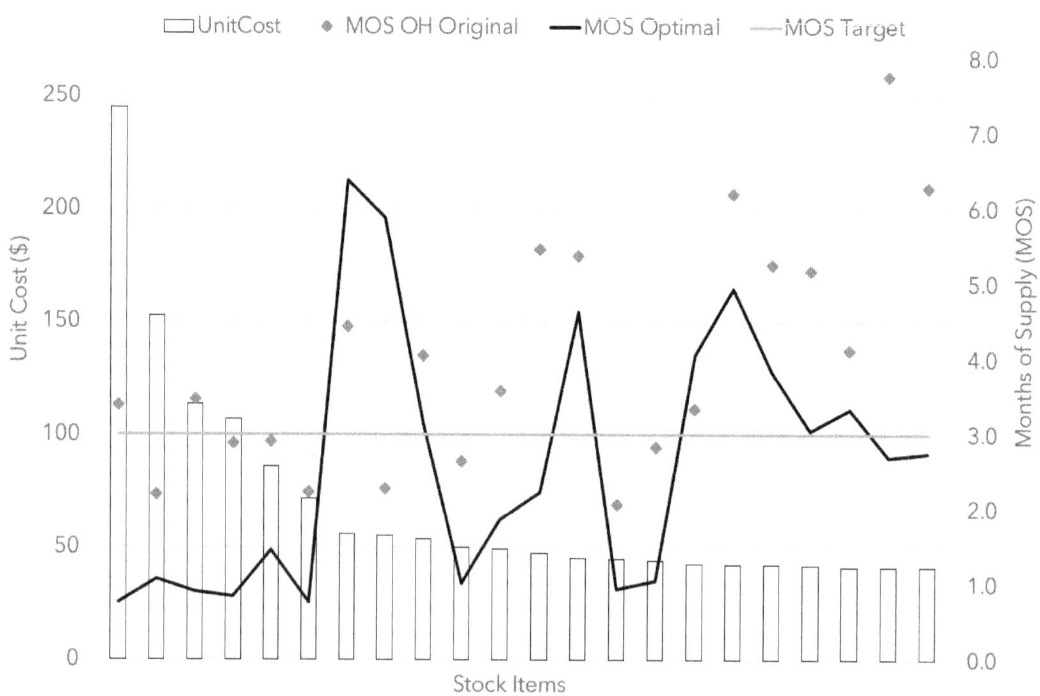

Figure 13-6: Comparison of original and optimal MOS for 22 most expensive SKUs

Change Management

Enough about inventory analysis and mechanics. How did the SpatialPrint team execute to achieve the predicted reduction?

The planners and the executive team were initially nervous about the idea of reducing inventory by 50%. Anyone with significant career experience has probably seen a new initiative implemented with vim, vigor, and faith in a bright future only to find out that the light at the end of the tunnel is actually the light of a freight train that runs right over all the plans and high aspirations.

When the SpatialPrint team voiced their reservations, our response was not "No, you must have faith! You can reduce inventory by 50%; let's go!" With operations science, the response is quite simple: "No problem; we understand, and it's not our role to tell you what to do. We can only tell you that, for example, if you do A, you will get B, or, if you do C, you will get D. How about you start with small reductions and gradually reduce the inventory until you are comfortable with the mechanics and predictability, and then you can reduce more aggressively as suits you?"

That was acceptable to the team, which is what they did, as illustrated in Figure 13-7.

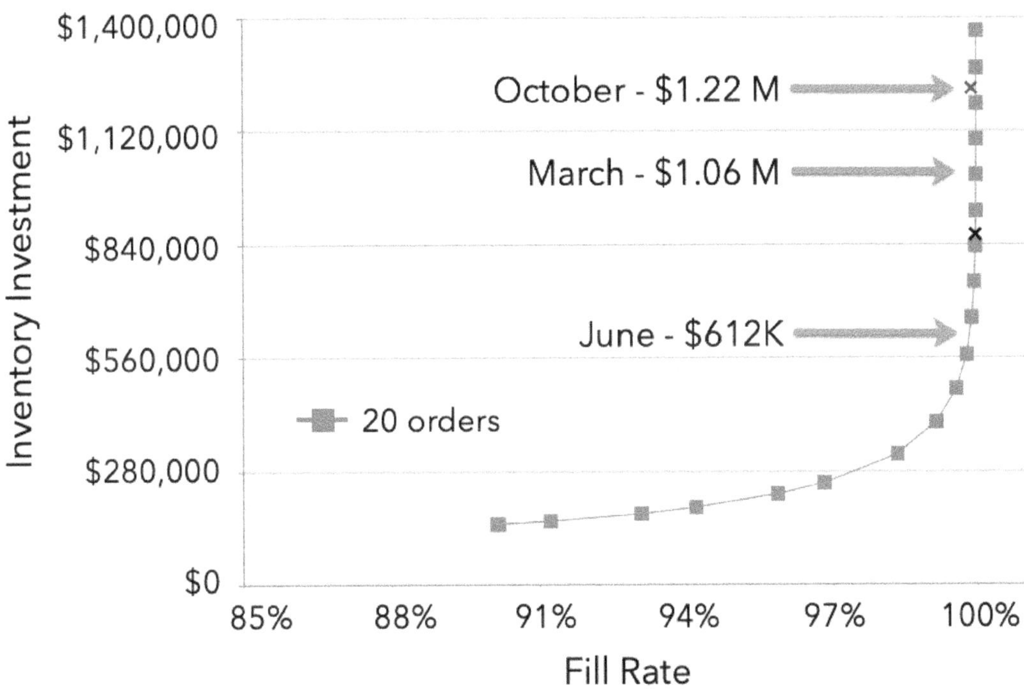

Figure 13-7: Inventory reduction management

Note that from October to March (six months) inventory was reduced only about $200K. In addition to a conservative inventory reduction approach, the team experienced an inventory bloom. An inventory bloom is the phenomenon where implementation of optimal inventory policies results in an initial *increase* in on-hand inventory. This is important to understand for managing expectations when embarking on an inventory optimization effort.

An inventory bloom is the result of the following situations:

Situation 1. For SKUs where optimal average on-hand inventory is higher than actual average on-hand inventory, inventory is replenished to get to optimal levels.

Situation 2. For SKUs where optimal average on-hand inventory is lower than actual average on-hand inventory, inventory must be consumed to get to optimal levels.

If inventory for SKUs in situation 1 is replenished faster than inventory for SKUs in situation 2 is consumed, on-hand inventory will increase temporarily. This effect is predictable and can be quantified and managed if necessary.

For SpatialPrint there was a small effect because they had so much excess inventory, but for other organizations the inventory bloom can be significant. It is helpful to quantify the inventory bloom and set expectations so the team sees the plan through to optimization.

Once SpatialPrint gained confidence in the process, inventory was reduced by about $400K, from March to June (3 months), to get to the goal of a 50% inventory reduction.

With the previously mentioned changes to their environment due to the optimal policies and structured practices in an operations science framework, final results were as follows:

- Average on-hand inventory: $565,000
- Fill rate: greater than 99%
- BTB: 0.66 days

These results were excellent, so the comprehensive optimization was worth the investment. However, for completeness, let's check to see what would have happened when we use the quick analysis technique of comparing reorder point to RTD. Recall that the point of the analysis is to compare reorder points in the system to average demand during replenishment time and then check the safety-stock buffer to see if the implied safety stock makes sense. Because the analysis is quick, it does not provide the predictive analytics for inventory investment, fill rate, and BTB. It does provide projected inventory reduction with a level of comfort that there is still enough inventory based on an adjusted safety stock buffer. Let's check some projected inventory reductions with different safety-stock choices.

The quick analysis showed that there was indeed a massive amount of safety stock built into the initial reorder points. Adjusting to a 21-day safety-stock level for all parts showed a $200K inventory reduction opportunity. It bears emphasis that this approach is adding 21 days of demand of inventory to replenishment time demand to set reorder point. This is not setting a target of 21 days of on-hand inventory. At high fill rates (~0 back orders),

$$I = \frac{ROQ}{2} + s$$

average on-hand inventory = (reorder quantity/2) + safety stock. One could implement these new policies and monitor fill rate to make sure customer service remains high. If high fill rates hold, the safety stock could be adjusted to 14 days of demand. In this case, the quick analysis results in a $360K inventory-reduction opportunity, but once again one would need to monitor fill rate to ensure customer service remains high.

Operations science intuition will provide quick decisions that are directionally correct and can provide great benefit. That said, detailed operations science–based predictive analytics do better to quantify buffer choices and policies needed to execute them. For starters, one need not wait to see what happens with iterations of conservative policy changes. It is possible to go straight to optimal policies that achieve desired results. It is true, as mentioned in the change management discussion, that a few iterative tests can be helpful to gain organizational confidence, but once that happens no further iterations are needed. The organization can quickly realize results by going straight to the optimal policies. One other advantage of using operations science–based analytics models is that leaders can experiment with the models instead of experimenting with the operations. The intuition-only approach requires testing the plan on the actual operations, which sometimes does not work out—this is the "test it and see if it breaks" approach. With operations science analytical models, the model must be validated, which requires checking against actual performance, but, once a model is validated, it can be used with confidence to achieve desired results.

In SpatialPrint's case, given the strategic importance of the 99% fill rate and the 50% inventory-reduction opportunity, the comprehensive optimization was ultimately the best path, but there are often significant benefits to starting with a quick low-tech analysis based in operations science.

In conclusion, Table 13-3 shows how the operations science operations management approach from Chapter 8 was applied by SpatialPrint.

Operations Science Application Summary Table: SpatialPrint

ELEMENT	DESCRIPTION
BUSINESS STRATEGY	Since the company is a sole-source supplier of consumables, its customers have contractual stipulations that SpatialPrint maintain 100% availability of material. Without that, customers will develop material sources elsewhere. And, more importantly, lack of service on consumables may open the door for customers to consider other sources for additive manufacturing machines. The challenge in this case was to implement a strategy that would optimize consumables inventory without jeopardizing service reliability.
OPERATIONS STRATEGY	
DEMAND	Driven by customer utilization of additive manufacturing machines, which drives demand for consumables stock inventory. Expectation was for product to be in stock and available.
TRANSFORMATION	Figure 13-1 shows transformation. Raw materials are baked, compounded, blended, extruded, and packaged before being placed into finished goods inventory.
VARIABILITY	Demand variability is caused by a wide variety of utilization levels at a wide variety of customers.

ELEMENT	DESCRIPTION
	Transformation variability results from batch sizing, scheduling, and timing of information system update.
BUFFERING OPTIONS: CAPACITY	Increased capacity could reduce replenishment times for inventory.
BUFFERING OPTIONS: INVENTORY	Inventory is the critical buffer in this case, with a goal of reducing inventory levels while maintaining a very high level of customer service.
BUFFERING OPTIONS: TIME	Expectation was for immediate availability of product, so time was not much of an option.
TOOLS AND APPLICATIONS	Operations science of inventory with reorder points and reorder quantities were useful for developing and implementing optimal inventory policies. CONWIP and virtual queue were used to manage production.

Table 13-3: Operations science framework summary

---

[i] "WeWork receives non-compliance notice from NYSE," Reuters, April 18, 2023, https://www.reuters.com/business/wework-receives-non-compliance-notice-nyse-2023-04-18/.

[ii] "Every Product Every Interval," Lean Enterprise Institute, accessed June 10, 2025, https://www.lean.org/lexicon-terms/every-product-every-interval/.

PART FOUR

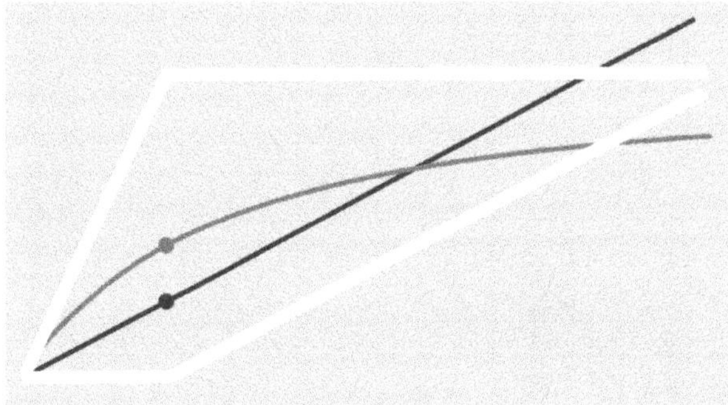

Stories of Operations Science in Action

The true test of any science is how well it works in practice. In this section, leaders from a variety of industries share their experiences applying operations science in manufacturing, services, and projects. Their stories demonstrate the universality of the science—whether the goal is to produce airplanes on time, improve lab services, or optimize construction projects, the same principles apply. Each contributor offers an insider's perspective on how operations science shaped their decisions, helped improve performance, and revealed new ways of thinking about problems. These accounts bring the science to life, showing how the "science of working better" helps leaders and teams across industries.

Operations Science and IT (Chapter 14)

Configure to Order (Chapter 15)

Contract Manufacturing (Chapter 16)

Operations Science in Clinical Laboratories (Chapter 17)

Operations Science and Project Management (Chapter 18)

Operations Science in Shipbuilding and Ship Repair (Chapter 19)

Operations Science & HealthCare Service Delivery (Chapter 20)

14

Operations Science and Information Technology

Uwe Göhring

Uwe is founder and president of bigbyte software systems inc. (bigbytesoftware.com). He is an SAP-designated mentor and has worked in supply chain optimization for more than 25 years.

Introduction

Since the dawn of the computer age 70 years ago, practitioners of supply chain and operations planning have repeatedly proclaimed a revolution of improvements in efficiency through digital (computer) technology. I have been working with planning systems for over 25 years. With the advent of advanced automation, intelligent execution, and digitalization, I have seen tremendous acceleration of digital technology adoption. However, I have not seen major strides in achieving consistent, successful results through automation or intelligent execution. Gains in efficiency have been hit or miss. In some cases, digitization has contributed to performance failures. Yes, there is undoubtedly much more automation in today's planning and scheduling systems, but the question must be asked: Did we automate effective processes, or did we just slap an advanced IT system on top of the old, outdated, and sometimes incredibly inefficient operations? In my experience, the latter has occurred most often, which immediately puts current quests for intelligent execution in doubt and, by association, raises serious reservations about the future efficiencies of such operations.

Starting from *Orlicky's Materials Requirements Planning* through Enterprise Resource Planning Systems(ERP) and Advanced Planning Systems to today's artificial intelligence and machine learning, software providers like SAP, Oracle, I2, Kinaxis, and many others have been seen as the enablers of digitalization. IT systems have become the hot new ingredient for improvement projects to make operations more efficient and to gain competitive advantage. Implementation of these systems usually requires incredibly expensive projects for both the acquisition of software functionality and the engagement of implementation consultants. The projects are usually very lengthy, realized benefits are unclear, and unsatisfactory outcomes have been the norm.

One might blame these bad outcomes on vendor-provided software, but that would be too easy an answer. Vendors offered functionality to support all known methods and methodologies, new and old. Software to support proven business methods like the theory of constraints, Lean, and the Toyota production system have been offered. Any new fad that had the slightest chance of becoming the next big thing was developed as a new offering. Eventually, as the competition between these software vendors grew exponentially, each one of them tried to beat the others by "functionality overload." As a logical consequence, functionality wasn't being offered to create the best possible system; functionality was developed for whatever the customer community requested.

This buildup in functionality went as far as consultants developing their own new business methods. Software vendors jumped on the new method immediately and offered it as part of their suite of functionalities, no matter whether it brought any increased efficiency, e.g., carbon footprint planning. The prevailing sales pitch became "We can do that, too." The following list is a sample of offerings:

- Make-to-order (MTO) versus make-to-stock (MTS)
- Carbon footprint and sustainability-based planning
- Push versus pull supply chain strategies
- Capacity-driven versus demand-driven planning
- Centralized versus decentralized planning
- Single-sourcing versus multi-sourcing procurement
- Lean, agile, and hybrid supply chains
- Supplier risk scoring and multicriteria supplier selection
- Batch sizing and consolidation rules
- Demand forecasting models
- Bullwhip effect mitigation strategies
- Back-ordering versus stock-out avoidance
- Dynamic versus static lead-time adjustments
- Real-time versus periodic replanning
- Collaborative planning, forecasting, and replenishment (CPFR)

Is this the fault of the software vendor? I think not; they're doing exactly what they should be doing to make money—providing what the customer requests, which is just good business. A major problem, however, is that what is often requested is not based on sound operations science. In fact, it rarely is.

The modus operandi has been to buy and implement standard software packages. But wait—no two operations are exactly alike, so the next step requires spending huge amounts of money and effort on customization so the software will optimize a particular company's operations. Also—and this is very important—the operations that software is being customized to control are themselves often set up to be inefficient. For example, an operation might be set up to minimize unit cost. Operations science shows that this usually drives highly dysfunctional behavior. Spending huge amounts of money and effort for software customization to accomplish that goal is an expensive way to ensure the company continues to shoot itself in the foot, though it may be able to do so much faster and with better reporting.

Now consider a different scenario. First create an optimized model of operations and then enable it with standard software. Of course, this depends on the quality of the model. If the model is a quick fix and ignores the laws of operations science, then we only continue to ask the software vendors to accumulate and offer functions and features that rarely result in efficient solutions.

If, however, we approach continuous improvement with sound operations science, we will achieve better results. It's like designing a better racing car. Adding a turbine jet engine to a VW Beetle does not automatically create a winner. One must consider the engine options available, the car frame available, and the overall automobile system performance for the given course of competition. This illustrates one important fact of life that is a prerequisite for good outcomes in managing operations: The operations design must be based on valid science. I can't design a car without knowing mathematics or the science of mechanical engineering. And so it is with operations. We can't ignore operations science, make unsubstantiated claims of what, for example, causes inventory to increase or decrease, and then expect that everything will work out.

The following is a glimpse into my experiences with clients, illustrating how operations science offers a transformative force in driving efficiency and continuous improvement and how, sometimes, the science alone does not satisfy.

A large commercial aircraft manufacturer needs increased production

At a large global manufacturer of civilian and military aircraft, we were tasked to help ramp up production to meet increased demand for the most popular type of aircraft. The company had to increase the delivery rate from 48 units per month to 63 units per month. The executives wanted to investigate optimization efforts first, before committing to large investments in resources and added capacity. Consequently, we suggested doing an analysis applying operations science principles.

We used flow benchmarking with flow performance curves to obtain information to assess the current situation. Little's law provided the pieces needed to define the controls that would help us increase efficiency and deliveries without large investments in capacity. As you have seen earlier, Little's law relates cycle time with work in process and throughput:

$$WIP = TH \times CT$$

The formula may be rearranged algebraically in three different forms. In practice, given some amount of capacity resource and some level of variability in demand and process time (the performance envelope), WIP is the independent variable, as shown on the flow performance curves, and cycle time is the lagging dependent variable, so the real-world representation of Little's law looks like

$$CT = WIP/TH$$

Also, it's generally simpler to calculate cycle time by counting items of WIP in a flow and the rate of items out of the flow (throughput) than it is to track each item from start to finish.

For the subassembly under review, we counted a current WIP level of €2.9 million presently on the shop floor (on average there were 107 subassemblies in process at any given moment in time at €27,000/subassembly). We used the final assembly demand for subassembly demand to tie the two together. This enabled the factory to output 48 final assemblies per month. Using Little's law and 22 workdays/month resulted in an average calculated cycle time of 48.4 days to make one subassembly:

$$Cycle\ Time = \frac{107\ subassemblies}{48\ subassemblies/month} = 2.2\ months$$

$$Cycle\ Time = (2.2\ months)\left(\frac{22\ days}{month}\right) = 48.4\ days$$

Now it was time to determine the effectiveness of the current policies. For that, we generated the flow performance curves as shown in Figure 14-1 and 14-2. First, we built the capacity envelope, knowing that the maximum available capacity to make the subassemblies was 88 pieces per month, which, in fact, was the bottleneck rate (BNR), the maximum output we could expect from the currently configured assembly line. We used 2 days for the raw process time (RPT), which represented actual work content.

In the diagrams, the marginal case is represented by the red line with white circles. Obviously, the more variability introduced into the calculation, the further the marginal case deviates from best case. We calculated an implied RPT based on the actual performance (48 subassemblies with 107 subassemblies in WIP) as described in Chapter 11. As a result, the current performance was equivalent to a process with a one-month raw process time. This means that even though the fastest time to complete a subassembly based on the actual process times was about two days, the fastest time with the current process was one month (22 days). This is the Actual Throughput line in Figure 14-1.

Note actual performance is much worse than the marginal case line indicating that variability or poor control or both are causing much wasted capacity.

Now, we know that if we continue with our current configuration and introduce the new rate of 63/month (see Figure 14-1), we need to have an average of 225 assemblies in the system to achieve the required output on average. However, *on average* means 50% of the time, and that means we don't achieve required output. As a practical matter achieving 63/22 = 2.9 subassemblies/day every day in pursuit of 63 assemblies/month, would require much extra capacity and much lower variability or both. The automobile moving assembly lines can manage constant daily output, most of the time, because they set the speed of the line to desired daily output and thereby set demand variability (arrivals) to zero. For the de-coupled subassembly process here, we would not expect 2.9 subassemblies every day, but we could set the WIP level higher to, for example, 260 subassemblies to ensure hitting the daily rate of 2.9 more often. In practice, the process might produce 5 subassemblies on some days and 0 on some but, on average, the process will produce 63 subassemblies/month. This analysis gave us a very good indication of where we were, where we needed to be, and it certainly motivated the managers responsible to do something.

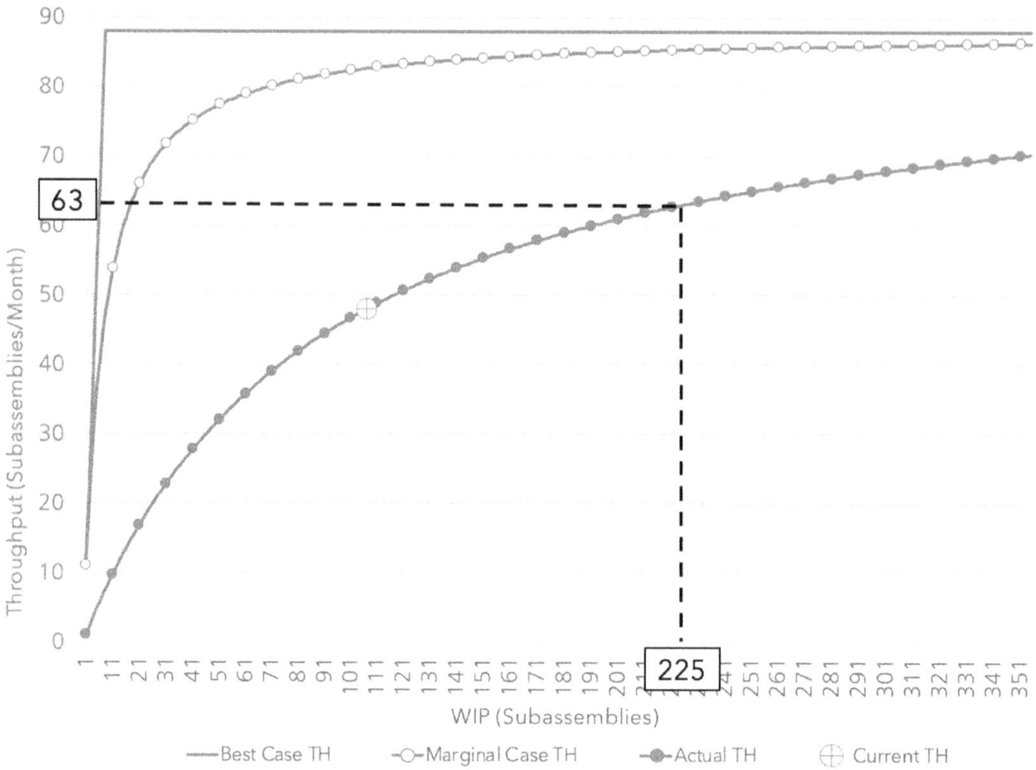

Figure 14-1: Subassembly flow performance curves

From a financial standpoint, this inefficient use of capacity has major implications. If we translate the WIP levels to euros and compare the existing financial investment (€2.9 million) to that required for achieving the desired output under current conditions (€6.1 million), a massive financial investment in working capital is required, about double the working capital as in the existing scenario. See Figure 14-2.

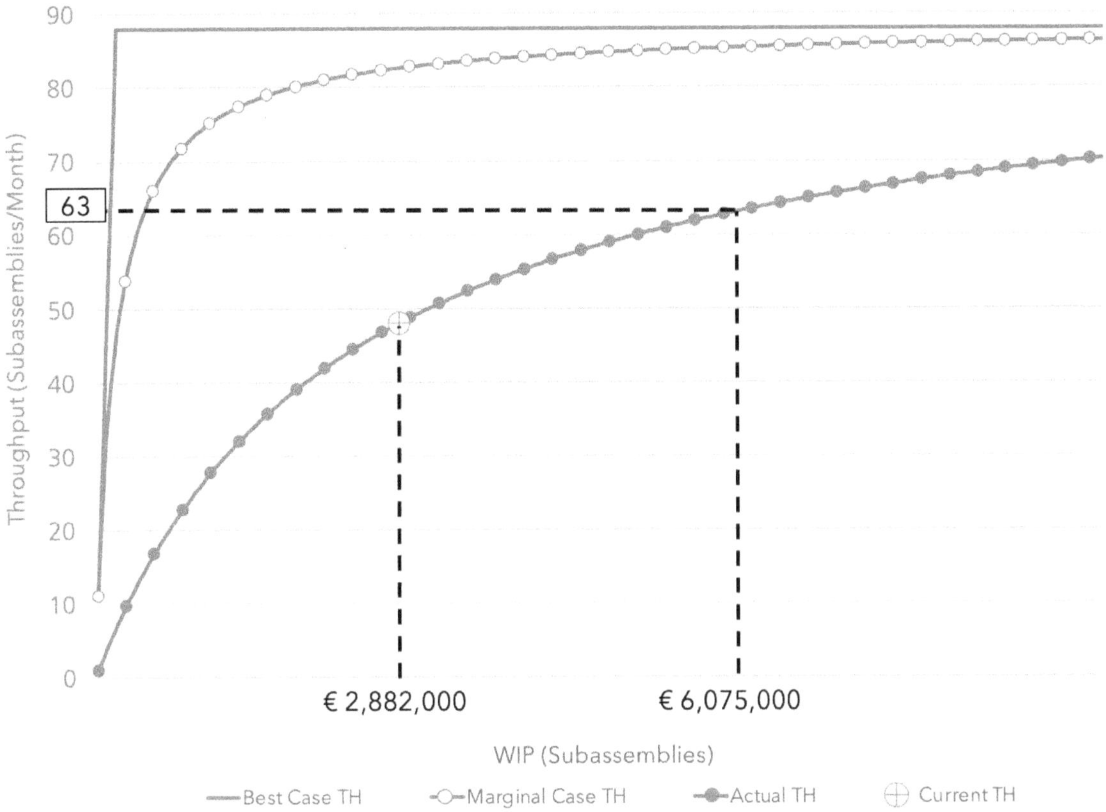

Figure 14-2: Flow performance curves with working capital

Looking at the effects of reducing variability, a process with the moderate variability of the marginal case would only require about 20 pieces of WIP (€540,000) to achieve a throughput of 63 subassemblies/month. So, reducing variability was a massive opportunity for financial savings. The level of variability implied by the current performance was so far greater than marginal case variability as to be strongly indicative of self-inflicted variability caused by poor control rather than variability from wildly variable process times.

Next, we considered how to improve the situation. It is very hard to refute that reduction in variability would improve the situation. An improvement in use of IT for planning, scheduling, and monitoring would reduce variability and certainly increase flow and move us up on the diagram, increasing the rate without the need to increase the WIP. Often, it is the scheduling practices that are sources of massive amounts of unintended variability. What we needed to do was now clear: focus on activities which decrease the amount of self-inflicted variability caused by poor scheduling and control and, if possible, reduce external variability in either demand or supply.

Sadly, while variability is common in daily life, people do not understand its behavior, its sources, and the operations science of its relationship to the buffers. Everywhere we look, variability dominates our processes, influencing outcomes in ways we often fail to recognize. From fluctuations in demand and lead times to variations in supplier performance and workforce productivity, variability introduces uncertainty, inefficiencies, and bottlenecks.

Yet organizations frequently focus on eliminating waste and streamlining workflows with IT without addressing the root causes of inconsistency. Without managing variability, even the best-designed IT practices will struggle to achieve predictable and sustainable performance improvements.

In the end, for the subassembly process, we don't know what happened. The manager of the process could not buy into the concepts of the analysis. This is not uncommon. When faced with an evaluation that shows such massive performance problems, managers can react defensively. This is especially true if the manager has been in charge of the entire process for a long time. Rather than focus on the opportunity, the reaction is to protect against any perception that they might have been doing a poor job. In that case, the result is a rejection of the analysis and a search for an evaluation that is less dramatic.

Addressing Unpredictable Cycle Times

In another instance at the same company, we achieved significant process improvements by once again leveraging operations science as the driving force. They were running production lines for manufactured panels, brackets, fairings, and other components eventually used to assemble pylons, the critical structures that attach aircraft engines to the wings. The lines had been experiencing delays, leading to longer-than-expected cycle times and difficulties meeting demand. When faced with inefficiencies in a production line, experience often drives us toward seemingly obvious problems—inefficient workstations, bottlenecks, or underutilized resources. But sometimes, the root cause of a persistent problem remains hidden in plain sight. This was precisely the challenge we faced when optimizing the production lines for these components.

Teams had already attempted various process improvements on the shop-floor efforts to balance the line flow and buffering the bottleneck, but the delays persisted. Spreadsheets were created to devise a schedule, and complex calculations generated capacity utilization proposals and load profiles, which were used to come up with better sequences, but the delays persisted. Eventually, someone suggested employing SAP software for capacity planning and workflow scheduling to get out of the spreadsheets in the hopes of shortening cycle times.

You have read and heard the phrase "hope is not a strategy" many times over, but here lies the deeper root of the problem. Managers jump to conclusions about what will solve the problem without ever analyzing the problem or validating the solution.

Instead, they take a software vendor's word that the sole implementation of one of the world's best software packages will do the trick. We should use IT offerings to automate the process where it makes sense. However, to simply implement an integrated software package can only deliver the anticipated results with lots of luck, and the implementation is usually anything but simple. In this case, we needed to know the source of the problem first before we could develop a sustainable solution with a predicted outcome. And no one knew what the root cause was.

That said, we felt compelled to use operations science. And, yet again, it was operations science that brought us insights and irrefutable evidence. Since the operations control problem manifested itself as delays in the delivery of the final product, we wanted to know where the delay or delays occurred. It was already known that cycle times for the machining of pylon components were long, so we agreed to investigate the routings and process times on the shop floor.

Previous initiatives looked at the times maintained in the routings, which, when added up, provided the raw process time. However, adding up process times provided no information about the time lost when orders were waiting in queue.

The operations science provided us with a structured approach to problem solving. First, we knew the throughput and were able to count the WIP on the shop floor, so we were able to calculate the actual cycle time. This was quite an eye opener. Raw process time from the routings added up to 12 hours (because the shop floor was open 8 hours a day this translated to 1.5 days). However, when taking queue time into consideration, as Little's law does, the calculated cycle time came out as 95 days!

Wow. How can that be? Well, if you know the principles, laws, and corollaries of operations science, you know that WIP on the shop floor is "visible" cycle time. The more WIP in front of the work center, the longer the waiting or queueing time of the order to be processed. This realization shifted our focus. Rather than adding more capacity or changing routings, we first reduced WIP levels, while maintaining throughput, and improved flow. Once we identified the excessive WIP as the primary issue, we investigated the specific stations and processes contributing to the buildup of unfinished orders, unconsumed material, and idle workstations.

A powerful technique for doing this is the VU analysis, as explained in Chapter 12. The VU analysis uses readily available information, cycle time and process time, to calculate VU values. High VU areas are prioritized as potential high-leverage areas for improvement.

A deep dive into the production steps revealed

- certain workstation operations were batching work unnecessarily, leading to long wait times between process steps
- a key inspection step was overloaded, forcing components to sit idle before moving forward
- missing raw materials forced orders to get stuck on workstations and blocked flow, and
- production orders arrived at machines that couldn't handle them because there was no operator and materials were missing

By targeting these issues, we implemented changes such as building flow lines with capacity buffers, moving from batch to flow by reducing transfer batch size, and improving material availability with rule-based replenishment. The result? Orders started to flow, and there was a seemingly magical increase in order throughput. The cycle time went from 95 days down to 20 days. That was a good result, but not good enough.

Now it was time to use the IT system for additional improvements. The standard software that had already been implemented contained all the functionality we needed. The functionality simply needed to be applied in a manner that would support our goal of further WIP reductions. This was accomplished by flattening bills of materials (BOM), building line hierarchies, and balanced routings for takt-based scheduling, and implementing effective replenishment policies and practices to ensure high availability of purchased parts on the lines.

With today's common planning and scheduling software, many companies have the needed functionality to accomplish effective planning and scheduling. Very many times companies never achieve the performance they are capable of due to ignorance of operations science and a misunderstanding of the mechanics of their planning and control systems.

Eventually, the cycle time went down to 5 days, which was still significantly higher than the raw process time of 1.5 days. However, if you have a good understanding of manufacturing systems, you know that raw process time is impossible to achieve in the real world. Attaining raw process time, zero inventory, and maximum throughput (equal to maximum available capacity) is not possible because variability exists. And this knowledge is another pillar of the teachings of operations science. Without it we would have looked at the wrong spots to further optimize. Continuous improvement becomes a very frustrating undertaking without the concepts of operations science.

The savings were substantial. Beyond reducing cycle times by an average of 90 days, we also significantly lowered WIP levels on the shop floor, resulting in a major financial impact.

I can only imagine the consequences if we had skipped the operations science analysis and jumped straight into capacity planning with a software package. Most likely, the solution would have been built around the existing work centers and BOM structure, leading to an excessive number of scheduling points across the shop floor. The software might have offered a graphical scheduling tool for capacity leveling, but planners would have faced the near-impossible task of constantly coordinating and rescheduling hundreds of individual orders each day.

Ultimately, whether you rely on Excel spreadsheets or the most advanced ERP or APS system, the key is to address the root causes of a problem. ERP systems won't help you identify those root causes; they only assist in implementing a solution. If the underlying issues remain unresolved, no system will deliver meaningful improvements.

Tackling Bi-Modal Inventory Swings: Insights from a Global Automotive Parts Manufacturer

Here's an example where inconsistent mental models, lacking any foundation in operations science, can lead to major problems. At a major automotive parts manufacturer, we encountered severe bi-modal inventory swings, alternating between excess stock and critical shortages. This issue was particularly visible at one of their spare parts distribution centers, but the root cause traced back much farther upstream.

The case involved distribution of aftermarket parts. They produced shock absorbers at a large manufacturing plant in Poland, supplying a distribution center (DC) in Belgium. On the surface, it seemed like a straightforward supply chain. However, the reality was anything but. The DC in Belgium was drowning in inventory. Shock absorbers that no one wanted were stacked high in the warehouse, tying up working capital and space. Yet, at the same time, the shocks that customers did want were routinely out of stock. The result was a strange and frustrating situation: Executives couldn't decide what to complain about more—the extreme inventory holdings or the consistently poor customer service levels.

Because customer service was seen as the top priority, leadership issued a blanket directive: Keep at least two months of inventory on hand to improve availability. It sounded like a sensible response, but it only addressed the symptoms, not the underlying issue. Making matters worse was the way the manufacturing plant in Poland operated. The plant manager, guided by global leadership, was incentivized to focus on unit cost. The mantra was clear: We must utilize all our resources to the maximum and engage in low-cost manufacturing. In practice, that translated to keeping the production lines running at full capacity, churning out the cheapest shocks to lower the average unit cost.

Unfortunately, these "efficient" production runs had little to do with what the market actually needed. The plant overproduced shocks that were cost effective to make but slow moving, as there was little or no demand for them. Meanwhile, there was no time, or capacity, left to manufacture the parts that were in high demand at the DC. This created an enormous disconnect between supply and demand, and the more the plant pushed for "efficiency," the worse the situation became.

We had to attack the problem from both ends simultaneously. Simply increasing inventory or tweaking the production plan wasn't going to cut it. We needed a system-wide solution, based in sound operations science, that attacked the root cause. First, we tackled the demand side by calculating the appropriate buffer requirements for the DC. The goal was to size those buffers so they could absorb as much demand variability as possible, smoothing out the spikes before they hit the plant. This was not a guesswork exercise; it was a deliberate, data-driven process.

We used a tool to segment the entire finished goods portfolio into nine distinct classifications. This segmentation was based on demand behavior, variability, and criticality. Once we had these classifications, we developed specific inventory and replenishment policies tailored to each group. These policies were guided by carefully constructed decision trees that considered lead times, service level targets, and variability factors.

On the manufacturing side, we worked to break the cycle of unit-cost–based decision-making. Instead of prioritizing unit cost, we aligned production priorities with actual demand needs driven by the DC's consumption and the segmented buffer strategies. The plant began producing what was needed when it was needed, not just what was cheapest to make.

Within a relatively short time, the bi-modal inventory swings stabilized. The DC no longer faced excess inventory of slow-moving products while suffering shortages of high-demand items. Customer service levels improved dramatically, and working capital tied up in unnecessary inventory was released. In fact, inventory levels in the DC stabilized at almost 50% less than the previous levels, while service levels rose by 12 percentage points to 97%.

But the biggest success wasn't just better inventory balance. It was the shift in mindset across the organization. By applying operations science, we replaced inconsistent mental models with a common, organization-wide understanding of how to holistically manage inventory, production, and customer service. .

If we had simply relied on an ERP system to fix the problem, we would have failed. ERP systems are great at executing transactions and automating workflows, but they don't think, and they certainly don't resolve systemic issues rooted in misaligned priorities or flawed decision-making. No ERP system can replace the critical thinking and the global approach that operations science provides. It was operations science that led us to the root causes of the problem: conflicting objectives, inconsistent mental models, and a manufacturing strategy that prioritized cost over customer needs. And it was operations science that guided us to a solution that worked for both the plant and the distribution center.

For any organization grappling with similar challenges, I strongly recommend taking a closer look at the root causes instead of relying on the IT provider's quick solution: Technology is just a tool. Without the right thinking and decision frameworks behind it, even the most sophisticated systems can lead you straight into trouble. Operations science is the key to understanding complex systems and making decisions that deliver real, sustainable results.

Avoiding Costly Investments in Additional Capacity Through Simple Process Adjustments

At a large global manufacturer of electrical components, we faced a common but costly challenge: inefficient use of existing production resources. Despite having a modern manufacturing facility with capable equipment, machine utilization was very low, at less than 40%. At the same time, a massive production backlog was growing, causing major delays in customer service and grinding down customer satisfaction.

Faced with these problems, leadership jumped to what seemed like an obvious solution. They put numbers together to invest heavily in new machines, thinking, If there's a backlog, we must need more capacity. They concluded their decision without further analysis. Unfortunately, without a deeper understanding of the real problem, they were about to commit a very large sum of money to a solution that wouldn't have addressed the true cause of the problem, and they would have continued underutilizing the new resources as well.

Someone in the organization had different ideas and asked us to take a closer look. What we found through a thorough, operations-science–based, root-cause analysis was both revealing and all too familiar: The real problem wasn't capacity. The issue was an inefficient and chaotic planning and scheduling process that caused existing resources to be underutilized. Despite having machines that could meet demand, orders weren't flowing through the plant in an orderly or effective way. Instead, they were getting stuck, delayed, or waiting for materials. The result was idle machines, frustrated production teams, and unhappy customers.

Our operations science analysis uncovered several key factors driving this inefficiency in the IT system:

- **Unnecessary complexity in the BOM:** Extra levels in the BOM added needless complications to scheduling and sequencing. Typically, when IT leads an implementation, material master records are created for every step along the way of a routing. This is usually driven by accounting wanting to accumulate cost at every step for detailed reporting. It is important to know an order's progress to planned completion, but that is different from having to transact a job in and out of every step in a process. One idea in support of detailed costing and complex BOMs is that they allow better control over costs and, therefore, profitability. This is an illusion because these efforts are noise generators and tend to complicate, i.e., add variability, to a process, which increases cost and reduces delivery performance. See the Chapter 11 discussion on contribution margin at the constraint for more insight. Much more important is that orders flow through the process to meet demand. If the IT system supports flow to achieve desired cycle time and meet demand with minimal cost, desired profitability will occur, which accounting can then report.
- **Incorrect scheduling parameters in the ERP system:** The wrong scheduling type was being used, leading to poorly planned orders that didn't align with real-world constraints. This happens when the IT department (or the SAP consultant) leads the implementation and doesn't employ the principles of operations science. Too often the setup of master data and the decision of what scheduling type to use are driven by what the software supposedly requires to work effectively. It is often what has been implemented before, at another company, or what we learned at a seminar, or a training program given by the software vendor.

- **The bullwhip effect running wild:** Variability, caused by deterministic planning in orders and lack of coordination between departments resulted in wildly fluctuating demand signals, amplifying uncertainty and confusion across the system. All too often, manufacturing systems are implemented without consideration of variability. The principles, corollaries, laws, and rules from operations science are ignored, and experience from other (often failed) implementations dictates decision-making and the need for quick resolutions.

We then mapped out the causal relationships that were driving the inefficiencies. Logical Thinking Process[i] helped us to identify not just the visible problems but also the deeper, systemic issues. For one, there was poor master data management, leading to incorrect BOM structures and material planning. This ended up in self-made interruptions of the flow, because after each processing step a goods receipt was posted. To continue, orders had to be released and a goods issue posted.

Additionally, the lack of a rule-based replenishment system resulted in frequent material shortages. Raw materials were ordered exclusively on deterministic expectations from a very unreliable production plan. Nobody followed the production plan anyway; the production team met in the early mornings to discuss what to put on the line that day. For the purchasing department this meant that anything with a lead time longer than a few hours had to be brought in on a wild guess.

And to make matters worse, there wasn't any control over the release of production orders. Imagine, as a supervisor, you're receiving many production orders every morning only to find out that most don't have components available. Then, when you locate an order with components and want to release it to production, you find out that it is scheduled to run on a machine that is currently busy with another job.

Once we identified the root causes, the solution became a little more apparent: It wasn't about buying more machines; it was about properly using the existing ones. And it wasn't about buying a new IT system, it was about properly using the one we already had.

Here is what we did. First, we flattened the Bills of Materials. By reducing unnecessary BOM levels, we reduced the number of scheduling points and the accompanying need for excessive goods receipts and goods issues. We were able to connect the processing steps into a flow along one routing, from raw material through finished goods. This greatly simplified the scheduling process and provided a more streamlined flow of materials and orders. It gave the planners better visibility and more realistic cycle times.

For better material availability, we implemented rule-based replenishment. We quit using deterministic expectations which never materialized and took a variability-focused approach using buffer planning and decoupling points. This helped stabilize material supply and eliminated one of the main sources of delay.

Eventually we leveraged the ERP system the right way. We configured the ERP's planning and scheduling tools to level orders based on real material and capacity availability. We introduced a collective material availability check, ensuring that only feasible orders—those that had both materials and capacity—were released to the shop floor, and we employed rate-based scheduling for an even flow along the production lines.

The result? A significant reduction in chaos and confusion. Orders began to flow smoothly through the factory, and machine utilization jumped *from less than 40% to close to 80%.* The backlog shrank rapidly, customer service levels improved, and, perhaps most importantly, leadership realized that the expensive investment in new machines was unnecessary.

In this case, operations science, combined with the logical thinking process, provided the framework to identify the real issues and implement a sustainable solution. Instead of throwing money at new machines, the company invested in smarter processes, better planning, and more disciplined execution.

Operations Science Application Summary Table: Shock Absorbers

ELEMENT	DESCRIPTION
BUSINESS STRATEGY	Provide shock absorbers on demand from stock inventory. Achieve lowest unit cost.
OPERATIONS STRATEGY	
DEMAND	Individual customers ordering from stock
TRANSFORMATION	Manufacturing facility in Poland.
VARIABILITY	Demand variability caused by many different customers ordering shock absorbers. Transformation variability was largely self-induced by running at a high level of utilization to ostensibly achieve efficiency. Poor service led to frequent schedule changes and frequent expediting which increases variability.
BUFFERING OPTIONS: CAPACITY	No capacity buffer was used because the company was focused on running at high levels of utilization.
BUFFERING OPTIONS: INVENTORY	Tremendous amounts of inventory used but it was the wrong inventory.
BUFFERING OPTIONS: TIME	Time was not an option because customers expected product be in stock when they ordered it.
TOOLS AND APPLICATIONS	Product segmentation Implemented optimal inventory policies Synchronized production more closely to demand

Conclusion

Across all these real-world cases the common thread is obvious. Inconsistent mental models and a lack of operations science can lead to major inefficiencies, misguided decisions, and ineffective use of large IT investments.

In each scenario, leadership initially reacted in ways that are all too familiar in many organizations. They focused on surface-level symptoms, whether by issuing blanket inventory directives or preparing for significant capital investments in new machines, without understanding the underlying systemic causes. If we had relied solely on technology, such as ERP systems, or defaulted to gut instincts, these companies would have invested heavily in solutions that didn't solve their problems.

Instead, by applying operations science and structured root-cause analysis, we were able to diagnose and address the real issues. The solutions were not about adding complexity or expensive technology. They were about simplification, alignment, and applying operations science to design robust, reliable processes. Flattening BOMs, implementing rule-based replenishment, segmenting inventory with clear policies, and using ERP systems the *right* way allowed our clients to stabilize operations, improve service levels, and maximize the use of existing resources.

Technology, such as ERP systems, is only as effective as the thinking behind it. Operations science provides the foundational framework needed to design systems and processes that work in the real world, where uncertainty, variability, and conflicting priorities are everyday challenges.

Before defaulting to expensive investments or relying on technology to "fix" problems, organizations need to ask the following:

- Do we understand the system dynamics at play?
- Are we applying operations science to guide our decisions?
- Have we identified the true root causes, or are we chasing symptoms?

By focusing on these questions and using disciplined methods like the logical thinking process, organizations can avoid costly mistakes, improve performance, and build sustainable competitive advantages.

People don't go to work thinking, "I hope we have a day full of firefighting, stress, and confusion." Unfortunately, when those days occur, that stress often gets carried home. The operations science approach not only provides process performance improvements, it also greatly reduces stress and confusion for the people working the process, which is a major benefit in and of itself.

---

[i] H. William Dettmer, *The Logical Thinking Process: A Systems Approach to Complex Problem Solving*, 2nd ed. (Milwaukee, WI: ASQ Quality Press, 2007)

15

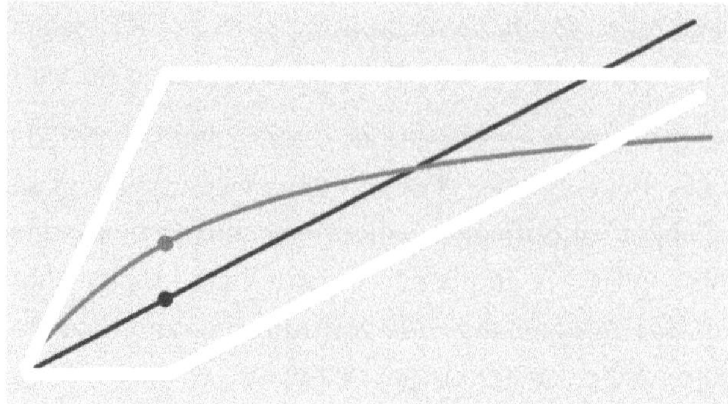

Configure to Order: The Vortex Flowmeter Team

The Vortex Team

Jeff Bell

I first experienced application of operations science theory to practice in the late 1990s with the vortex flowmeter team in Emerson Electric's Rosemount Flow Division. I managed the vortex flowmeter business unit. My colleague Dale Nugent managed the magnetic flowmeter business unit. We had recently graduated from the Master of Management in Manufacturing (MMM) program at Northwestern University. We were students of Spearman and Hopp during the time the *Factory Physics* textbook was being written. We wanted to bring the concepts to our work at Emerson's Rosemount Flow Division. Mike Severson, featured multiple times in this book for his work as an operations scientist for different organizations, began his operations science journey on the vortex team. Sue Thomas was the production supervisor leading the vortex team to execute tactics and install controls to support operations strategy. The vortex team agreed to hold a weekly lunchtime study session with the *Factory Physics* textbook to see if they could get deliveries under control in a high mix, engineer-to-order business. Sue often invited leaders from the shop floor to participate and provide perspective. As mentioned, a deep dive into technical calculations is not required on the journey to better operations performance. Thus, the team started with the basic concepts. They extended this learning by inviting Professor Kate McKone-Sweet from the University of Minnesota to teach a large group of colleagues in the vortex business unit on the principles of operations science in an on-site session. After a few years, Sue, Mike, and a few leaders from the floor attended an executive session on factory physics taught by Wally Hopp at Northwestern University to reinforce and extend the concepts they were applying. This is the story of how the team, armed with the science of operations, improved deliveries and regained the confidence of the sales force to drive profitable growth in the Rosemount Vortex Division.

Business Strategy

In the late 1990s, the Rosemount Flow Division consisted of two very different products in terms of positioning in the market. This example primarily focuses on the Rosemount vortex flowmeter product and business. The Rosemount vortex flowmeter significantly differentiated itself from the competition based on technical attributes. The integral-cast meter body greatly reduced leak potential, and the sophisticated electronics minimized noise in flow readings. It was the only vortex meter on the market that worked in certain applications. As stated on the Emerson Electric website, "Rosemount 8800 Series Vortex Flow Meters offer world-class reliability with a gasket-free, non-clog meter body that eliminates potential leak points, resulting in maximum process availability and fewer unscheduled shutdowns."[i]

The meter was in great demand despite long lead times and inconsistent deliveries. Unfortunately, since the Rosemount vortex meter was only a small portion of the salesperson's offering of Rosemount pressure, temperature, and flow instruments, many salespeople did not promote it due to customer relationship concerns from the inconsistent delivery performance of the vortex product.

Emerson had the opportunity to realize the growth potential of the technologically differentiated Rosemount vortex meter by consistently meeting delivery expectations, which would help the sales force promote the advantages of the product. This required that the vortex team meet delivery promises. The salespeople would not promote the product until they could trust the lead times. Thus, the business strategy for the differentiated vortex flowmeter product required that lead time promises for delivery were met.

The vortex flowmeter product line shared the high fixed cost and difficult-to-move asset of the flow lab with the magnetic flowmeter product line. Dale Nugent, the newly named director of the magnetic flowmeter business, recalled,

> The magnetic flowmeter was in what was essentially a commodity market. Customers made the purchasing choice for magnetic flow meters based on cost, lead time, and "pull through" with other Rosemount products purchased for new installations. To help with the cost part of the decision, the company invested in a new plant in Mexico to produce all magmeters. The transition, like most construction projects (see Chapter 18), was behind schedule and affecting deliveries. We developed a quick-ship program called MAGIC—Mags Impressing Customers—to keep the sales force engaged. The MAGIC program also greatly improved business unit margins with its quick-ship expedite pricing.

The entire vortex product line was produced in Rosemount's facility in Eden Prairie, Minnesota, and the final assembly of magnetic flow meters for MAGIC was produced in the same facility. Until the Mexico transition was complete, Rosemount magmeters were also assembled in the Eden Prairie facility. This caused a capacity crunch because all products shared the very expensive flow lab for testing and configuration. Flow meters are configured based on technical specifications required by specific customer applications and provided through the sales force. Thus, each shipment was essentially unique, although the meters contained many common components. The challenges the vortex team faced to execute the business strategy focused on

1. how to manage a wide variety of final product configurations requiring calibration while still providing high on-time delivery and high-quality products to customers,
2. how to convince the customers and the sales force that the first goal had been achieved so they could be confident in buying and promoting the product, and
3. how to manage the high fixed cost and difficult-to-move asset of the flow lab.

Operations Strategy: Demand and Transformation

Demand

Demand for vortex flow meters could be placed into two groups, new plant installation projects and emergency repair projects. In a new plant installation project, there was typically significant visibility to demand and an ability to schedule based on lead times. That said, it was critical in these cases that on-time delivery be high so projects would stay on schedule. In emergency repair projects, short lead times were critical because plants essentially could be shut down if a meter needed replacing. In both cases, demand was engineered/configured to order so an order had to be placed with correct configuration specifications prior to it being configured for shipment.

Transformation

The production process was “make-to-order”, with a flow consisting of welding, hydrostatic testing, sensor production, transporting the meter to the flow lab, calibrating the meter in the flow lab, configuration, testing, transporting the meter back for final assembly, final assembly, final inspection, packaging, and shipping. The transportation piece was necessary because the flow lab was in a separate building from the rest of the operation.

Buffering Options

The team had to manage the portfolio of buffers (time-inventory and capacity) in the context of the two market segments it served. The new plant installation segment could have a time buffer but the time quoted for the project needed to be accurate so planned build schedules could be maintained. The emergency repair segment did not want a time buffer. Finished-goods inventory was not feasible since vortex meters were an engineer-to-order/configured-to-order product. Thus, the team understood it would be critical to manage variability, the capacity buffer, and their associated impact on the time buffer.

Finally, there was an opportunity to leverage common components for mass customization with an inventory order interface prior to customizing based on configuration and flange size that fit the customer’s specifications. This allowed the team to manage variability and provided some flexibility in how much of the time buffer the customer would see on certain higher-demand line sizes.

Tools and Applications

Capacity Planning with a Focus on the Bottleneck

As discussed in Chapter 4, the bottleneck constrains the output of the operation. Recall that the bottleneck is the highest-utilized station. Struggling to keep up with demand, the team agreed to apply operations science to identify the bottleneck.

To demonstrate how the team found the bottleneck with quantitative specifics, this example references data in Exhibits 6 and 7 of the case study "The Rosemount Vortex Flowmeter Plant" by Professor Kate McKone-Sweet.[ii] The case study is available at the Harvard Business Review Store (https://store.hbr.org/product/rosemount-vortex-flowmeter-plant/BAB045). Dr. Wally Hopp created a spreadsheet of hints for his classes to use with the data when they worked on the case. We thank him for his permission to use these models in this chapter.

The data for the flow lab are shown in Table 15-1, with utilization calculated using Dr. Hopp's spreadsheet:

CASE DATA

Line	Scale	Product	Mean Demand/ Week	Std Dev Demand/ Week	Average Mix	Process Time (min)	Setup Time (min)	Run Time/Week (min)	Total Min/Scale	SCALE USE Util
1	A	18-24 M	6.668	1	0.01	48	9	380.08	2070.4	48%
1	A	6-8 V"	70.014	5	0.09	17	4	1470.29		
2	A	14" & 16" M	6.668	5	0.01	12	4	106.69		
3	A	12" M	6.668	5	0.01	13	4	113.36		
3	B	8" & 10" M	6.668	1	0.01	35	8	286.72	2690.5	62%
4	B	4" M	16.670	1	0.02	43	7	833.5		
4	B	6" M	10.002	1	0.01	22	7	290.06		
4	B	4" V	80.016	7	0.1	12	4	1280.26		
5	C	2" M	16.670	1	0.02	54	6	1000.2	4981.0	115%
5	C	3" M	16.670	1	0.02	41	6	783.49		
5	C	2" V	116.690	7	0.15	10	3	1516.97		
5	C	3" V	120.024	8	0.15	10	4	1680.34		
6	D	1.5" M	10.002	1	0.01	40	6	460.09	2817.2	65%
6	D	1" V	123.358	8	0.16	8	3	1356.94		
6	D	1.5" V	100.020	7	0.13	7	3	1000.2		
7	E	0.5" M	10.002	1	0.01	23	6	290.06	1210.2	28%
7	E	1" M	16.670	1	0.02	22	6	466.76		
7	E	0.5" V	56.678	3	0.07	5	3	453.42		
		Total M	123.358							
		Total V	666.800							

Table 15-1: Case data for flow lab operations. M = magmeter V =vortex

Scale C, with 115% utilization, was definitely the bottleneck! The team started to monitor throughput (TH) at the bottleneck and found it to be far less than capacity. In fact, when they walked by the flow lab, they often found Scale C was not being utilized because there were no meters ready for Scale C at the flow lab! How could that be the case? It turns out that meters were moved only periodically from the production area in the other building which created "move batches" that starved the bottleneck. This optimized efficiency for the "move resources" but suboptimized the entire process because it starved the bottleneck. Mike Severson also remembered,

> We learned the weld team had been setting up for large process batches of meters so they could minimize setups. While this helped optimize welding results, it suboptimized the entire process

> because it meant critical meters for Scale C were waiting in welding behind large batches of less critical meters for lower-utilized scales. Once again, this suboptimization starved the bottleneck.

Sue Thomas said,

> The team immediately began moving meters more frequently to avoid starving the bottleneck. This, in conjunction with prioritizing meters in weld that went on Scale C in the flow lab, solved the starving problem, but utilization was still over 100%, and capacity could not keep up. The flow lab did not typically operate on the weekend, so overtime was instituted to increase capacity and help the situation. The team also invested in equipment to be able to run multiple meters at the same time on a line, which increased capacity and TH.

Recall that the MAGIC program for magmeters was by definition an expedite program so those items always got moved to the front of the line. Dale Nugent, the leader of the magmeter business unit said,

> Unlike the rest of magmeter production, MAGIC production was not slated to move to Mexico as the United States quick-ship program needed to avoid delays at the border. Clearly, the expedite feature of MAGIC was a disruption that required management. The MAGIC units were sold at a premium and vastly improved the profitability of the business. It was worth the effort to manage the disruption.

Since MAGIC units were appropriately prioritized, the vortex team found that weekend overtime increased magmeter throughput on a weekend but did not have a big impact in terms of vortex meters completing over the weekend. The company was not going to invest in a flow lab expansion when the current flow lab had plenty of capacity for vortex and MAGIC flow meters only. The rest of the magmeters were soon planned to be out of the facility. As will be discussed Chapter 18, projects are often significantly behind schedule so "soon" can be a really long time. Ultimately, capacity improvements and the completion of the Mexico facility lowered utilization below 100%, and the team was able to achieve feasible production rates.

With the flow lab situation improving greatly, the bottleneck shifted to welding, as shown in Table 15-2, with case data for nonflow lab operations. Dr. Hopp's spreadsheet in Table 15-2 shows the analysis. The good news in this case is that hiring another welder and adding a station were pretty straightforward solutions to the problem. Sue Thomas remembered,

> We added capacity to welding by investing in automation, and that really helped.

CASE DATA FOR NON-FLOW LAB													ADJUST FOR FAILURES		Adjust for Yield	Adjust for	Station Capacity			Demand	Utilization
	No of stations	Raw Process Time (hr)	Std Dev (hr)	Effective Hours/Day	Yield	Setup Time (hr)	Setup Std Dev (hr)	N_S	MTTF (hr)	MTTR (hr)	Batch Size (in)	Batch Size (out)	A	t_e	t_e	t_e	per hour	per day	per week	Vortex Demand per Week	
Welding	5	0.262	0.150	8	0.99	0.03	0.01	1	40	0.17	1	1	0.996	0.263	0.266	0.296	16.9	135.2	676.2	666.8	98.6%
Hydro	1	0.019	0.030	8	0.96	0.15	0.01	5	40	0.17	1	1	0.996	0.019	0.020	0.050	20.0	160.4	802.0	666.8	83.1%
Sensoring	1	0.025	0.006	8	0.95	0	0	5	40	0.17	1	1	0.996	0.025	0.026	0.026	37.8	302.7	1513.6	666.8	44.1%
Config	1	0.030	0.006	8	0.85	0	0	1	40	0.17	1	1	0.996	0.030	0.035	0.035	28.2	225.7	1128.5	666.8	59.1%
Assembly	1	0.023	0.006	8	0.99	0	0	1	40	0.17	1	1	0.996	0.023	0.023	0.023	42.9	342.9	1714.5	666.8	38.9%
Inspection	1	0.015	0.006	8	0.99	0	0	1	99999	0	1	1	1.000	0.015	0.015	0.015	66.0	528.0	2640.0	666.8	25.3%
Packing	1	0.015	0.006	8	1	0	0	1	99999	0	1	1	1.000	0.015	0.015	0.015	66.7	533.3	2666.7	666.8	25.0%

Table 15-2: Case data for nonflow lab operations

WIP Caps, Due-Date Quoting, and Virtual Queues

Orders were released to the floor with an ERP system based on standard lead times in the sales catalogue as well as deterministic cycle times in production and the supply chain. In general, orders were worked on in a sequence of first in, first out, but as orders started to go late and WIP grew, this changed. The beginning of each day turned into an exercise of sorting for the expedites so the most strategic customers could be moved to the front of the line. Sue Thomas, the production supervisor, lamented,

> This was crazy! We spent the first two hours of every day looking for expedite meters instead of producing more meters.

The team agreed that instituting CONWIP control would help end the current chaos on the floor. They decided to start with a CONWIP level of 700, equal to about one week's throughput so the cycle time from meter start to meter finish would be one week. They left all the current expedites in WIP and then kept the first-in, first-out approach for the rest of WIP until they hit the CONWIP level. All orders not in WIP were placed in a virtual queue in a first-in, first-out manner. Sue Thomas remembered the organizational change effort required to plan and execute the CONWIP system, stating,

> There was initially a lot of pushback on CONWIP. We explained what we were doing by referencing the textbook *Factory Physics* and asked people on all levels to trust us and give us time to catch up so we could really understand the whole quote-to-cash process and not just individual tasks. We succeeded in training customer support and getting their buy-in, so they were able to deliver to promise within agreed upon parameters. Once we caught up, it was much easier to stay caught up and use the reservations spots for occasional expedites of high priority to the business.

CONWIP ended the early-morning expedite-sorting window, enabling valuable resources to be focused on capacity to produce meters instead of spending time finding expedite meters. The team began to achieve planned shipment levels and was able to reduce lead times. Predictive control using Little's law and due-date quoting was now possible because the team capped the WIP and routinely achieved planned TH.

With predictive control of the time buffer, the team could now use it strategically. They knew that in many cases, the customers wanted the technically differentiated product and were willing to wait for it as long as the time quoted for delivery was correct. The overdue backlog was about two weeks' worth of work. The total backlog including overdue was five weeks' worth of work and growing. Based on the size of the backlog and the expected TH, the operations team wanted to communicate a standard lead time of 10 weeks to the sales force. This provided a time buffer that enabled the team to deal with variability in regular demand and expedite demand. It also helped efforts to reduce the chaos on the floor. When the marketing team questioned the size of the buffer, the operations team used the lead time vs. service level charts discussed in section 9.5.4 of the *Factory Physics* textbook to provide data-based options for lead time and service level. The business team, including sales and marketing, was now one team, making choices for the business together instead of in silos. The production team worked with the sales force to agree that expedites would take two weeks, which enabled expedites to move to the front of the virtual queue but not be released until the CONWIP system allowed room for them to enter production. This plan provided a time buffer so the vortex team could meet commitments on expedites and standard lead times. Even though these lead times seemed long (who wants an expedite of two weeks?!) and competitor standard lead times were three weeks, recall that the most important strategic issue for long-term growth was the confidence of the sales force. In most cases, on-time delivery mattered more than a short lead time because the customers were willing to wait for the technically differentiated product. In terms of the sales force, they knew they could get the orders based on differentiation but did not want to spend time explaining missed deliveries or risk damaging their overall relationship with the customer because of delivery problems with vortex products.

Sales Force and Product Line Collaboration

Jeff Bell

As director of the vortex business unit, I hit the road with marketing to collaborate with the sales force on this plan. We even renamed the business unit's marketing efforts as field-centered marketing, with 100% focus on collaboration with the salespeople in the field. In addition to working together to create a lead-time/on-time plan, the business funded a technical specialist completely dedicated to helping the salespeople make the technical differentiation sale. It was important to win the confidence of the sales team. Sales force collaboration helped stabilize confidence for vortex deliveries with the sales force. It also catalyzed growth because the sales force felt like the vortex business unit really listened to them in terms of what they needed to win. The technical specialist was a huge win. Along with reliable and reduced lead times, the sales force got continual reinforcement that the team was working together. It really helped produce positive mindshare in the sales channel.

Finally, the time buffer also helped the magmeter product line. As Dale Nugent, director of the magmeter business, explained,

> The reality was, we had to deliver MAGIC to our customers with an incredibly quick response. When it came to the shared flow lab, MAGIC magmeters needed to go to the front of the line. The vortex strategic time buffer approach enabled what was needed in the market for both product lines.

Inventory/Order Interface for Mass Customization

As the team started to think about how to reduce lead times through continuous improvement, they wondered if they could include a strategic approach in addition to grinding out improved TH and yields. In one of their weekly lunch sessions, they discussed potential for an inventory/order interface with common components enabling mass customization. Many of the components in vortex were common despite the fact that line size and configuration combinations made stocking finished goods virtually impossible. An Emerson sister company, MicroMotion, had a program called flanges on last, where meters were built up to the point that they needed the specified flange size. What could work in the vortex situation? Mike Severson did a Pareto analysis of demand and noted,

> While we ship literally thousands of different SKUs, a Pareto analysis shows that the 80/20 rule basically holds. It is not a surprise when the majority of what we ship are two- and three-inch meters. Perhaps we could build to stock before configuration of these meters and configure to order. This worked great to level load production capacity and reduce the time needed to get these units from customer order to configuration, because most of the work was already done.

Cycle Time Components

As mentioned before, reducing transfer batches by delivering flow meters to the lab every hour instead of every day significantly reduced starvation at the bottleneck flow lab. Ultimately the production floor was moved next door to the flow lab so the transfer batch could be reduced to a size of one. When the meter was done in production, it could be put right into the flow lab. This was a key enabler to reaching two-day cycle times for the whole process.

The team was successful in meeting quoted lead times and rearranging priorities in the virtual queue to be responsive to strategic customer needs. They achieved significant reductions in lead times by focusing on the bottlenecks and started a quick-ship program for 2- and 3-inch meters. There was just one remaining issue that was not improving—a stack of orders at the configuration station that could not be produced because their orders had nonexistent or unfeasible calibration data. These orders waited until a salesperson provided the needed configuration data, which often took weeks or even months. The team waited to work on this issue until they had gained the confidence of the sales force with several years of collaboration. At that point, the team worked with sales force leadership to agree that if an order came in with missing or infeasible calibration data, the meter would be shipped to the customer as standard configured, with a note that the salesperson would configure the meter in the field. Because salespeople wanted to go to the customer to sell and not to install or configure meters, this provided great incentive to make sure orders came in with needed configuration data. This source of variability and delay was significantly reduced.

Controls and Measures

The team knew operations strategy execution required controls and measures to support the strategies and tactics discussed above. The team realized it needed to go beyond the corporate focus on being able to ship any order in five days or fewer. While this was a good slogan, it did not really match the needs of different business segments. Large, new plant installation projects could handle longer lead times but required that the meters showed up on time. Five days was not needed and would have required a huge capacity buffer to meet this artificial need. For expedites, the team thought five days was not short enough for a goal. They wanted to go for a goal of two days to better serve their customers who needed to get lines up and running in emergencies. Ultimately, after a few years implementing not only the previously mentioned strategies and tactics but also many continuous improvement initiatives, the team routinely achieved two-day cycle times. The team was now in the wildly different position of choosing lead times based on market strategy instead of operations constraints. Competitor lead times were three weeks, so they chose to make the standard lead times two weeks and expedite lead times two days with an expedite charge. This made use of an available time buffer to increase profit with expedite pricing and allowing the market-based time buffer to enable tighter management of costly capacity and inventory buffers.

Of course, lead time is a choice. The outcome is percent on-time and average days late. The team reported those but really focused on the leading indicators by counting WIP and measuring TH to make sure CONWIP levels were maintained and planned TH achieved.

Business Results and Lifelong Friendships

Jeff Bell

The vortex team leveraged operations science to align strategies, tactics, and controls to regain the confidence of the sales force and engage them to drive profitable growth for the product line. Equally, if not more important than the business results, are the lifelong friendships and subsequent business relationships that developed from the experience. Mike Severson and I worked together after this experience in many different business situations. We still work together today at Arc Precision and on consulting engagements where we teach and deploy operations science. Dale Nugent became an investor in Arc Precision, a company founded to apply the principles of operations science in contract manufacturing; this company is featured in the next chapter. Sue Thomas worked with us at Arc Precision to help recruit and develop shop floor talent. I personally think operations science fostered these relationships. We are confident that we are competent with our shared approach and have a lot of fun working together. It is rewarding to make a difference by achieving results today as well as developing people and processes that will drive results in the future.

Operations Science Application Summary Table: Vortex Flowmeters

Table 15–3 provides a summary of how the vortex team used the operations science framework and management process to achieve results:

ELEMENT	DESCRIPTION
BUSINESS STRATEGY	The opportunity was to realize the growth potential of the technology differentiated Rosemount vortex meter by consistently meeting delivery expectations to help engage the sales force to promote the product.
OPERATIONS STRATEGY	
DEMAND	Demand for vortex flow meters could be grouped in two segments—new plant installation projects and emergency repair projects—with different response requirements but both requiring compliance with promised lead time for delivery.
TRANSFORMATION	The production process was made to order with a flow consisting of welding, hydrostatic testing, making the sensor, transporting the meter to the flow lab, calibrating the meter in the flow lab, configuration, testing, transporting the meter back for final assembly, final assembly, final inspection, packaging, and shipping.
VARIABILITY	Demand variability was caused by the wide variety of final product configurations requiring calibration with a wide range of delivery requirements.

ELEMENT	DESCRIPTION
	Transformation variability was caused by process batches, move batches, and frequent changes in sequence/schedules due to changing priorities, based on responses to poor delivery performance from customers, salespeople, and management.
BUFFERING OPTIONS: CAPACITY	It was critical to manage the capacity buffer and its associated impact on the time buffer by finding and addressing bottlenecks. The team successfully addressed bottlenecks in the flow lab and welding.
BUFFERING OPTIONS: INVENTORY	Finished-goods inventory was not feasible since vortex meters were an engineered/configured-to-order product. There was an opportunity to leverage WIP inventory with common components for mass customization with an inventory order interface prior to "customizing based on configuration and flange size that fit the customers specifications."
BUFFERING OPTIONS: TIME	The new plant installation segment could have a time buffer but needed the time quoted for the project to be accurate so planned build schedules could be maintained. The emergency repair segment did not want a time buffer. The salespeople were reluctant to promote the product until they could trust the lead times. Thus, the business strategy for the differentiated product enabled time buffer management but required that lead time promises for delivery were met.
TOOLS AND APPLICATIONS	Capacity planning with a focus on the bottlenecks Optimal process and move batching WIP caps, due-date quoting, and virtual queues Assemble to order for mass customization

Table 15.3 Operations Science Framework Summary

---

[i] "Rosemount 8800 Series Vortex Flow Meters," Emerson Electric Co., accessed July 10, 2025, https://tinyurl.com/3usw8rrb.

[ii] McKone, Kathleen (2003). *The Rosemount Vortex Flowmeter Plant*. BAB045. Boston, MA. Babson College licensed for publication to Harvard Business School Publishing. https://store.hbr.org/product/rosemount-vortex-flowmeter-plant/BAB045

16

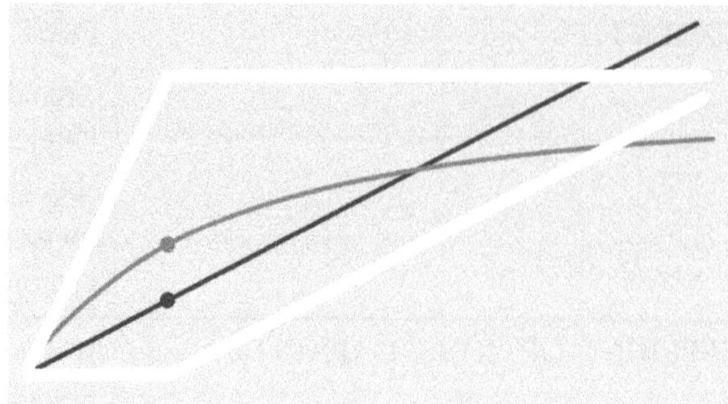

Contract Manufacturing: Arc Precision Medical Device

The Arc Team Starts a Company

Arc Precision (Arc)[i], a precision machining company that supplies components and assemblies for medical device companies, was started in 2008 by Jeff Bell (coauthor of this book) and Arc's current president, Jim Lind. (For full disclosure, note that all three authors are investors in Arc.) Early in his career Jeff developed an ambition to be an entrepreneur in manufacturing. That ambition became a reality when Jeff met Jim Lind to start Arc. Jim spent his career in precision machining with significant experience in precision-machined components for medical devices. As discussed in the last chapter, Jeff met Mike Severson who helped him apply the factory physics framework in the vortex business unit. Ultimately, Mike joined the team with another engineer, Andy Kolkind. Arc had a vision to drive profitable growth by leveraging operations science across the supply chain to support business strategy in an underserved space in contract manufacturing for the medical device industry. Arc was started in Minnesota's Twin Cities area due to Jim's connection to the high-quality labor force serving the Minnesota-based hotbed of medical device innovation, known as Medical Alley.

Business Strategy

Arc's initial business strategy was to serve emerging medical device companies, defined as late-stage funded through $500 million in revenue, because this was an underserved market. Large contract manufacturers focused on major OEMs in the medical device space like Medtronic, Johnson & Johnson, and Abbott Labs. Because they prioritized their capacity for the big players, they had difficulty effectively serving the smaller companies in the emerging medical device segment. Poor service from the large contract manufacturers caused the emerging medical device companies to turn to the highly fragmented base of capacity in small machine shops that weren't well equipped to scale and support growth in a highly regulated environment. Arc's approach was to differentiate itself for the emerging medical device companies, relying on its use of operations science to successfully scale and support the growth of its customers. Arc focused on providing precision-machined components to medical device companies that serve the orthopedic extremity and minimally invasive surgical markets. Customers in this market require rapid and dependable customer service that supports both new designs and on-time delivery for new and repeat products. Arc's command of operations science made it possible to meet these needs.

In the medical device market, the early stage of a device design is key for a contract manufacturer such as Arc. Arc employed operations science to dependably accommodate the early-stage product development cycles of rapid prototyping and design stabilization. As a result, Arc established strong customer performance and became an approved supplier for devices as they were rolled out on the market. Since these devices are strictly regulated by the government, the company's dependable quality and service created strong partnerships, making Arc a preferred supplier over the life of the device.

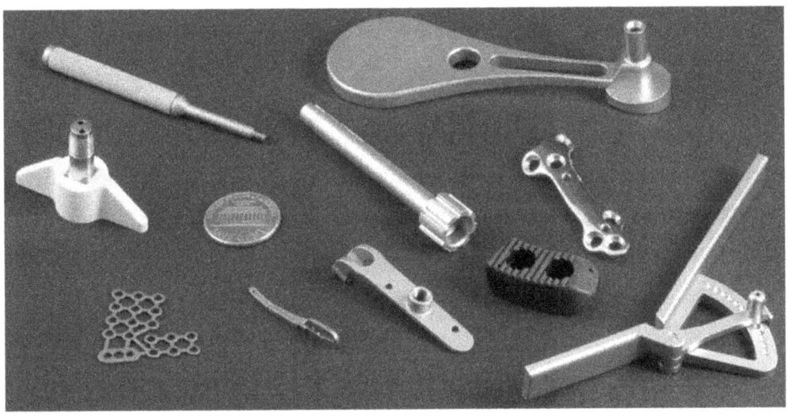

Figure 16-1: Sample of parts manufactured by Arc. Penny for scale. Source: Arc Precision

The Arc team faced several operational challenges in their new endeavor. They needed to manage a wide variety of demand types while still providing a high rate of on-time delivery and high-quality products to customers. They needed to meet this challenge while steadily growing the company with needed profit and cash flow.

Operations Strategy: Demand and Transformation

Demand

Arc demand is driven by OEM's requirements for components to build their medical devices. Safety-stock levels can be used for legacy repeat components. Most other components are make-to-order. Additionally, customer R&D engineers will often step in with quick-turnaround requests for prototype production as part of new product development.

Transformation

Raw materials (stainless steel, titanium) are procured and machined to customer specifications. Individual machined parts may be shipped directly to customers, or part assemblies may be created and then shipped to customers. Parts typically have some type of finishing requirements (electropolish, laser marking), and assemblies often involve welding. Arc's capacity includes CNC equipment and skilled personnel. This capacity is primarily in-house, though it may be outsourced as needed.

Transformation involves applying specific capacity resources at the right time in the right process. As Jim Lind, co-founder and president of Arc, stated,

> Owning a machining business is not just about hiring machinists and buying machines. The most effective way to manage the business must take into account the machinist, the machine, AND the product. Not every machinist has the skill to run every operation for a product on every machine required.

Transformation also requires supply chain management of engineered materials that often have long, variable lead times and the aforementioned finishing requirements from suppliers, which are often capacity constrained.

Buffering Options

Arc has a dilemma that many companies face: How to manage the portfolio of buffers to respond to highly variable demand for newer products without negatively affecting on-time delivery and cost for products with consistent, predictable demand? One way to do this would be to have lots of unused capacity sitting around, but that is expensive and not viable for a growing company trying to conserve cash.

Instead, Arc collaborates with its customers on visibility to demand to help reduce variability and enable Arc to level load capacity with time and inventory buffers. Through contract negotiations, Arc also provides incentives for customers to allow Arc some flexibility on building ahead to level load capacity. While attempting to manage demand through these methods, operations strategy also includes specific control techniques to plan and monitor production so that Arc meets its commitments to its customers based on feasible production goals for its employees. Specifically, WIP control is paramount. In a company such as Arc, where there is a high mix of many different parts, physical WIP is difficult to monitor in terms of parts. Instead, Arc has to work with WIP in units of hours. This allows the same operations science benefits of WIP control and does so in a high-mix environment with much variability in part type and volume. Arc also buffers variability by building capacity slots into its schedule to react to quick-turnaround requests from highly valued customers. Ultimately, Arc's operations strategy focuses on dynamic management of the portfolio of inventory, capacity, time, and demand variability to enable great customer service and value for its customers.

Supply Chain Collaboration

Arc grows successfully when it meets the service needs of its medical device customers while optimizing its high fixed cost of capacity in terms of CNC machining centers and skilled labor. Arc's medical device customer base now consists of both emerging medical device companies with new innovative technologies and larger companies who develop and acquire new innovative technologies. Arc was able to expand its original strategy to include new technologies at larger customers as well as those at emerging medical device companies. Arc's customers deliver these technologies with specifically designed instruments that enable surgeon/patient success in orthopedic extremity, trauma, and minimally invasive surgeries. Arc provides its customers the machined components and assemblies for these instruments, which are produced with an ISO 13485 medical device quality approach.

Arc works with its customers throughout the product life cycle, starting in the design phase with design for manufacturing assistance. It then produces prototypes and small pilot runs for clinical studies and validations. In this phase of the product life cycle, customers need Arc to respond quickly with capacity as designs and plans evolve. As products are approved and launched, the product life cycle moves to a production phase that can leverage the entire portfolio of buffers.

Supply chain collaboration focusing on visibility to demand enables planning throughout the product life cycle. Collaboration significantly reduces planning variability because it enables Arc and its customers to be on the same page about requirements. This allows both parties to benefit in terms of needed responsiveness and lower capacity costs. In the phases of design, prototype, and clinical study, it is not possible to use an inventory buffer for responsiveness because designs and plans are constantly evolving. As we will see, operations science and associated tools can be used to develop a reservation system using CONWIP concepts, due-date quoting, and virtual queues so that customers who collaborate can get quick turnarounds on manufactured product as designs evolve. Once a product is launched and in production, it is possible for Arc to use an inventory buffer to reduce cost, lead time, and inventory. In both cases, the key to success is true supply chain collaboration with a focus on visibility to demand, as shown in Figure 16-2 which Arc uses with customers to explain the reservation system:

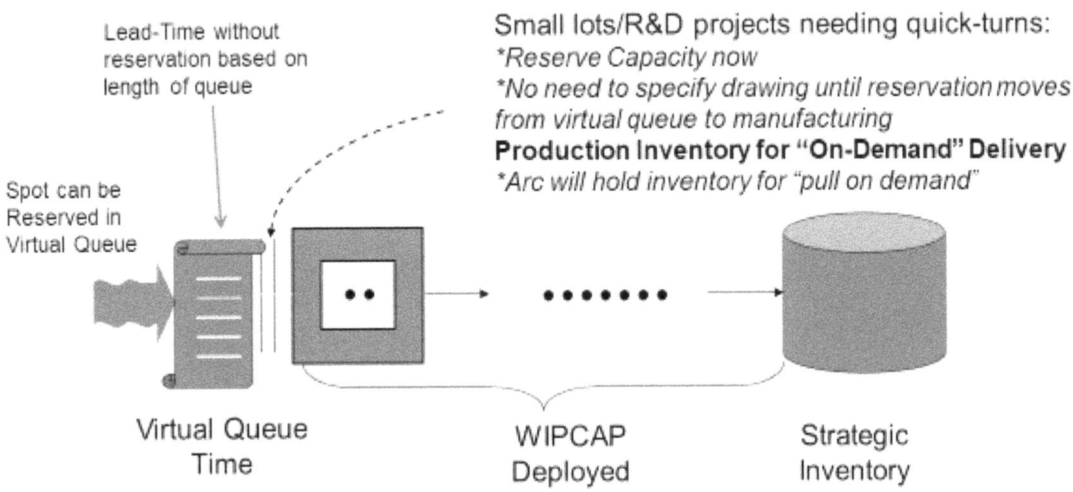

Figure 26-2: Supply chain collaboration and optimal use of buffers Source: Arc Precision

Arc highlighted a medical device OEM, Skeletal Dynamics[ii], as being best in class with supply chain collaboration. Tom Norman, chief operating officer, and Adrian Castro, director of supply chain management, empower Skeletal's teams to collaborate across the supply chain. Luis Dominguez, senior supply chain analyst, wrote the following about the customer perspective from medical device manufacturing:

> Supply chain collaboration between customer and supplier is not only vital, it is by far the most important factor when it comes to choosing whom to do business with. A great business partnership involves both parties being transparent with each other about their goals and having a responsive and honest communication channel. From a demand-planning perspective, being able to trust a supplier to deliver items by their agreed promised date significantly reduces variability of incoming supply. A major payoff and end goal of establishing this type of collaboration from the customer's perspective is to reduce inventory costs.
>
> As we know from the science of operations, reduced variability results in reduced buffers. Thus, it's safe to reduce inventory safety stock of items manufactured by suppliers who consistently score well in their level-of-service (LOS) scorecards. LOS scorecards are reviewed monthly by the supply chain team to ensure suppliers are meeting desired benchmarks. The starting point to establishing the best LOS scorecard is to have accurate data to measure performance, which includes percent of on-time delivery, percent of orders delivered in full, and percent of nonconformances (NCRs) per lot delivered. Once established, it's important for both the supply chain team and supplier to be in accordance with how these key performance indicators (KPIs) are measured, and the supplier should understand this enough so they can validate the data themselves and put in effort to improve, especially when the scorecard determines how business is allocated among suppliers.
>
> The first step to ensuring consistent LOS is by having weekly open-order reports and periodic meetings to discuss them. Data in the open-order reports from both supplier and customer should be identical, and both parties should put in effort to reconcile any differences (it's essentially their gradebook!). Periodic, and sometimes even weekly, meetings to discuss the reports are critical in maintaining relationships and understanding the workflows of each company. Suppliers should always be proactive enough so that customers are never caught off guard by any issues. However, a hallmark of a top supplier is when there are fewer weekly open-order meetings, if any at all, due to the open-order reports showing every delivery on time and with no issues to discuss.

Adrian Castro, director of supply chain management at Skeletal, remarked:

> The collaboration is truly paying off. It's rewarding to see the positive impact reflected in the supplier scorecard, which clearly demonstrates that our efforts are making a meaningful difference.

Tools and Applications

Capacity Planning

Arc's business has inherently high variability in demand and transformation. The company makes a strategic choice to target about 80% utilization in the long run to provide a buffer for this variability. The company also uses two tools to help plan capacity in the face of significant variability in demand.

As its top priority, where possible, Arc works with its customers to reduce variability with blanket orders and inventory agreements so it can build inventory and level load capacity. Building inventory helps level load capacity, provides responsiveness on legacy components, and helps create openings needed for quick response on new prototype orders.

Second, when level loading is not possible, Arc uses flex capacity so it can run effectively at long-term utilization higher than 80%. Flex capacity can include overtime and outsourcing.

Arc measures current utilization quarterly, by comparing hours of throughput (TH), measured in a throughput tracking spreadsheet, to work hours available with current staffing on a two-shift basis. Arc plans staffing for a utilization level of about 80%. Arc also measures open machine capacity by comparing hours of TH to hours available on a three-shift, seven-day basis. This helps inform the need for longer-term capacity expansion relative to machines and building space.

CONWIP, Due-Date Quoting, and Virtual Queues

Arc uses the tools of CONWIP concepts, due-date quoting, and virtual queues presented in Chapter 6 to manage the time buffer and communicate with customers. In each work area, Arc develops a weekly schedule with a CONWIP level in terms of hours of work. The company also tracks TH and schedule adherence in each work area. Since CT= WIP/TH, Arc knows it will hit planned lead times if it achieves planned TH. TH and schedule adherence are reported in daily emails, tracked weekly, and tracked with a four-week moving average.

Mike Severson, Arc's operations scientist who helps the Arc team use the tools stated,

> In applying the operations science concepts to Arc operations, one concept we use is due-date quoting. It seems that, based on science, it would be a no-brainer. However, practice from customer buyers and the noise variability generated by their internal planning/ERP systems requires some finesse to maintain good customer service and keep Arc production under control.
>
> The lead time any company quotes to its customers should depend on the utilization level of its resources. There is a very strong relationship between utilization and cycle time. The science shows that as utilization with low variability in processing time approaches 100%, cycle times explode. With high variability in processing times, as in an engineer-to-order environment, cycle times blow up well before 100% utilization, often around 80% utilization. In my experience across industries, it is fairly common for companies to quote a fixed lead time to customers because that is what is in their ERP system or what sales has dictated, or for other reasons out of an individual manager's or planner's control.
>
> At Arc this often results in modification of due-date quoting to due-date negotiations. Despite our best efforts at providing customers with visibility to our capacity, a customer's purchasing agent will occasionally call us with a request like "I know you said that you could provide 50 sets of surgical instruments per month, but I need to increase that to 150 per month as soon as you can." Usually with repeat parts, we have been tracking the customer's demand for years, so this is not a complete surprise but still is not feasible with capacity alone. Collaboration with customers is key. Where possible, we work together on long-term blanket orders so we can use an inventory buffer to level load capacity. When the demand signal is too late or too large to use an inventory buffer ahead of time, we collaborate on the customer's real needs to keep production running and ship partials to keep the supply chain flow meeting true production needs. Often, we find that customers were expediting to safety-stock levels, which defeats the purpose of safety stock to manage supply chain variability. Customer partners who collaborate with us get better service, better inventory turns, and lower costs because we work together to manage the buffers.

Arc's effective execution of operations science tactics hinges on its ability to control WIP and achieve TH. Arc uses Process Behavior Charts discussed in Chapter 8 to track schedule adherence. Process Behavior Charts acknowledge a feasible goal and the natural variability around results. Some days things go better than planned, other days they go worse than planned, but it is important to understand process performance to quote response times and meet them. Arc's release plan for WIP and achievement of TH targets with good schedule adherence enables predictive control of the time buffer. This allows Arc to use due-date quoting to communicate lead times. Of course, as process performance improves in a statistically significant manner, the company increases its goals for TH, which reduces response time. Finally, Arc uses the concept of virtual queues to reserve spots so it can strategically respond to the development needs of customers. Virtual queues also act as a capacity trigger to enable proactive capacity increases.

Assemble-to-Order

For one customer, Arc makes different surgical kits that are made from largely common components. For component inventory that could be assembled to order, the customer could make a smaller commitment but needed to allow a time buffer that enabled the assemble-to-order process. The customer received both the benefit of reduced inventories and shared cost reductions as Arc was able to level load capacity based on this supply chain collaboration.

Business Results and Supply Chain Relationships

Arc leverages the science of operations to deploy tools that enable needed customer service while achieving utilization of fixed-cost resources needed for profitability. The science of operations also enables supply chain collaboration based on a common understanding and language for making decisions about demand and responsiveness. Over the years, this collaboration catalyzed development of deep relationships with Arc's customers at all levels of the organization, from the executive suite to engineering to operations. These relationships drive profitable growth and success across the supply chain in support of product development, product launches, and on-going production for medical devices.

Operations Science Application Summary Table: Arc Precision

Table 16 – 1 provides a summary of how the Arc team used the operations science framework and management process to achieve results:

ELEMENT	DESCRIPTION
BUSINESS STRATEGY	To deliver customer service throughout the product life cycle for emerging medical device companies and technologies while enabling profitable growth and return on assets at Arc.
OPERATIONS STRATEGY	
DEMAND	OEM medical device demand drives component demand to safety-stock levels for repeat product and make-to-order on new product. Additionally, customer R&D engineers will often step in with quick-turnaround requests for prototype production as part of new product development.
TRANSFORMATION	Raw materials (stainless steel, titanium) are procured and machined to customer

ELEMENT	DESCRIPTION
	specifications. Individual machined parts may be shipped directly to customers or part assemblies may be created and then shipped to customers. Parts typically have some type of finishing requirements (electropolish, laser marking) and assemblies often involve welding.
VARIABILITY	Demand variability caused by wide variety of components at various stages of the product life cycle. Transformation variability caused both long and variable lead times from suppliers and part complexity in production.
BUFFERING OPTIONS: CAPACITY	It was critical to manage the high fixed-cost capacity buffer and its associated impact on the time buffer to serve customers and grow profitably. Arc collaborates with its customers on visibility to demand to reduce variability and level load capacity with both time buffers and inventory buffers.
BUFFERING OPTIONS: INVENTORY	Blanket orders and assemble-to-order programs enable the use of inventory to help level load capacity. Collaboration across the supply chain is a key to success.
BUFFERING OPTIONS: TIME	Segmentation of demand across the product life cycle enabled a time buffer for repeat products in production but required fast response for product development iterations.
TOOLS AND APPLICATIONS	Capacity planning with a focus on the mix of equipment and personnel skills relevant to part complexity and routings Blanket orders enabling inventory to level load capacity and provide customer service CONWIP, due-date quoting, and virtual queues Assemble-to-order for mass customization

Table 16-1: Operations Science Framework Summary Table

---

[i] www.arc-precision.com

[ii] https://skeletaldynamics.com/

17

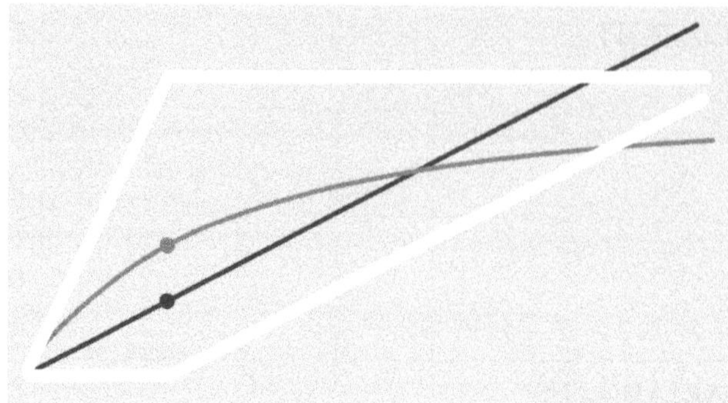

Operations Science in Clinical Laboratories: Lessons from the Pandemic

Dr. Frederick Strathmann has worked in laboratory medicine for over 15 years, holding leadership roles in academic, clinical, and industry settings. Board certified by the American Board of Clinical Chemistry in both clinical and toxicological chemistry, Dr. Strathmann serves as a part-time clinical laboratory director and clinical consultant for contract research organizations (CROs) operating under the Clinical Laboratory Improvement Amendments (CLIA) and is also the senior vice president of global business at MOBILion Systems.

Introduction

The COVID-19 pandemic threw the world into unprecedented chaos, with clinical laboratories at the forefront of the response. For many, routine test volumes dropped suddenly, impacting both hospitals and reference laboratories, leaving operational managers struggling to reassess capacity and plan for potential scenarios, and wondering where to turn for guidance.

Fortunately for me, almost one year prior, I had read a book by Mark Graban titled *Measures of Success* that provided numerous clues for monitoring a system and ensuring that overreacting does not become the norm. This chance reading led me to Mark's colleagues that included Dr. Mark Spearman and the pioneers of a scientific approach to operations management—operations science. Far from the conventional wisdom taught in my MBA operations courses, the operations science management approach provided a quantitative framework for assessing, planning, and ultimately improving how a clinical laboratory operates.

With the support and input of numerous colleagues on my operations teams, I had the good fortune to set off on a journey that provided invaluable insights and an enduring management philosophy that guides me through any situation. This chapter delves into the operational issues and lessons learned from running a clinical laboratory during this period.

As Figure 17-1 shows, diagnostic labs do not resemble classic manufacturing lines. Fortunately, it turns out that classic manufacturing is not unique in its behavior. I found that I could use the universal concepts of operations science to successfully manage lab operations without having to invoke manufacturing "best practices" with my associates. This avoids the "we're not a factory" cultural resistance that can be a significant obstacle to effective change management. I was able to create an operations management approach uniquely suited to our operations. This greatly eased the successful internalization of the concepts and adoption of the approach by lab personnel and by executive management.

Also included in this chapter are stories shared by numerous colleagues over the years. The stories shed light on the intricacies of utilization, capacity planning, Lean principles, the crucial role of buffers in absorbing variability, and the challenges associated with debunking engrained concepts in laboratory operations.

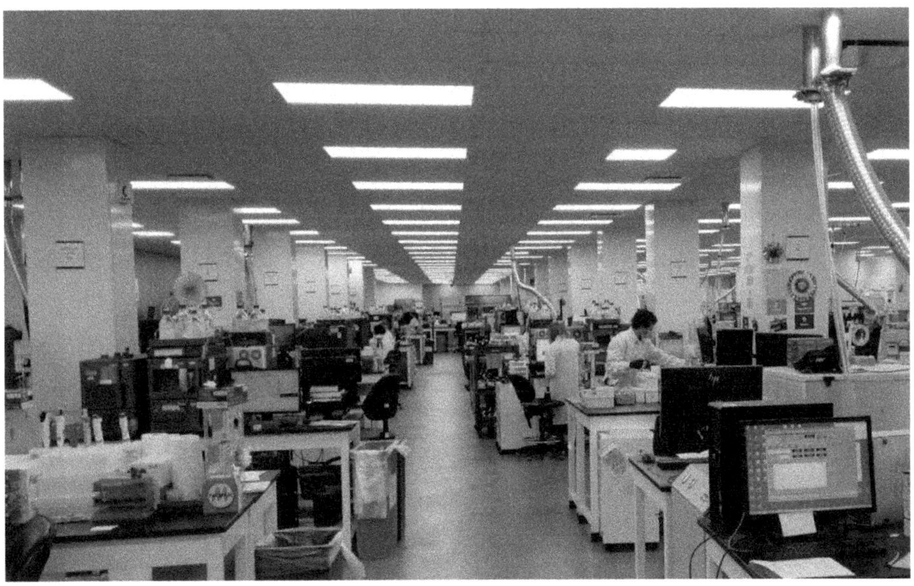

Figure 17-1: A typical diagnostic laboratory layout

Utilization and Capacity Planning

In the world of clinical laboratories, aiming for 85% utilization might seem like an efficient goal, and most CFOs and CEOs will drive their operations to achieve this as the ultimate benchmark. However, the target becomes a perilous pursuit in most laboratories because most are in the unfortunate situation of being inadequately staffed.

The Fallacy of High Utilization

Before the pandemic, the majority of labs grappled with long-standing staffing and resource issues that high utilization rates exacerbated. The constant drive to keep every piece of equipment and staff member at peak productivity left no room for error or unexpected surges in demand. A visual representation of what understaffed and overworked laboratories experienced is found in Figure 3-12: Cycle Time versus Utilization graph, from *Factory Physics for Managers*.[i]

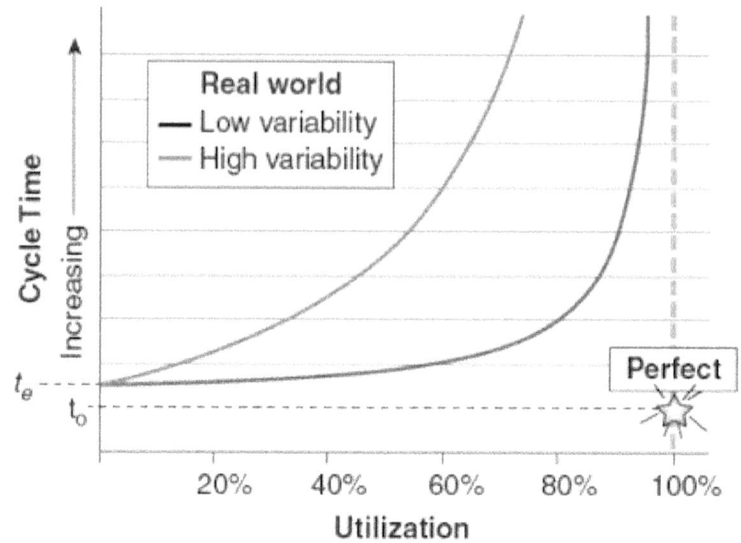

Figure 17-2. The corrupting influence of variability.

With utilization rates from low single digits to 75%, cycle time remains relatively stable. However, as utilization approaches 80%, an inflection point occurs that sends cycle time into the stratosphere, with a tiny increase in utilization. In laboratory terms, that means turnaround time suffers, and patient impact quickly becomes a critical concern. In the real world, where variability lurks around every corner, this cycle time versus utilization graph tells an even uglier story. As variability increases, this inflection point occurs at lower and lower utilization rates. For the typical laboratory being asked to achieve high utilization rates by uninformed decision makers, long hours and overworked staff are often the only answer.

Capacity Leaks and Where to Find Them

One critical oversight that complicates the utilization metric often occurs during capacity planning. It is the lack of accounting for indirect work when calculating staff utilization levels. Most budget reviews focus on the number of tests and full-time equivalents needed to meet the testing demand. The reality of the clinical laboratory, however, is filled with meetings, emails, physician phone calls, standard operating procedure updates, and numerous other tasks that can easily take a substantial portion of staff time. In our planning discussions, these became known as capacity leaks and were one of several reasons why historic capacity estimates never seemed to jibe with the reality of the operations leadership and staff utilization regarding workload. To combat this in our capacity planning, we revamped our approach to include direct and indirect tasks after a thorough cataloging effort. This exercise was not only valuable in providing a reality check on how much time away from the bench was required to adhere to regulations in a clinical environment but was also useful in identifying non-value-added work that could be reduced.

One additional capacity leak was a misalignment with hiring times and capacity needs. It is not uncommon for capacity assessments to be conducted retrospectively, using number of tests completed. In reality, this is a great way to see just how understaffed a lab truly is but does nothing to rectify the situation. Shifting to a forecasting model provides us with a better view of what was coming and what we needed to do to prepare. The cadence we set was to plan two quarters ahead to ensure recruiting, hiring, and training could be completed to align with the capacity needed in the laboratory. This proactive stance allowed us to better align our workforce with anticipated demand and identify bottlenecks in recruitment that were widespread during the pandemic. This was a major challenge since recruitment for employees was especially intense during COVID. It was challenging to get financial personnel to understand that hiring in January was required for a test-volume increase that might start in April—especially when those finance folks thought 30% turnover was expected. Little did they realize that excessive turnover is a symptom of a poorly staffed system falling victim to capacity leaks, among other issues.

Pandemic Realignment and Measures That Mattered

One of the greatest demonstrations of balancing capacity and demand was provided by the test-volume drop during the early days of the pandemic. In the prior year, the operational leadership was making requests for dozens of staff to make up the gap in capacity that was being uncovered by the more structured and quantitative operations science approach to capacity planning. A very hard pill to swallow for the executives was that it would take time and process improvement resources to chip away at this gap each year. We worked to improve our internal operations and reduce needed head count. Head count was reduced through natural attrition. We were then more responsive and profitable so we could grow without excess cost. (Side note: Improvement projects and professional development are another source of capacity leaks if project time is not properly budgeted.) One challenging budget period saw an operations executive battling for resources, stating, "Our capacity and our demand will eventually balance out—either because we get the resources we need or because we'll lose business!"

Ironically, the initial drop in testing volumes during the early stages of the pandemic briefly aligned capacity with operational demands in many organizations. Despite the chaos and uncertainty of the outside world, for a brief period the daily grind of the lab environment felt less stressful, and the natural capacity match provided a unique opportunity to streamline processes and address inefficiencies that had long been masked by overutilization.

Alarmingly, some misguided senior managers were scrambling to figure out what to do with "all this extra capacity" and a balance needed to be struck to ensure fiscal responsibility in a time of extreme turmoil. This mismatch in understanding was a stark reminder of the importance of maintaining a balance between utilization and capacity, especially in times of crisis.

Misapplied Lean Thinking

Lean principles, when applied correctly, can drive significant improvements in operational efficiency. However, the misapplication of these principles can lead to detrimental outcomes. For the less-experienced operations leader, the push to eliminate all work-in-process (WIP) in favor of single-piece flow created significant bottlenecks that even the most novice laboratory workers knew was doomed to fail. The idea was to streamline processes, but it quickly became evident that reducing all WIP to zero did not align with the realities of clinical laboratory work.

In reality, operations science shows that zero WIP does not align with the realities of any operation. Zero WIP means zero throughput, which means zero revenue.

Misguided efforts to limit WIP and cycle time (aka turnaround time, or TAT) ignore the interdependency of capacity, utilization, and fixed costs. A primary flaw in thinking that leads to misguided efforts is ignoring the highly likely occurrence of a step being delayed, i.e., variability. A sample that does not pipette quite as easily as the norm, a test order with written instructions that are a challenge to decipher, or a label that is smeared and cannot be read easily are all very common occurrences that can throw a wrench in a single-piece flow system. In the end, the trade-off for single-piece flow and shortened TAT is usually a tremendous drop in throughput. The loss of WIP removes a primary buffer of variability and starves the bottleneck, thereby decreasing throughput. Ultimately this increases the need for even more capacity and increases costs which are results completely opposite the results needed in an already understaffed environment.

Successful Lean implementation requires a deep understanding of work and process flows and the operations science governing their behavior. Rather than blindly minimizing WIP, a better focus is to comprehensively analyze operations. This involves moving away from merely shrinking batch sizes and instead addressing the root causes of inefficiencies in the context of an operations science framework. Operations science greatly strengthened our Lean efforts to foster a culture of continuous improvement and process understanding. We were able to quickly and efficiently implement Lean principles that genuinely enhanced our operational effectiveness.

The Role of Buffers

In the dynamic environment of clinical laboratories, buffers play a crucial role in absorbing variability. In operations science there are three key buffers: capacity, time, and inventory. Inventory buffers are less applicable in labs when considering test results as the finished product—it is not easy to store finished tests in a warehouse before samples arrive. For this reason, the lab must operate by using capacity and time buffers to dampen the impact that the inevitable variability in demand and output. Buffers are not something new that one would read about in *Harvard Business Review*: Buffers are a fact of life. When managed effectively using operations science they provided improved performance and profitability.

In laboratory operations, like in many environments, variability ensures that someone or something is always waiting—hence the emergence of buffers. Samples waiting on staff or instruments increase cycle time and impact the time buffer. Staff or instruments waiting on samples decrease utilization and impact the capacity buffer. Problems in the lab delay results and make physicians and patients wait for results—hitting the time buffer again. As discussed prior, it is a very common mistake for uninformed leadership to push management higher and higher on their utilization of staff and instrumentation. No executive likes to walk through a lab and see idle staff or idle instruments and will often use these anecdotal visits as evidence for continued belt-tightening. On the flip side, if instruments and staff are not waiting, then most likely physicians and patients will be.

Realistic Turnaround Times

Setting realistic turnaround-time expectations based on data is essential for effective laboratory management and optimal patient care. During the pandemic, many laboratories learned a hard lesson about the differences between turnaround-time distribution versus averages. By focusing on the ninety-fifth percentile, we provide a more accurate picture of our capabilities and can plan capacity accordingly. The key component is to ensure that the lead time, or what is referred to as our published TAT in lab medicine, is set to a number greater than our measured cycle time (i.e., the time it takes from arrival to the lab and result to the physician). Unfortunately, many labs overly focus on average TAT. and some even have made the critical mistake of setting their published TATs to the average of their historic TATs. While it is a great idea to use data to guide decisions, assuming a normal distribution of TATs, the story unfolds quickly since 50% of the results will be late. Going back to our discussion of buffers, it is essential that the laboratory establish lead times (published TATs) that are based first on clinical need and then set to align with economic realities of operations. This provides a framework where the lab is not pushed to turn out results faster than required, which can drive up capacity needs and costs without adding value.

One Chart to Rule Them All

One of the most effective tools we employed was a production-flow[ii] behavior figure, shown in Figure 17-3, derived from operations science. This visualization (referred to as Figure 3.19 in *Factory Physics for Managers*) helped our managers, many of whom lacked formal operations training, understand the relationships between different aspects of our workflow. By combining intuition with scientific operations principles, we fostered a more cohesive and informed approach to managing our lab.

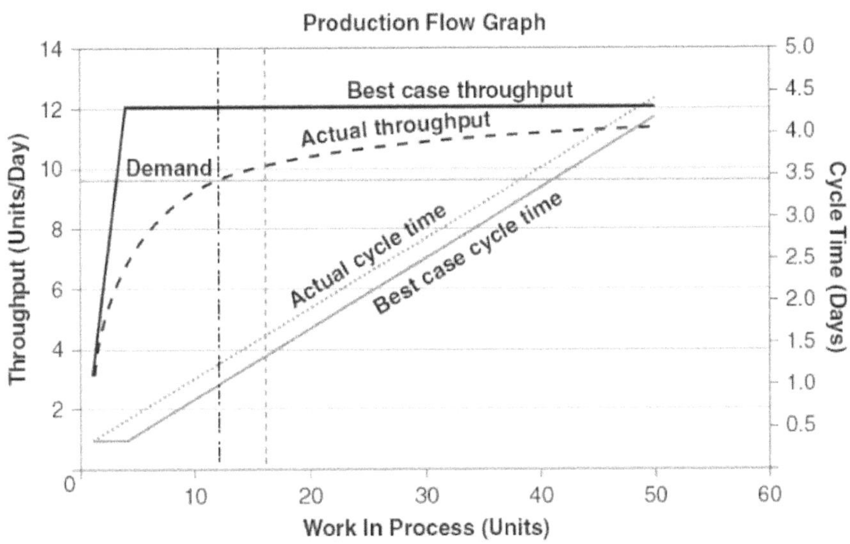

Figure 17-3. Visual representation of flow behavior[iii]

The Little's law equation

$$TAT = \frac{WIP}{Throughput}$$

quickly becomes a common topic of executive meetings for a struggling operations leader, so much so that at one point her CEO exclaimed, "You're becoming challenging and difficult to work with!" Little did that CEO know the only thing keeping the laboratory from going off the TAT cliff was the constant forced lessons on what the alternative would be if this warning were not heeded. In other words, laboratory TAT is a function of WIP and throughput. From a practical standpoint, this means that WIP is a leading indicator of TAT. As Dr. Mark Spearman likes to say, "WIP is visible cycle time.[iv]" A daily walk through the lab tells the informed manager of just how close that cliff might be.

Putting these concepts into language that laboratory leaders can appreciate gives rise to the following description:

> In a clinical laboratory, the number of specimens in process (similar to work-in-process or WIP) directly influences both turnaround time (the time it takes to complete tests) and the lab's overall throughput (the number of tests completed). Essentially, the number of specimens being processed is a key factor that controls the lab's efficiency. By adjusting this factor, you can influence both the time it takes to get results and how many tests the lab can handle. Keeping an eye on WIP at numerous steps in the workflow is a fantastic way to make the most out of daily reviews and early-morning standups trying to allocate scarce resources to where they are needed most.

In laboratory management, controlling the number of specimens in process is crucial. Whether it's described as a pull system, batch processing, or any other term, the critical point is that regulating the flow of specimens determines how well the lab performs. When managed properly, specimens in process can serve as a buffer, helping maintain a smooth workflow and reduce bottlenecks.

Additionally, bottleneck utilization is key to maximizing efficiency. This can be measured by comparing the demand for testing to the capacity of the lab's most limiting process, such as an instrument or a heavily manual process (e.g., homogenizing tissues prior to analysis) with the longest wait time. For example, if the demand is 9.6 tests per hour, but the instrument can only handle 12 tests per hour, the utilization rate would be 80%.

There's also an ideal level of specimens in process, where the lab operates at its most efficient. This optimal specimen load balances maximum throughput with the fastest possible turnaround time, similar to the performance curve seen in manufacturing. The goal is to avoid overloading the system, which can cause variability and a sharp drop in throughput and delay test results. While the exact optimal point can vary, the focus should be on keeping the specimen load within a manageable range to ensure smooth and timely operations. This is where historic data comes in handy and where future planning with simulated volumes can help identify problems before they occur.

Conclusion

The COVID-19 pandemic underscored the need for a balanced approach that integrated both intuitive understanding and scientific principles. At the same time, the lack of understanding of how operations science governs laboratory performance meant many labs were left to jump from fire to fire and relied on brute-force efforts and the accompanying high levels of stress to keep up with demand. Leveraging operations science concepts, we were better prepared to make the arguments we needed—even if those arguments were not always able to convince the powers that be who could be overly focused on the bottom line without contemplating the impact on patient care. Visualizing flow behavior allowed us to identify bottlenecks, optimize processes, and ultimately deliver better outcomes for our patients and stakeholders. The COVID-19 pandemic was a stress test for clinical laboratories worldwide. Through the lens of our experiences, this chapter has explored the critical aspects of utilization, capacity planning, Lean principles, and the importance of buffers. By embracing these lessons in operations science, we can build more resilient and efficient laboratory operations, better equipped to handle future challenges.

Operations Science Application Summary Table: Diagnostic Labs

Table 17-1 provides a summary of how the lab team used the operations science framework and management process to achieve results:

ELEMENT	INSIGHT
BUSINESS STRATEGY	To deliver test sample results quickly and efficiently to meet the needs of physicians and help achieve best possible patient outcomes
OPERATIONS STRATEGY	
DEMAND	Test samples. Origination can be for scheduled visits ordered by physicians as part of office or out-patient treatment or high priority quick turn samples for emergency analysis
TRANSFORMATION	Test samples are received and routed through the lab to receive required analysis. Technicians are directed to conduct analysis depending on test instrument availability and sample priority
VARIABILITY	Demand variability caused by wide variety of tests and demand arrival variability Variability in transformation due to conditions of sample labels, unclear test instructions, and skill levels of lab technicians
BUFFERING OPTIONS: CAPACITY	Understanding capacity utilization and "capacity leaks" is vital to determine resource utilization, establish strong control and desired performance
BUFFERING OPTIONS: INVENTORY	A service business so no stock inventory option. WIP (test samples) is a primary control parameter
BUFFERING OPTIONS: TIME	Managing WIP and utilization to achieve required turnaround time provides significant competitive advantage and improved outcomes for patients
TOOLS AND APPLICATIONS	Capacity Planning with a focus on the mix of equipment and personnel skills relevant to test priority, complexity, and routings. Bottleneck identification and control Little's law application, flow performance graphs, and WIP control concepts (CONWIP)

Table 17-1: Operations science framework summary

[i] Pound, Bell, and Spearman, *Factory Physics for Managers*, p. 75

[ii] Note this reflects the more manufacturing specific terminology in Factory Physics for Managers. In the general operations science context, this graph is called a Flow Performance Graph since there are flows everywhere whether in manufacturing, services, or projects.

[iii] Pound, Bell, and Spearman, *Factory Physics for Managers*, p. 87

[iv] Ibid, p. 82

18

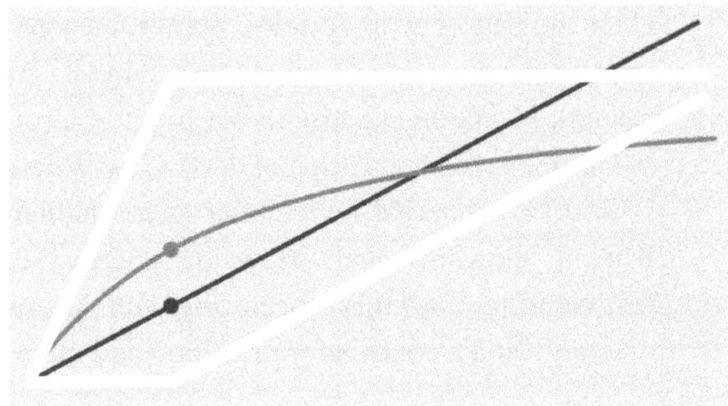

Operations Science and Project Management

Projects are everywhere. Their outcomes produce the buildings, roads, bridges, software programs, new products, telecommunication infrastructure, cybersecurity, and defense systems that are the foundation of modern civilization. On a more personal note, every time someone plans a vacation, manages a school science fair, or plans a wedding, they are doing project management. Projects are a collection of production systems, as we first described in Chapter 4's rebar cage example. Entire industries have failed to recognize this fact and its implications at the cost of massive, ongoing failure in project delivery.

Large projects, also called major capital projects or megaprojects, are very often spectacularly unsuccessful, with billions of dollars of cost overruns and completion months or years after the planned completion date. Dr. Bent Flyvbjerg has spent his career studying the performance of megaprojects. In a 2023 *Wall Street Journal* article, he stated the problem bluntly: "You shouldn't expect that [projects that build assets valued greater than $1 billion dollars] will go bad," he says. "You should expect that quite a large percentage will go disastrously bad."[i] By "quite a large percentage," he means 99.5% of megaprojects do not meet their combination of budget, timeline, and benefits targets. A 2017 McKinsey study estimated that, in engineering and construction alone, achieving the same level of productivity as in the general economy would free up $1.6 trillion dollars a year that is otherwise being wasted.[ii] Dr. Flyvbjerg's iron law of megaprojects is "Megaprojects will be over time, over budget, under benefits, over and over again.[iii]"

This carries over to businesses that build major capital projects. As Todd Zabelle writes in *Built to Fail*,

> At the same time, contractors undertake way too much risk for the return on investment they realize. Every contractor, no matter how large, is only one step away from bankruptcy.

> Over the past several decades, numerous contractors have failed or have avoided failure through being acquired, including Guy F. Atkinson (now part of Clark Construction Group), Morrison–Knudsen (Washington Group, then URS, then AECOM), the collapse of Carillion, the acquisition of AMEC by Wood Group, and the UK's mighty Laing, which O'Rourke purchased for £1 after three challenging projects. A few years back, the giant Korean shipyards that serve the energy sector also found themselves in need of restructuring… All this is occurring with the construction industry being a major consumer of energy and a contributor of carbon and other types of waste.[iv]

Operations science provides the knowledge and approach to break Flyvbjerg's iron law. This has been demonstrated many times. Ignorance of project behavior as described by operations science is a primary cause of chronic failure of megaproject management. While this fact is not widely recognized, there is good progress being made at the Project Production Institute[v] or PPI. PPI's founders recognized in the early 2000s that the existing mental models evidenced in common project management practices were flawed and did not provide a scientific explanation for project performance. See the PPI's website for a wealth of resources on operations science in project management and applications that have broken Flyvbjerg's iron law. This chapter explores operations science application in project management and will

- provide an overview of the relationship of projects to production systems
- demonstrate the application of operations science in engineering design
- review operations science implications to the common stage-gate approach for project management for both new product development and engineering and construction major capital projects and
- illustrate operations science application for a major capital project

Projects, Products, and Production Processes

A major obstacle obscuring project managers' vision on the true nature of their projects' behavior is the use of financial project control measures and Gantt charts as a proxy for control of project progress. As Arbulu, Choo, and Williams[vi] clearly describe, project controls are not the same as project control. Gantt charts and financial controls and measures are great for planning and reporting, but the widespread use of those tools to manage the progress of a project is like watching only the scoreboard during a sporting event to determine the outcome of the contest. The project management profession, as defined by the Project Management Institute, seems to see the successful delivery of projects through coordination of many entities (contractors, subcontractors, asset owners, governments) with only passing regard to the mechanics of how those entities get work done. From the sixth edition of the *A Guide to the Project Management Body of Knowledge,*

Operations management is an area that is outside the scope of formal project management as described in this guide. Operations management is concerned with the ongoing production of goods and/or services. It ensures that business operations continue efficiently by using the optimal resources needed to meet customer demands. It is concerned with managing processes that transform inputs (e.g., materials, components, energy, and labor) into outputs (e.g., products, goods, and/or services).[vii]

Another perception problem about operations science's role in project management is illustrated with the product process matrix[viii] (see Figure 18-1). First proposed by Robert H. Hayes and Steven C. Wheelwright in 1979 in the *Harvard Business Review*, the matrix shows the relationship between types of products produced and types of processes used to produce those products. Primarily, it focuses on the relationship between volume of demand for a product and the processes used to produce that product.

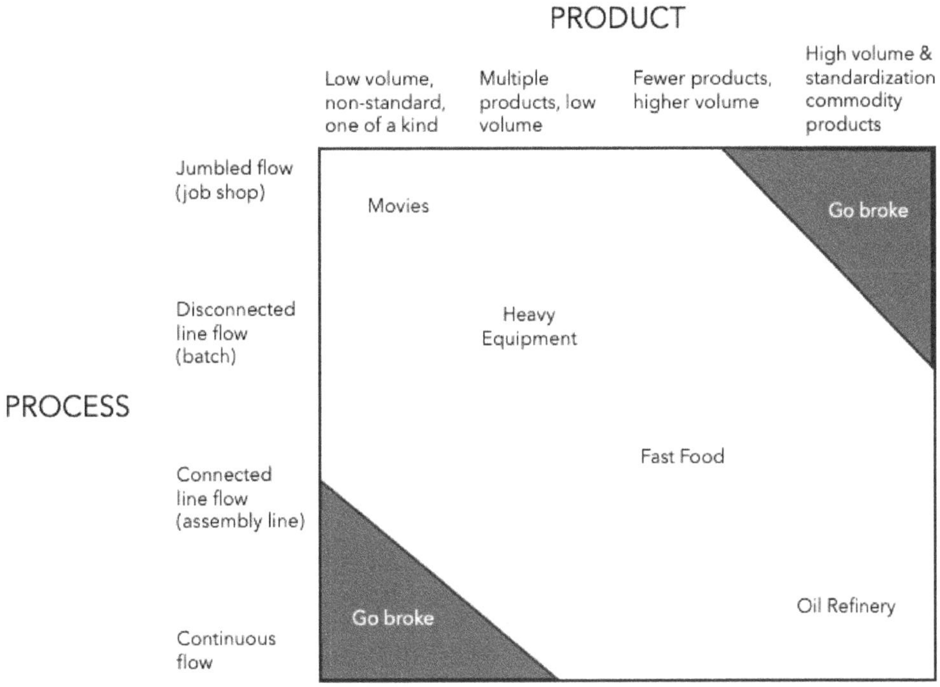

Figure 18-1: Hayes and Wheelwright product process matrix

This was an innovative and insightful look at the relationship between products and the processes used to produce them. In the bottom left corner, one would not try to use a continuous flow process to produce a one-of-a-kind product. While Steven Jackson gets credit for moving away from one-offs by filming all the *Lord of the Rings* films as one project, movie companies for the most part must build unique sets or create unique computer animation for specific movies. In the top right corner, no one tries to supply fuel for a nation's automobiles by making small batches of gasoline in a disconnected job shop flow.

While the product process matrix illustrates some fundamental relationships between product demand and products' production processes, the matrix does not provide a scientific description of these relationships or any insight into the best combination of buffers for a company to implement and control to achieve its goals. However, on a different level, the matrix is very useful for illustrating a powerful insight that project managers can use to their great advantage to establish predictable control of their projects.

A major challenge that project managers face is deciding how to best categorize and control the processes in their projects. Some managers are unaware of the implications of this challenge. In his 1993 textbook *Production/Operations Management: From the Inside Out*, Roger Schmenner fits projects into the product process matrix shown in Figure 18-2. The idea is that projects are a collection of jumbled resources for a one-off product. For example, there was only one Chunnel project connecting England and France, there was only one Big Dig project in Boston, and there was only one Burj Khalifa built in Dubai.

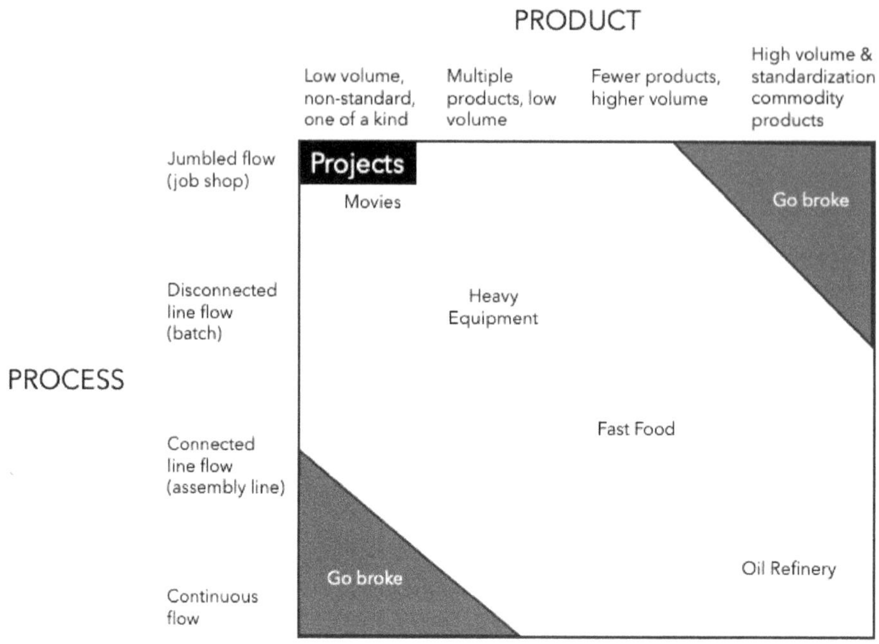

Figure 18-2: Projects in the product process matrix

Schmenner's classification of projects perpetuates the misconception that, since projects are one-offs with widely varying resources and stakeholders, they cannot be managed in a production approach, and each must be shepherded to a successful outcome. In other words, a project manager should be good at herding cats. The flaw is that projects are not simply a one-off product using jumbled resources. Projects are, in reality, a collection of products and production processes, as illustrated by Figure 18-3 and Figure 18-4.

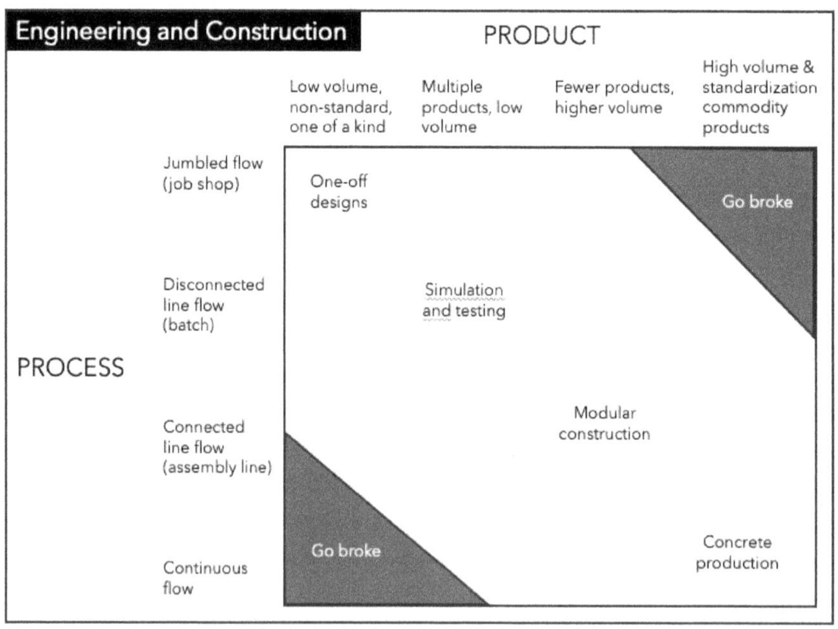

Figure 18-3: Engineering and construction projects are composed of many products and processes

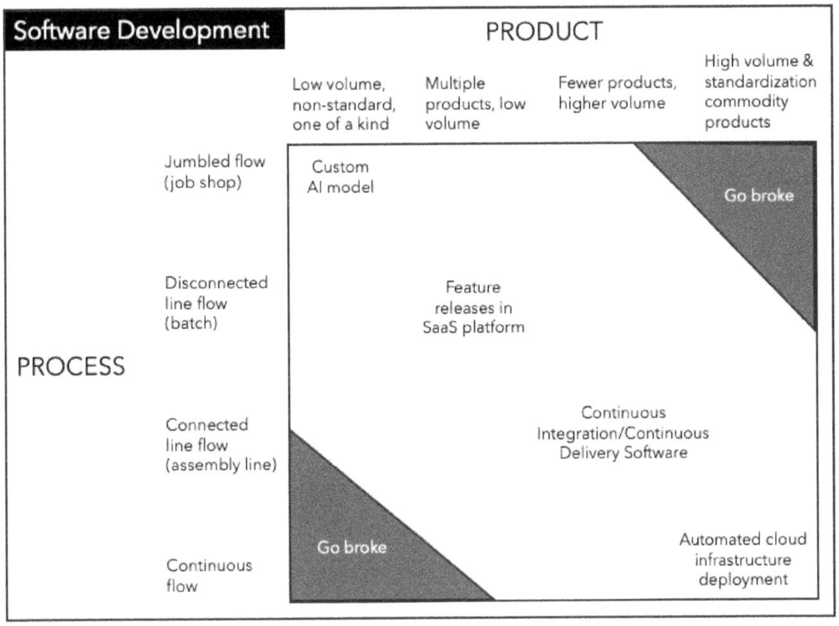

Figure 18-4: Software development projects are composed of many products and processes

Because of this fact, project managers can find themselves adrift when trying to control project progress without understanding operations science. First, they must see their projects as a collection of products and production processes, and then they must use the science describing the behavior of those production processes (physical or virtual) to establish and maintain full control of the project. This applies across all projects, even software development, as illustrated in Figure 18-4.

Project Production Management (PPM) is the development and application of modern production theories, principles, and methods to better understand, control, and improve project delivery. This includes the application of operations science, digital, and autonomous technologies.[ix] Of the following methods and techniques for project management and control, only PPM explicitly incorporates operations science to its fullest extent for effective project management.

Advanced Work Packaging

Agile

Baseline Variance Analysis

Critical Path Method (CPM)

Earned Value Management

Gantt charts

Industrialized Construction

Integrated Project Delivery

Lean construction

Modular construction

Program Evaluation and Review Technique (PERT)

Project Production Management (PPM)

Risk Analysis

Stage-Gate Control

Work Breakdown Structure

Without considering the buffering (capacity, inventory, or time) trade-offs for expected levels of variability within a project's various production processes and the relationships that govern those trade-offs, project design and management are almost certain to remain shackled to Flyvbjerg's iron law of megaprojects—over time, over budget, under benefits, over and over.

Operations Science in Engineering Design

A configure-to-order company, we'll call them ElectroPro, producing electro-mechanical devices requires extensive design input for configuration of the final product before release to production. Each order is essentially a design project. The challenge was that the design project management process was taking too long even after the company added capacity, which was hard-to-get, expensive engineering talent. Figure 18-5 illustrates the product design process at ElectroPro.

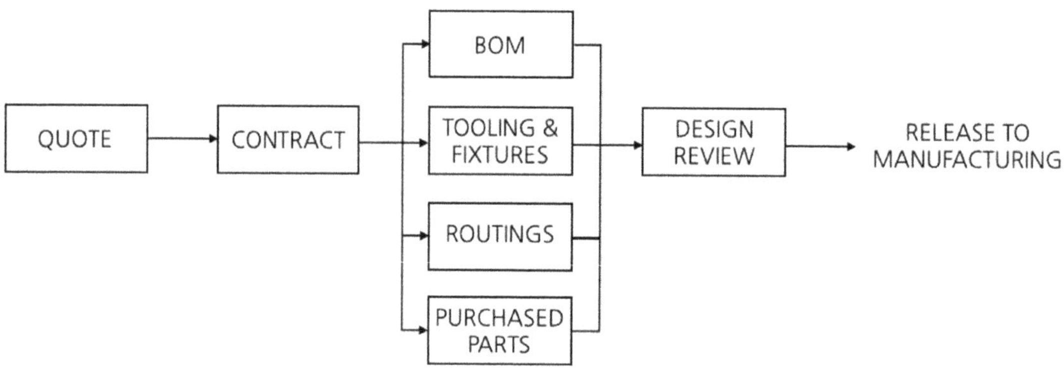

Figure 18-5: ElectroPro's product design process

The primary people populating this process (number of people in each position is given in parentheses) were as follows:

- Sales engineer (2)
- Design engineer (5)
- Purchasing (3)
- Project manager (1)

In any situation where there are people with deep experience in a process, the first step is getting to know them, their issues, their challenges, and their thinking on solutions. This is where operations science becomes very powerful, not just for mapping processes and identifying improvements—that is common for almost any continuous improvement approach. When the opening discussions occur in the context of operations science, people respond differently because the science reveals the mechanics of their work in a way that is easily digestible without sacrificing technical accuracy. The concepts of operations science provide new insight. At the same time, learning operations science concepts does not require acceptance through faith alone. Especially with technical types, such as engineers, this establishes great credibility. Too many people have experienced the "program of the month" approach where everyone is asked to chip in for the common good (of the company and its employees) based on subjective slogans such as "Reduce waste." Everyone is in favor of that, but what does it mean and how is it quantified?

Natural skepticism is to be expected with any new initiative. People generally don't like change. With operations science, the sequence of events for application is not much different than most other improvement efforts. The major difference is the operations science content and the design, implement, and control structure as introduced in Chapter 8. ElectroPro's version of this approach is illustrated in Figure 18-6.

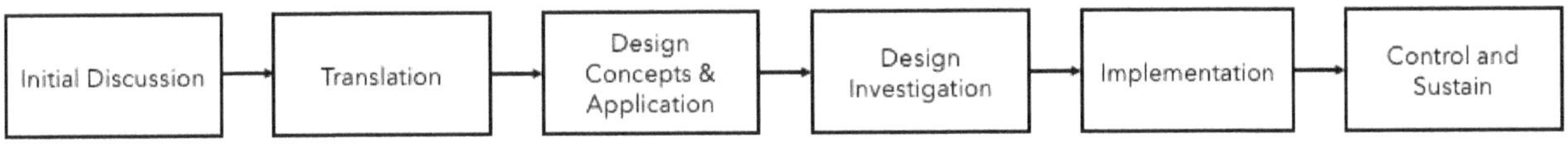

Figure 18-6: Operations science design, implement, and control for ElectroPro

Initial discussion is all about the business's strategy and how the operations strategy supports it. In ElectroPro's case, the product design process of Figure 18-5 was taking too long. It was at 5 weeks, and the ElectroPro's President wanted something closer to 2 weeks. The initial discussion also provides introduction of operations science concepts and discussion of the team's challenges in the context of the science to begin socializing the concepts and allowing the team to adapt the science to their work.

Translation is a step that should not be overlooked. The generic operations science concepts, which apply everywhere, usually won't be successfully applied anywhere if they are not translated to the team's day-to-day terminology. If the team refers to cycle time as design time, flow time, order completion time, or whatever, training in operations science concepts must be delivered using the team's terminology. Telling everyone, "When you see cycle time in the material, just think 'design time' since that's what you call it," is a surefire way to lose support for the effort.

There are four primary actions for working with project teams who have no previous experience with operations science. These people are usually stuck in the classic project management mindset of looking at Gantt charts produced by some software such as Primavera and at reporting measures such as earned value. If the team is like most project management teams, they are commonly overwhelmed with constantly changing priorities and tasks.

Starting with any project team requires a basic introduction to operations science. Once the concepts are translated to the team's terminology, the training sessions take about 1.5 to 2 days, depending on the industry.

Then, very quickly, the focus moves into design, implement, and control of their process using operations science. As with ElectroPro, the process was already in place, so the exercise became an operations science design review of the existing process. For building a new asset in engineering and construction or for creating a new cybersecurity software system, operations science is a very powerful guide for designing project delivery processes to deliver the project on time, on budget, with all predefined benefits—more on that shortly. In any case, as we did with ElectroPro, the four primary actions required are as follows (We will introduce primary action four, commitment reliability, shortly as it is another control component of the design, implement, control sequence.)

1. Determine capacity and capacity utilization

   Show the team how to calculate capacity, and then calculate the team's capacity utilization. Time available is usually pretty easy: It is the time the engineers, purchasing personnel, and project manager work each week. Time used is trickier as that requires the team to determine its WIP, which is the next primary concept.

2. Determine the team's WIP

   This concept tends to draw a blank stare when it is first introduced to project personnel. They have work they are working on and tasks to do, such as making calculations, creating prototypes, drawing up specifications, and making drawings. They think of this as their work; it is not WIP like in a manufacturing process. Except, it is *exactly* like WIP in a manufacturing process. Engineers don't typically have physical work like cutting metal to make a part, a task that requires time (setup and machining) and resources (a mill and a machinist). They have virtual work or knowledge work requiring time (creating a CAD drawing) and resources (a computer and an engineer).

3. Establish WIP control

   Engineering work on projects requires much coordination. There are many interdependent tasks. Engineering managers want to keep their engineers busy. There is a tendency to load up work so engineers are constantly busy whether the work is needed or not. This makes things worse by causing longer cycle times. WIP control is a vital action required to prevent WIP explosions and long task durations, or cycle time.

These are primary actions required for the design component of operations science application. Once a team understands the basic relationships of operations science, maps its process, understands the WIP in its process, calculates its capacity in detail (but not in extreme detail), and establishes WIP control, they are on their way.

Let's look at ElectroPro's capacity utilization calculation. ElectroPro's primary work was configure-to-order work, so there were modifications for nearly every order they received. Their demand was divided into three main categories:

1. High-volume, low-cost designs, sourced internationally as they, supposedly, didn't need engineering resources
2. Modifications of existing designs produced at ElectroPro's headquarters factory
3. New designs produced at ElectroPro's headquarters factory

When faced with the idea of applying the science to all the complexity of their work, engineers can be overwhelmed because they think the same level of detailed specification applied to an ElectroPro solenoid, for example, would be required for using operations science to control their processes. In our experience with operations science, more data and more data faster do not mean better control. The initial analysis for ElectroPro process consisted of the three product categories listed previously. The next steps were to create a capacity matrix for the product development process. That required estimating times per task (WIP) per resource per product. This was detailed, but it was manageable. The matrix is illustrated in Figure 18-7.

	PROCESS TIMES (HOURS)											
RESOURCES ==>	SALES ENGINEER			DESIGN ENGINEER			PROJECT MANAGER			PURCHASING		
PRODUCT ==>	Existing Designs	Low-Cost Standard	New Design	Existing Designs	Low-Cost Standard	New Design	Existing Designs	Low-Cost Standard	New Design	Existing Designs	Low-Cost Standard	New Design
005 Quote	1.0	0.0	8.0	0.5	0.5	8.0	0.2	0.0	0.5	2.0	0.3	8.0
010 Contract	3.0	0.5	8.0	0.5	0.0	0.5	0.3	0.0	1.0	0.0	0.0	0.0
015 BOM	0.0	0.0	0.0	4.0	0.5	8.0	0.0	0.0	0.0	0.0	0.0	0.0
020 Design Analysis	1.0	0.0	3.0	4.0	2.0	24.0	0.0	0.0	1.0	0.5	0.0	1.0
025 Tooling and Fixtures	0.0	0.0	0.5	2.0	0.0	24.0	0.0	0.0	0.0	0.5	0.0	3.0
030 Routings	0.0	0.0	0.0	0.5	1.0	16.0	0.0	0.0	0.0	0.0	0.0	0.0
035 Purchased Parts	0.5	0.0	1.0	0.3	1.0	3.0	0.0	0.0	0.0	4.0	0.5	8.0
040 Design Review	1.0	0.0	3.0	1.0	2.0	4.0	1.0	0.0	3.0	1.0	0.0	4.0
WEEKLY PRODUCT DEMAND ==>	6.1	7.2	1.3	6.1	7.2	1.3	6.1	7.2	1.3	6.1	7.2	1.3
TIME USED (HOURS/WEEK) ==>	39.7	3.6	30.6	77.8	46.8	113.8	9.2	0.0	7.2	48.8	5.8	31.2
RESOURCE HOURS/WEEK ==>	73.9			238.7			16.4			85.8		

Figure 18-7: ElectroPro process design capacity matrix

This approach provided a workable assessment of the capacity of the process and pointed to opportunities for improvement.

In addition, ElectroPro applied the WIP control process of CONWIP to their process. There was a tendency to put any order that had a contract into the design process. Sometimes the contract had all the information needed for the design, but often it did not. The result was to stuff the design process with so much WIP that the engineers were always very busy, but they often felt like they were just putting out fires—always responding to the latest "hot" design. By defining the WIP and doing the flow performance calculations first described in Chapter 4, ElectroPro was able to manage WIP much more effectively, which reduced cycle times and much self-inflicted stress.

Within a few months ElectroPro was able to free up 20% of engineering capacity requirements with process changes; this was a large cost savings as the company faced increasing demand. Also, as seen in our earlier discussions about the relationship between capacity utilization and cycle time, this was a huge benefit to cycle time reduction. With the improved utilization management and a few other changes identified through the operations science analysis, ElectroPro reduced order cycle time (from order receipt to release for production) by 32%. A significant contributor to freeing up engineering capacity and reducing order cycle time was reducing the number of escapes.

An escape is any task in a design project that gets past a step in the process with incomplete information. This causes rework and destroys productivity. Operations science demonstrated very clearly the effect of rework on productivity, which led to a concerted effort to reduce escapes. Escapes were reduced by 71%. One contributor to escape reduction was better communication through implementation of commitment reliability tracking.

We mentioned four major actions to take when starting operations science application to projects. The first two, capacity calculation and WIP identification, are primarily operations science design tasks. WIP control is a control action. The fourth control action that is extremely effective for improving project management processes is commitment (or commit) reliability.

4. Implement commit reliability
   The process is straightforward.
   a. A team meets at the beginning of the week to discuss what they are going to work on. This is not new. Teams regularly meet to coordinate work. For engineers, usually once a week is frequent enough. For work on an active construction site, teams meet once a shift or, at least, once a day. However, we have been in large corporations where engineers coordinating work was very *laissez-faire*, and work coordination was not heavily structured. Engineers are typically smart and independent and can fall into the trap of just figuring it out as they go along.
   b. Team members commit to the tasks they will accomplish that shift, day, or week. This commitment is recorded. This often takes some getting used to. Note that the commitment is not to what a person will do with every minute of their time. The tasks committed are from the project's required task list, the project WIP. Each task should have some time content estimate, use general categories, for example, 0–2 hours, 2–6 hours, 6–10 hours, and over 10 hours. All tasks have a code of 0 – New when loaded in to the work queue. When a team member commits to a task, its status is changed to a 1 – Committed.
   c. The following shift, day, or week, team members report what they have accomplished. The results could be

      0 – Not completed

      2 – Completed

      3 – Completed, not committed (this is for tasks that dropped in during the week)

      4 – Cancelled

      5 – On hold

A few words about status code 3: Completed, not committed. This happens all the time. One plans the work week and then something that wasn't originally needed in the week suddenly takes priority or management makes something a priority. There is nothing wrong with that. Where it becomes a problem is when it is so frequent that commit reliability suffers. In that case, the process reverts to the similar problem of chronic uncertainty in planning as with low commit reliability for committed tasks. The result is the same: poor reliability, increased variability, and longer cycle times for the project.

A sample commit reliability form is shown in Figure 18-8. This is what it would look like at the beginning of the period (shift, day, week). At the end of the period, the status code and the cause code fields are updated, and the data are saved. The cause code is updated only if the task was not completed. Tracking cause codes is a great aid for identifying root cause and corrective action.

ENTRY WEEK	TEAM MEMBER	TASK DESCRIPTION	JOB CODE	STATUS CODE	HOURS	CAUSE CODE
24-May-20	Zeus	Finish filling out process times in utilization model	1-Implementation	1-Commited	0 - 2	
24-May-20	Zeus	Determine new review process	5-Design and Analysis	1-Committed	4 - 6	
24-May-20	Athena	Set resources and utilization levels	5-Design and Analysis	1-Commited	2 - 4	
24-May-20	Ares	Plan and implement SO backlog reduction	7-Training	1-Committed	> 10	
24-May-20	Athena	Determine optimal WIP control	5-Design and Analysis	0-New	2 - 4	
24-May-20	Ares	Implement escape tracking reporting	10-Audit	5-On Hold	> 10	

COMMIT CODE	DESCRIPTION
0	New
1	Committed
2	Completed
3	Completed, Not Committed
4	Cancelled
5	On hold

CAUSE CODE	DESCRIPTION
1	Design change
2	Material quote
3	Authorization delay
4	Waiting on customer
5	Supplier

Figure 18-8: Commit reliability tracking sheet example

As this process is repeated, it generates, among other things, a nice graph for evaluation of commit reliability (Figure 18-9).

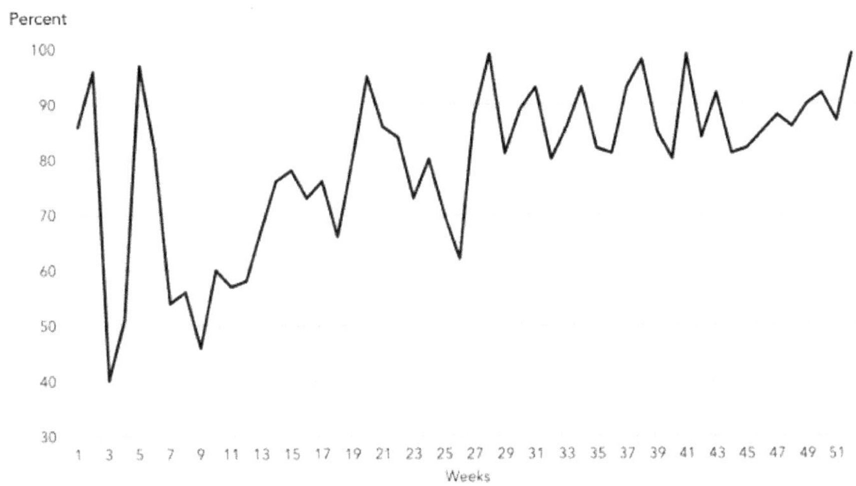

Figure 18-9: Commit reliability graph

Data from PPI over hundreds of construction projects showed that average commit reliability for one day was only 52% before the project production control (PPC) process and commit reliability were implemented. PPC is a technique of PPM, which organizes and enables control of production, or work execution. Gary Fischer, during his time as general manager of special projects at Chevron, said, "when we launch PPC on a project in construction, commitment reliability (meaning work planned for the next day versus actual work done) quickly jumps from 50 percent up to 80 or 90 percent. However, more importantly, we see reductions in the hours and time it takes to complete the work. For example, on one of our current projects, we are seeing 40 percent fewer hours and 60 percent less calendar time to complete work under PPC."[x]

Imagine being able to predict only 50% of what is going to be accomplished in one day, at the beginning of the day. This is very poor forecast accuracy and a huge source of variability. As the reader now knows, operations science shows that more variability requires more buffering. On projects, this buffering shows up as longer durations to deliver projects and the associated extra cost. As projects go late, contractors typically throw more money and resources at the project to shorten delays.

A few more observations about commit reliability:

1. 100% commit reliability is rarely the goal. There is generally too much variability in projects to expect that to happen with regularity. For a construction project, looking at one-day commit reliability, a goal of low to mid 90% range is great. For engineers working on design work and reporting one-week commit reliability, a lower result should be expected, low to mid 80% range is great. This is because engineering work is, by its nature, full of unknowns. Setting an unrealistic goal for commit reliability drives counterproductive behavior. People will work to achieve the measure, or just ignore it, rather than doing the required work and using the measure to record results for useful evaluation.
2. Commit reliability is basically forecasting, and the operations science laws of forecasting apply. Forecasting further into the future results in less accuracy.
3. The commit reliability process is part of production control for projects. One wouldn't run a factory very well without a schedule and without measuring on-time delivery to schedule. The project management profession has basically ignored operations management and operations science and relied on financial performance reporting and high-level schedule attainment reporting, rather than applying operations science to design and control the work (production) that goes on in projects.
4. Commit reliability makes a well-structured framework for coordinating work. It can be started in about one week.

Operations Science and Stage-Gate Control

A nearly universal practice in project management is the separation of projects into stages of development and then use of decision points, or gates, to make a go/no-go decision about the entire project or about moving to the next stage. This is a natural control structure that has been around as long as there have been projects because of the nature of the work. As Arbulu, Fischer, and Hartung point out, "During the 1960s, NASA integrated a phase review process that consisted of five stages with a required review at the end of each stage. As such, it was arguably the first generation of a governance process that presaged the stage-gate processes we see today."[xi] Robert Cooper popularized the stage-gate name when he published *Winning at New Products* in 1986. The Stage-Gate® trademark is owned by Stage-Gate International, which Cooper co-founded.

Stage-Gate at Procter & Gamble

The success of the stage-gate process at Procter & Gamble (P&G) is described by Cooper and Mills.[xii] They illustrate P&G's stage-gate process as in Figure 18-10.

FIVE STAGES

DISCOVER	DESIGN	QUALIFY	READY	LAUNCH
Promising Consumer Proposition	Integrated Business Proposition	The Initiative	Prepare Market Launch	Execute Market Entry

FOUR GATES

KEY DECISION	Staff it?	Design complete? Start implementation?	Criteria met? Launch plan agreed?	Ready for launch?
MILESTONE	Project Establishment	Project Commitment	Launch Plan Agreement	Launch Authorization

Figure18-10: P&G's SIMPL™ process—an idea-to-launch Stage-Gate® model

The stage-gate process has frequent project evaluation checkpoints (gates), allowing teams to evaluate project viability in terms of technical and market feasibility. In the words of Dr. Cooper, "It's really a way of taking any project—not just a new product project, but any project that is done over time—and breaking it into a series of manageable, bite-sized chunks, with decision points along the way."[xiii] This meters spending on projects to stop spending on projects that are not meeting necessary criteria, filtering out less-promising initiatives and focusing resources on the most viable ones. The approach is in opposition to the method of approving a project and committing all the money for its completion at the beginning.

At first glance, stage-gate sounds good and often works to reduce WIP in research and development to speed remaining projects. It is a best practice that works in the right situations. It worked great when P&G used it in the retail household care segment. According to Bob McDonald, P&G's vice chairman of global operations, "This work brought us a new discipline to manage our innovation programs and yielded a major increase in the in-market success of our initiatives. We aligned our organization on how to use Stage-Gate® success criteria, and portfolio and resource management to deliver better innovations for the consumers we serve."[xiv]

P&G was not able to bring the same success to development of pharmaceutical drugs. It bought Norwich Eaton Pharmaceuticals in 1982. P&G's stage-gate process gained momentum during the late 1990s in its cosmetics business. However, P&G sold its pharmaceutical business in 2009, well after stage-gate had been introduced in P&G. It seems that the development process for new products and brand extensions in consumer goods wasn't transferable to produce blockbuster drugs. It may be that the stage-gate process had little or nothing to do with P&G's inability to develop blockbuster drugs: Having a stage-gate process does not guarantee success.

Different businesses require different portfolios of buffers and variability. The concept of setting specific gates during project delivery can work against the most efficient management of the production processes within a project. For a quick look at that, we return to major capital projects.

Stage-Gate in Major Capital Projects

Stage-gate as conceived and described in Figure 18-10 is not well suited to major capital projects. Stage-gate was created for the management of new product development projects as mentioned in the case of

P&G. Major capital projects have many different characteristics than new product development, such as creating a new type of lipstick. Table 18-1 compares major capital projects and new product development.

ITEM	MAJOR CAPITAL PROJECTS	NEW PRODUCT DEVELOPMENT (CPG, manufacturing)
PROJECT DURATION	Long (5–15 years)	Short (6 months–3 years)
REGULATORY REQUIREMENTS	Highly regulated (permits, compliance, safety)	Moderate (safety, quality standards)
INVESTMENT SIZE	Very high ($100M and higher)	Lower ($1M–$100M)
MARKET DYNAMICS	Slow moving, long-term demand stability	Fast changing, competitive landscape
RISK TOLERANCE	Low tolerance for failure, high risk aversion	Higher tolerance for failure, iterative approach
INNOVATION SPEED	Slow, sequential development process	Fast, agile, iterative innovation cycles
STAKEHOLDER INVOLVEMENT	Owners, governments, investors, communities, contractors	Customers, retailers, marketing teams
PRODUCT LIFECYCLE	Decades, with continuous upgrades	A few years to a decade, requiring rapid innovation
DECISION-MAKING COMPLEXITY	Complex, multiphase decision-making	Simpler, market-driven decision-making

Table 18-1: Comparison of major capital projects and new product development[xv]

The original stage-gate process envisioned rapid iteration of ideas and creation of designs to take them through a series of gates where weak or insufficient ideas would be weeded out. That is possible in some aspects of major capital projects such as early design, but the general idea of creating and evaluating a prototype of a building, or drilling platform, or a tunnel is impractical. That will become more practical with CAD versions of major assets. Even so, there is much work done in engineering and construction between design, fabrication, and installation in the physical world. There are often major feedback loops between design, fabrication, and installation when the physical world does not turn out as planned in designing and constructing an asset.

Consider the Sydney Opera House (Figure 18-11). It was originally estimated to finish in 1963 at a cost of $7 million. It finished in 1973 at a cost of $103 million. It is an extreme case but illustrates well the back-and-forth that occurs between construction work on the site, construction engineers, and workers, and architects. Major modifications were required at the beginning of construction due to inaccurate understanding of site geology. The podium was completed before the shape and weight of the famous shells of the roof was known. The podium had to be modified several times as the roof structures were added.[xvi]

Figure 18-11: The Sydney Opera House required multiple iterations and modifications during construction.[xvii]

Stage-gate, as a recognized technique, was not around when the Sydney Opera House was built, but the project illustrates the type of back-and-forth that goes on in major capital projects, which weakens the utility of stage-gate in engineering and construction. More recently, stage-gate has been adopted for many major capital projects but, as Dr. Flyvbjerg pointed out, major capital projects remain almost always over time, over budget, and under benefits, so stage-gate has not been a fix.

As Arbulu, Fischer, and Hartung describe,

> Stage gate is a business decision-making and governance framework that overlays the actual work that must be done to create a capital project asset. Contrary to Cooper's principles and in accord with his admonishment, it has evolved as a rigid and inflexible approach which makes it difficult or impossible to cope with the variability that inherently resides within all projects.[xviii]

We support their proposed continuous flow approach, recognizing the production flows in projects and tying those flows to well-defined milestones rather than governing gates whereby work accumulates in large batches, a primary cycle time driver, for approval. This doesn't eliminate gates, but all work is not calibrated to gates. Arbulu, Fischer, and Hartung describe the continuous flow approach as follows:

1. Concepts are selected and then defined through mapping the allowable design space and identification of constraints arising from technical, regulatory, and governance requirements. This mapping must also clearly identify the business drivers. For example, in the energy sector, the driver is typically return on investment; for health care, it is driven by the need to provide health care to a growing population; and for data centers driven by expansion of AI and cloud computing, the drivers are 5G networks and edge computing.

2. The allowable design space is expanded by addressing technical constraints through science and engineering, regulatory constraints through advocacy and negotiation, and governance constraints through modeling and cost benefit analysis.
3. Milestones and conditions of satisfaction are rigorously defined to maximize overall flow of the work.[xix]

Operations science starts from first principles, describing

- the effect of variability when trying to synchronize demand and production,
- the required buffers (capacity and time-inventory) for synchronization, and
- the relationships between the buffers and variability.

Production could be site construction of a building, coding for a software program, or new product development. Without understanding these concepts and applying them to the production work (physical or virtual) that goes on in projects, project delivery will continue to be over budget, over time, and under benefits.

Fortunately, there are numerous successful applications of operations science to major capital projects. We close out the chapter with a couple of examples from a major capital project.

Operations Science in Major Capital Projects

As discussed in Chapter 4, the basic elements of operations science can be combined to represent extremely complex production systems such as an entire construction project. This enables one to use operations science to plan and execute projects with better results than traditional project management. This has been repeatedly demonstrated in practice with great success.

Traditional project management focuses on developing a plan and then tracking it with schedules and lagging reporting measures (i.e., CPM, Gantt charts, schedules, due dates, and assigned responsibilities, earned value analysis, etc.). Gaps in schedule completion are identified with the hope they will be addressed. One major issue is that sometimes the original schedule is a wish list that does not consider capacity and queuing (time delays).

In a 2021 *Management Business Review* article, Spearman, Choo, and Zabelle describe PPM, an approach that explicitly incorporates operations science into project management practices. They explain,

> PPM continues to use a high level schedule but in a different way. The overall schedule of the project provides the "demand" for the production system. The understanding is that if the needed rates can be maintained, the required dates will be obtained. The only way to know this is to understand the capacity requirements of the production system. Project management focuses on dates and a schedule that tells what should happen while the PPM focus is on rates and what can happen.[xx]

In Chapter 3, we introduced an example of a construction project as a production system. That production process is shown again in Figure 18-12. These data are taken from an actual project. To complete the project, 108 pipe spools, each with its supporting piling and footing, must be installed in 18 months (548 days). The pilings are custom made and must be engineered, then fabricated and delivered before they can be installed at the construction site. The supply flows, consisting of engineering, fabrication, and delivery, feed the site installation flow.

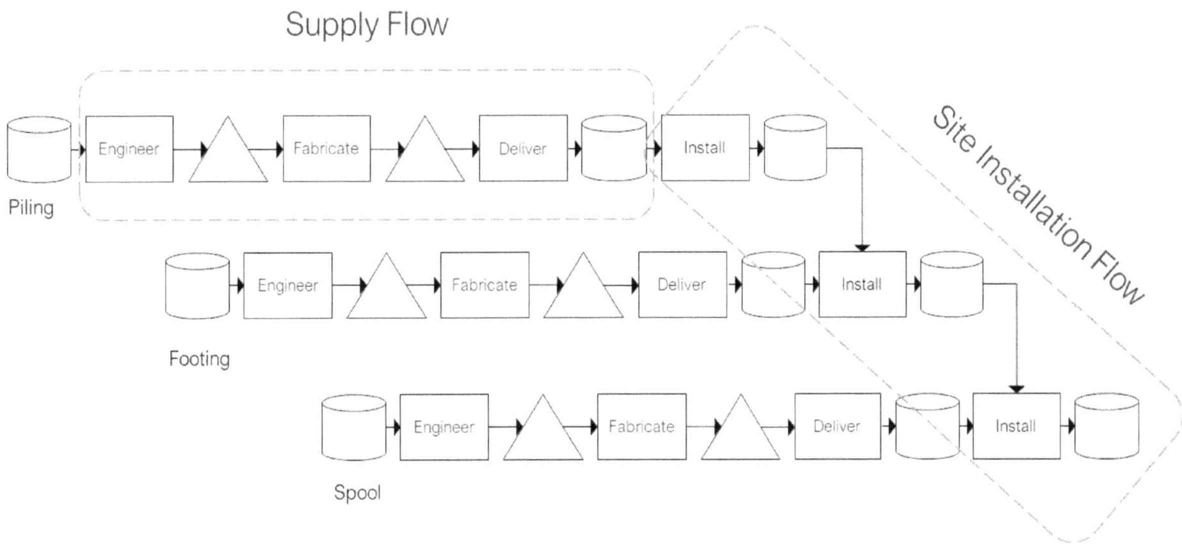

Figure 18-12: Basic production flows in a portion of a construction project

A major reason that projects are deemed unsuitable for operations science and production management is that they are defined as "a temporary endeavor undertaken to create a unique product, service, or result."[xxi] The key words here are *temporary* and *unique*. If one refers to the product process matrix in Figure 18-2, *temporary* and *unique* would seem to fit in the upper left corner of the matrix. So, Schmenner's classification of projects in Figure 18-2 would appear to be correct. However, this is an oversight of time horizon and scale.

A Formula One (F1) race car team will spend millions of dollars to ensure that they save micro-seconds on a three-second pitstop in a highly choreographed production process. Project managers, owners, and workers often do not think the same production concepts apply to a project that designs, fabricates, delivers, and installs 108 pilings in 548 days. The F1 teams go to 24 or so races a year (not 108) in a repeatable process. Each one of the 108 pilings might be unique, but the process of engineer, fabricate, deliver, and install is not unique; It is repeated 108 times.

An F1 race is a temporary endeavor; courses and conditions are unique for each race. Successful F1 teams are relentless in designing, implementing, and controlling the processes to design, build, deliver, and race their cars. Even F1 teams are temporary endeavors, much like construction companies, as Zabelle noted earlier. Those who do not recognize the immutable process of engineer, fabricate, deliver, and install for projects are leaving their success to chance. Without the foundation of operations science to design, implement, and control the process, owners, project managers, and workers put themselves at a severe disadvantage. F1 teams spend hundreds of millions per year orchestrating the production process to deliver their cars and drivers through the race circuit, but the costs and benefits pale in comparison to the $10 trillion opportunity cited earlier for major capital projects when applying operations science to their projects. Following is further description of operations science in application on a major capital project.

Rebar Cages for a Major Capital Project

The following story illustrates the application of operations science to a civil infrastructure project, valued at over $7 billion,[xxii] built in the early 2000s. This is an edited excerpt from Spearman et al.[xxiii] It describes the rebar component fabrication for the poured concrete columns as analyzed in Chapter 4. See Figure 18-

13 for a detailed production process map for poured concrete columns. Rebar cages is one segment of that process.

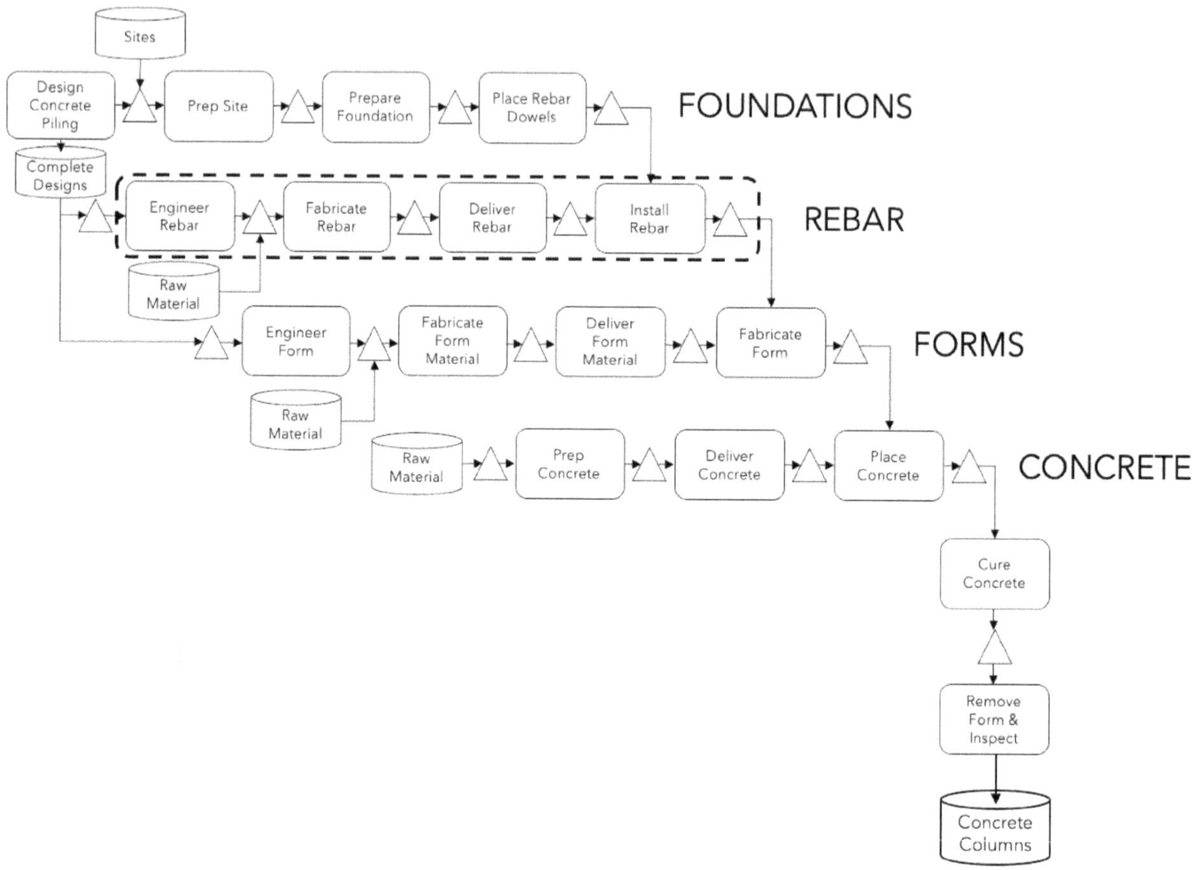

Figure 18-13: Production process map for poured concrete columns

The story provides an excellent illustration of how construction professionals typically configure production systems without considering the operations science of their systems and describes the unintended consequences that arise. It clearly depicts fundamental operations science relationships among production throughput, inventory, and reliability.

Here we build on the initial discussion in Chapter 4 with detailed financial implications and discussion of stakeholder perspectives.

The subject project production system is for the fabrication, delivery, and installation of foundation pile (rebar) cages. The rebar fabrication facility (the fab) has a theoretical capacity of 12 cages per day, while the site can install only 10 cages per day. The cost per day is $77,373 for the fab and $30,757 for the site. The project was scheduled to run 250 days, which should provide sufficient time to install 2,050 cages, since the system would, theoretically, be capable of producing 2,500 cages in that time. But when the time came for installation to begin, the owner was unaware that the design of the production system was not capable of hitting its target of 12 cages per day and was on heading for a massive WIP buildup and long delays that would end up with a $8.5 million overrun on a $27.1 million project. Note the rebar cage and column installation is only one in hundreds or thousands of projects within a major capital project.

Because there are different contractors for the production and the installation of the cages, the production system ends up being designed such that an inventory buffer would continuously grow between fabrication and installation until all pile cages were consumed (fabricate 12 cages/day, install 10 cages/day). The fab contractor gets paid for fabricating rebar cages. With no other incentive, the fab contractor will build cages as fast as possible because that means getting paid sooner. Left unattended, at the rates of 12 and 10 respectively, the WIP on-site would grow to around 370 cages before decreasing.

But there were other, more significant, problems as well. After the first two weeks, it was evident that while the capacity of the fab was low (10.8/day instead of 12), the installation site was running woefully under capacity (7/day instead of 10). Upon a review of the production processes, the new estimate of the completion time became 318 days, adding 68 days and another $7.4 million to the cost. Moreover, the differential production rates would cause excessive WIP and rehandling, adding another $1.1 million to the cost. If allowed to continue at the demonstrated rates, WIP would overwhelm the site with not 370 cages but over 700 cages being stored on-site (Figure 18-14).

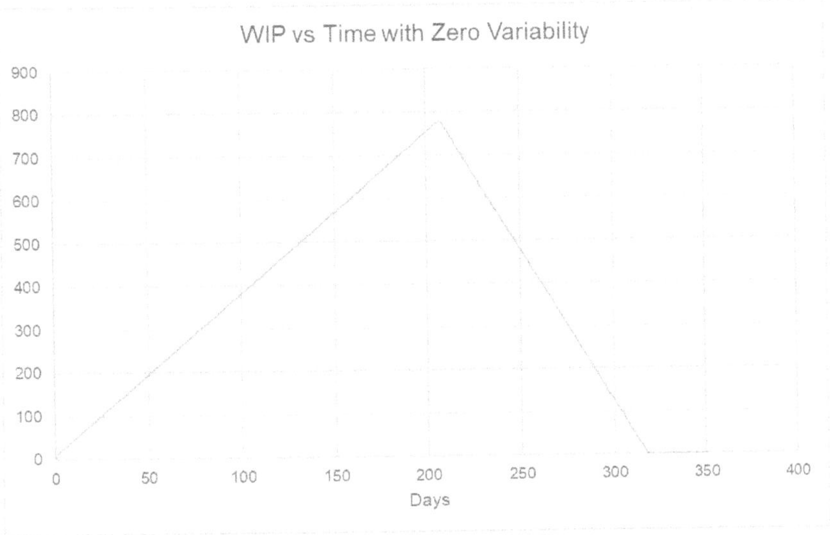

Figure 18-14: Rebar cage WIP buildup on-site

The photographs in Figure 18-15 depict the offloading and temporary storage of the cages and the additional cost associated with handling (equipment and labor), laydown/storage, preserving, and reloading to deliver to the point of installation.

Figure 18-15: Rebar cage handling and storage

ıy project production systems, WIP can be limited as a buffer between fabrication and installation ommodate variation in fab production while not starving the site. In this case limiting the buffer was htforward: The amount of WIP on-site was restricted. By limiting WIP storage to 20 cages using WIP control, the total rehandling cost dropped dramatically from $1.1 million to less than $93,000 no increase in number of days to completion.

a quick side note on WIP control, it can be a very important safety issue. As seen from Figure 18-15, rebar cages were large and heavy. Minimizing the amount of WIP meant fewer moves of WIP for rage and handling, which also meant fewer opportunities for safety incidents.

nce WIP control was installed, the more pressing problem of increasing capacity was addressed. This as accomplished by considering all the steps involved in installation and optimizing the entire process rom receipt of the cage to its placement in the excavation to the final pour of the concrete. Involving the nstallation team was key to the success of this effort. In the end, the productivity increased from a range of 62%–80% to 87%–95%, with an average capacity of 9.1 cages per day.

At this point, a more complete model was needed that included the cutting of the rebar as well as the delivery to the site. Although neither of these caused any capacity limitation, the times involved did affect the cycle time as well as the WIP. The production process model was constructed as shown in Figure 18-16.

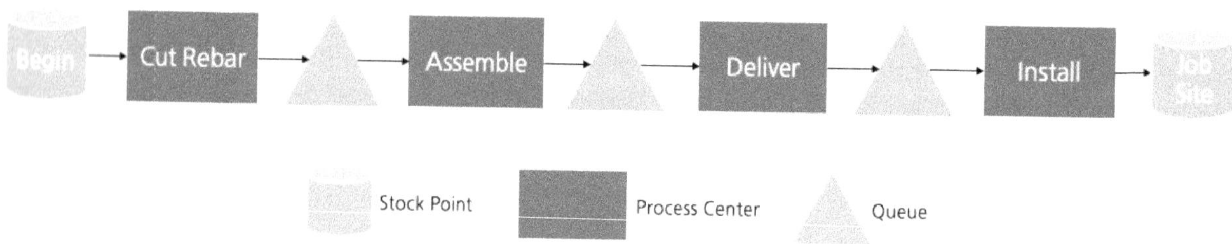

Figure18-16: Rebar cage production process

The limiting operation of the entire flow was installation at 9.1 cages/day. Cycle time and throughput versus WIP were plotted and compared to the ideal case using flow performance curves (Figure 18-17).

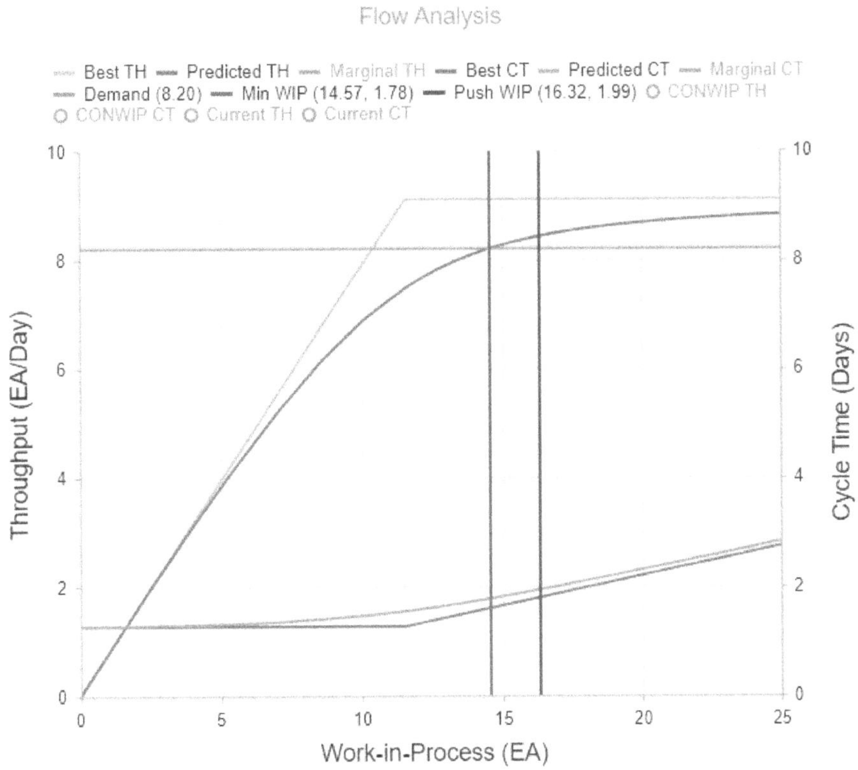

Figure 18-17: Performance curves analysis for rebar cage production

Operations science showed that if WIP was limited to around 20 cages, throughput would be 8.67 cages per day, which would finish the project well within the 250 allowed days. Also, the cycle time from cut rebar through site installation was only a little over 2 days.

An important note is that an operating throughput of 8.67 cages per day is a 95% utilization level (8.67/9.1), yet cycle time is only 2.3 days. This is an illustration of the power of operations science in practice, very short cycle times with high utilization of expensive resources.

This project production design allowed the project to be finished early (although the contract provided for a minimum of 250 days) with around $52,000 in rehandling cost. The total cost went from a forecasted $35.6 million to an actual $27.1 million, a savings of $8.5 million.

Conclusions

Despite all the confusion and protestations, production methods can and should be used to manage major capital projects to ensure that such projects are successful. This case has shown how the use of production techniques such as WIP limitation, high utilization of bottlenecks, inventory control methods, and frequent monitoring of the actual production are essential to managing any construction project.

The idea of using a detailed schedule to control a complex project by observing when milestones will be missed and then addressing the problem is almost always too little, too late. Indeed, while the schedule should describe what should happen, the production system determines what can and will happen. This does not mean that schedules are not useful. They provide the demand estimates for the production system that delivers the project. But they do not predict what will happen when.

Summary

To establish and maintain full control of a project, project owners, managers, and other stakeholders must see their projects as a collection of products and production processes and then they must use the science describing the behavior of those production processes (physical or virtual).

Operations science starts from first principles describing

- the effect of variability when trying to synchronize demand and production,
- the required buffers (capacity and time-inventory) for synchronization, and
- and the relationships between the buffers and variability

Production could be site construction of a building, coding for a software program, or new product development. Without understanding these concepts and applying them to project production work (physical or virtual), project delivery will continue to be over budget, over time, and under benefits.

Operations Science Application Summary Table: Rebar Cages

Table 18-2 summarizes how the rebar delivery team applied operations science to achieve excellent performance.

ELEMENT	DESCRIPTION
BUSINESS STRATEGY	Timely delivery and installation on budget.
OPERATIONS STRATEGY	
DEMAND	Project schedule of 250 days to produce and install 2,050 rebar cages.
TRANSFORMATION	Fabrication, delivery, and installation of foundation pile (rebar) cages.
VARIABILITY	Variability in the use of capacity for both production and installation. Weather was a source of variability, also.
BUFFERING OPTIONS: CAPACITY	Rebar production, storage, and transportation equipment and labor. Site installation equipment and labor.
BUFFERING OPTIONS: STOCK INVENTORY	Rebar cage components.
BUFFERING OPTIONS: WIP INVENTORY	Rebar cages, site installation tasks.
BUFFERING OPTIONS: TIME	250 days as dictated by schedule.
TOOLS AND APPLICATIONS	Capacity calculations, CONWIP control, and flow performance analysis to improve capacity output and responsiveness. Commit reliability for task coordination.

Table 18-2: Operations science application to rebar cage production

---

[i] Ben Cohen, "99% of Big Projects Fail. His Fix Starts With Legos." *Wall Street Journal*, February 2, 2023, https://www.wsj.com/arts-culture/books/lego-megaprojects-bent-flyvbjerg-big-things-11675280517.

[ii] Filipe Barbosa, et al., *McKinsey Global Institute: Reinventing Construction: A Route to Higher Productivity* (McKinsey and Company, February 2017) https://mck.co/4kHwpJa.

[iii] Ben Cohen, "99% of Big Projects Fail. His Fix Starts With Legos." *Wall Street Journal*

[iv] Todd R. Zabelle, *Built to Fail* (Forbes Books, 2024), chapter 1, Kindle. Section 1.2.3.4

[v] www.projectproduction.org

[vi] Roberto Arbulu, H. J. James Choo, and Mike Williams, "Contrasting Project Production Control with Project Controls," Project Production Institute, December 16, 2016, https://projectproduction.org/journal/contrasting-project-production-control-with-project-controls/.

[vii] Project Management Institute, *A Guide to the Project Management Body of Knowledge (PMBOK® Guide)*, 6th ed. (Function, 2023), Kindle.

[viii] For more in-depth discussion of the product project matrix and projects, see Mark L. Spearman and H. J. James Choo, "Rethinking the Product-Process Matrix for Projects," Project Production Institute, December 9, 2018, https://projectproduction.org/journal/rethinking-the-product-process-matrix/.

[ix] "Project Production Management," Project Production Institute, https://projectproduction.org/glossary/project-production-management/.

[x] "PPI Featured in *The Voice* Magazine," Project Production Institute, August 26, 2019, accessed March 2025, https://projectproduction.org/ppi-featured-in-the-voice-magazine/.

[xi] Roberto Arbulu, Gary Fischer, and Jack Hartung, "Stage Gate—What Are the Implications?" *Journal of Project Production Management*, 5 (2021), https://projectproduction.org/journal/stage-gate-what-are-the-implications/.

[xii] Robert G. Cooper and Michael S. Mills, "Succeeding at New Product Development the P&G Way: A Key Element Is Using the 'Innovation Diamond,'" *Visions*, 29, no. 4 (2005): 12, https://www.five-is.com/wp-content/uploads/2013/12/Cooper_Mills_2005_Succeeding_at_New_Product_Development_the_PG_way.pdf.

[xiii] Robert Cooper, "Stage-Gate in a Nutshell," Planisware, YouTube, June 11, 2015, 4 min., 23 sec., https://www.youtube.com/watch?v=FNlja0mplDI.

[xiv] Cooper and Mills, "Succeeding," p. 10.

[xv] Table generated by ChatGPR, March 18, 2025. Prompt used was "Create a table showing the differences between major capital projects and new product development in consumer goods or manufacturing and the suitability of stage gate for each item listed."

[xvi] "The Podium," Google Arts and Culture, accessed March 2025, https://artsandculture.google.com/story/the-podium-sydney-opera-house/xwURWKxhIgpfKg?hl=en.

[xvii] Architecture Stock photos by Vecteezy, https://www.vecteezy.com/free-photos/architecture

[xviii] Arbulu, Fischer, and Hartung, "Stage Gate—What Are the Implications?"

[xix] Ibid.

[xx] Mark L. Spearman, Hyun Jeong Choo, and Todd Zabelle, "Project Production Management: The Key to Delivering Large Capital Projects on Time and Under Budget," *Management Business Review*, 2, no. 1 (2022): 55-66,

https://www.researchgate.net/publication/383466560_PROJECT_PRODUCTION_MANAGEMENT_The_Key_to_Delivering_Large_Capital_Construction_Projects_on_Time_and_Under_Budget.

[xxi] Project Management Institute. A Guide to the Project Management Body of Knowledge (PMBOK® Guide)–Sixth Edition (Function). Kindle Edition

[xxii] All values are in 2008 dollars.

[xxiii] Spearman et al., "Project Production Management."

19

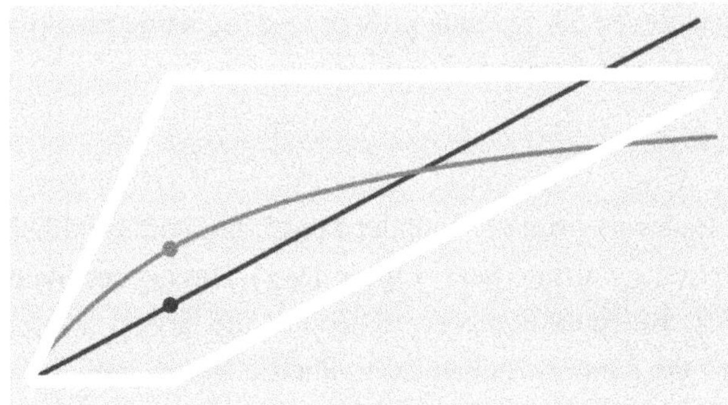

The Maritime Industry: How a Small Business Used Operations Science to Fuel a Sea Change in Ship Repair

In 2018, Pacific Shipyards International (PSI) was like many ship repair companies. It operated in a fast-paced, high-variability production system where "heroes" were recognized, a bias to action was rewarded, and the ability to adapt to constant daily changes at the deck-plate level drove a high level of expediting and the associated chaos. Over-budget and behind-schedule projects were common in the industry and occurred occasionally at PSI. Business was steady and, while PSI was one of the larger Hawaiian subcontractors for both commercial and government work, PSI's management was not satisfied with the status quo.

PSI began a massive five-year transition effort that included buying a new dry dock and, at the same time, restructuring and developing new internal business and operations practices to accommodate growth. When the US Navy switched from a single-source, cost-plus contracting structure to a fixed-price, multi-supplier approach, PSI found itself with a massive opportunity to accelerate growth. To do so successfully, it would have to manage a huge increase in work content as a prime contractor. This opportunity also produced an enormous challenge: To win and deliver the new business, PSI's original five-year transition plan had to be completed in two years.

Operations science concepts provided this small shipyard and maritime community a practical, scientific foundation to help successfully transform itself in 24 months. PSI tripled its revenue in the first two years and achieved 5X revenue growth to-date, while breaking records in customer service, cost, and on-time delivery performance.

Pacific Shipyards International

PSI traces its origin to founder Fred Loui, who established Pacific Refrigeration in 1944 to service America's wartime fleet (Figure 19-1). Having received commendation from the Navy for its services, Pacific Refrigeration grew to become the largest commercial ship repair operation in Hawaii by 1954 when the name was changed to Pacific Marine and Supply Company, Ltd.[i] Eighty years later, the Loui family remains at the helm. Steven Loui, current chairman of the Pacific Marine and Supply Company, has expanded his father's maritime vision, and PSI has become the largest privately owned shipyard in Hawaii, with two of the largest dry docks in the state, as well as driving leading-edge maritime research through its subsidiary, PacMar Technologies. Under Steven's decades of leadership, an entrepreneurial spirit has been deeply embedded across the family of companies—emphasizing strategic investment in infrastructure, technology, processes, and people. His long-term commitment to growth and innovation is matched by a deep sense of loyalty to employees and a steadfast focus on purpose. Guided by these values, he has led the corporate team and a distinguished board of advisors in making proactive, forward-thinking decisions aimed at fixing tomorrow's problems and positioning the company to sustain its strong growth.

Figure 19-1: PSI builds upon an 80-year legacy supporting the Hawaiian maritime industry

The Maritime Industry

Shipping sustains the world's economy—without it, international trade would come to a standstill. Despite major advancements in other forms of transportation, a surprising 90% of all goods are transported by ship. In Hawaii, the transport of cargo by rail or interstate trucking does not exist, so a thriving and reliable maritime capability is vitally important to the Hawaiian Islands and the quality of life of its residents because 98% of all goods shipped into Hawaii are transported by sea.

The US maritime industry is "responsible for almost 650,000 jobs and over $150 billion in annual economic output," according to a recent study conducted by PricewaterhouseCoopers for the Transportation Institute.[ii] In addition to transportation, the maritime industry also encompasses activities like shipbuilding, repair and maintenance, port operations, marine engineering, and critical support to national defense. The US ship repair industry supports commercial and government customers within major maritime regions along our coastlines, including the Hawaii Ship Repair Industry.[iii]

Impending Change

PSI's transformation began with the arrival of a new, state-of-the art dry dock (Figure 19-2) capable of supporting a range of mid-sized US Navy vessels to the largest vessels of the Coast Guard, Army Core of Engineers, and National Oceanic and Atmospheric Administration, as well as Hawaii's vital interisland commercial fleet. While this new Hawaii-based strategic asset immediately caught the attention of ship owners, including Navy officials, PSI's CEO, Iain Wood, understood that this "field of dreams" approach would only take them so far towards their vision. They also needed to strengthen core processes, build-up expertise within PSI, invest in technology, and strengthen the Hawaiian maritime industrial base.

Figure 19-2: Translated as "healing the ships," Ho'ōla I Nā Moku *is PSI's newest and largest floating dry dock*

Across the industry, ship repair projects are commonly delivered behind schedule and over budget. A large driver of this is the complexity of the projects and the inherent growth work[iv] that emerges during a repair project. PSI's ship repair system was loosely defined and largely dependent on "tribal knowledge," the experience and talent of the people involved.

PSI performed the majority of its government ship repair work as a subcontractor to a single defense contractor that had been awarded all private-contractor Navy ship maintenance work in Hawaii for the preceding 15 years. Between 35% and 45% of ship repair work in Hawaii is subcontracted, higher than other maritime regions.

Just as PSI was positioning the new strategic dry dock asset in Hawaii, the Navy announced it was transitioning its long-running acquisition process from single-source, cost-plus to a multisource, fixed-price business model. So not only did the Hawaiian maritime industrial base need to make changes, its largest customer, the US Navy, was about to uproot 15 years of core contract management processes. To mitigate transition risk, the Navy shifted a portion of maintenance work from the private sector back to its own shipyards, which provided more flexible capacity for modulating the effects of demand variability. The change also created a capacity buffer for ramp-up demand during the transition period. PSI made a gutsy decision to pursue a prime contract to strengthen the Hawaii Maritime Industrial Base, at the potential risk of losing significant subcontract work. Little did anyone anticipate that national ship repair performance records would be set right out of the gate on this first fixed-price ship repair project with PSI as the prime contractor.

Self-Assessment

PSI's transition story began with an in-depth, and sometimes painful, review of current state to establish their operational baseline. PSI's senior management team hired Hepinstall Consulting, a consulting group that specializes in shipyard improvement, to perform an independent deep dive into every aspect of PSI's operations, including assignment of daily production work on the deck plate, bill payment processes, and correction of quality defects. A 360-degree current state analysis was performed to shine a light on PSI's deeply rooted practices and norms and better understand operations from an internal, external, transactional, organizational, and human-factor perspective.

Hepinstall and PSI leveraged the operations science framework in working through PSI's transition. Hepinstall brought the expertise in operations science to PSI, though PSI didn't call it their operations science[v] framework. That is a substantial benefit. Employees don't have to be told they are starting a new initiative—the science strengthens and accelerates existing initiatives. Working together, Hepinstall and PSI designed operations practices that could be implemented and controlled to provide PSI with the paradigm shift needed for successful response to maritime industry changes.

This current state analysis identified the need to

- reduce process variation through standardization and improved flow
- shift PSI's ability to control projects instead of managing project controls
- shift from an accounting-based model to a work-delivery model
- improve the shipyard's operational discipline capability
- implement enabling technology
- strengthen the safety/environmental/quality capability with more structured processes

- elevate the role of production throughout the organization

PSI's willingness to self-assess at such an immersive level was a pivotal step in their continuous improvement journey.

Setting a Vision

The gap analysis validated the need for change. The data collected and demand forecasts produced also demonstrated how quickly PSI needed to improve in key areas and drove the need to accelerate the improvement timeline. As another defining moment, PSI made a strategic decision to accelerate the program, realizing that the pace of change is not always driven by the organization's immediate capacity to implement change. To focus the improvement journey, CEO Wood collaborated with his senior management team and board members to inspire the organization with the objectives of a clear vision:

> At Pacific Shipyards, we are laser-focused on our customer's needs, and that shapes the minds, attitudes, and actions of our workforce and strategic partners. Pacific Shipyards is a locally owned and operated full-service shipyard, with long-standing roots to the region that span decades. We are recognized for the technical proficiency of our workforce and our ability to achieve consistent world-class quality performance. Our processes are aligned, controlled, and repeated across programs to enable high OPTEMPO[vi] performance so that our customers can always rely on our ability to respond to their planned and emergent repair requirements.
>
> Our people are our strongest asset and drive the success of the company. Our empowered workforce embodies a culture of commitment, integrity, transparency, and personal accountability, and demonstrates strong work ethics and the desire to achieve continuous improvement. Our management team and support functions are motivated to ensure exceptional customer value, provide resources, remove obstacles, and enable high-productivity performance.

The vision was further distilled into detailed components for customer engagement, workforce, plant, and process. Significant work was required to convert the PSI vision into a working operations strategy.

Constructing Operations Strategy in Support of Business Strategy

Ship repair organizations commonly develop strategies at the program project level. A program typically includes multiyear contracting strategies and multiple awards of contracts, for example, a regional indefinite delivery, indefinite quantity (IDIQ) surface fleet ship maintenance program. A ship repair project called maintenance and modernization availability[vii] is performed at the specific ship level, for example, USS Daniel Inouye (DDG 118) SRA. Project capture strategies are created to increase the attractiveness of PSI's value proposition. Upon award, resource strategies and repair execution strategies are developed, describing how the project team will approach a specific ship repair project. During the transition period to fixed-price contracts, PSI wanted to be able to understand and control operations across programs and be able to set strategic direction over a longer-term period, three to five years.

A central tenet of successful operations science application is that business strategy drives operations strategy. While PSI's business strategy had always been to provide expert ship repairs and good customer service by investing in a skilled workforce and leading technologies, Wood realized that much of PSI's approach was informal, using standard project management practices and cost "controls" such as Gantt charts and earned hours. As discussed in Chapter 18, these and many standard project management techniques are controls only in the sense that they provide cost and schedule attainment information in hindsight. These techniques have their place and are often included as contractual requirements. However, an effective operations strategy requires treating a project as a production system and applying operations science concepts to productively manage variability. Then operations strategy can be wielded as a powerful competitive weapon in support of business strategy.

A primary drawback of traditional project management control techniques is that they ignore the effects of variability. For example, the sum of durations for tasks on the critical path for a Gantt chart is projected to be the duration of the project. In practice, the projected critical path at the beginning of the project, and during most of its duration, does not correspond to the actual critical path the project follows. A major benefit of operations science is it provides a scientific lens that enables businesses to view their projects as production systems. When viewed through the operations science lens, variability and its effects are easy to see and can be addressed to achieve predictable and profitable project performance.

Operations science concepts guided PSI personnel, from the board of directors to the frontline management and workers, on how to shape their operations strategy and control their operations performance to achieve the business results they wanted.

In operations science, operations strategy is the plan, or set of plans, to design, implement, and control demand, inventory, capacity, time, variability, and cost. The first step in operations strategy is defining the performance envelope a business will target. The performance envelope includes primarily

1. products and services a business will offer,
2. targeted average and variance of demand for those offerings, and
3. targeted average and variance of durations for tasks required for total project duration and expected replenishment times for vendors supplying goods and services to the business.

The primary reason for understanding the performance envelope is to understand demand and what capacity utilization (efficiency)[viii] should be targeted. Many ship repair businesses target highest possible efficiency for their people. In other words, keep everyone working as much as is possible to ensure they produce as much as possible per hour. That in itself is a worthy goal, especially in the face of labor shortages. However, without good understanding of operations science, this approach often devolves into keeping people as busy as is possible. *Busy* means doing anything, whether it is productive or not. Operations science shows that, in the face of variability, it is often more productive to idle a resource to wait for work that is needed in the appropriate sequence rather than giving a resource work that clogs the process or occupies the resources so they cannot start on a needed task when it becomes available. Additionally, many cost-plus contracts encourage a burn-and-earn approach; in other words, as soon as work gets done, the contractor or subcontractor can bill for it. This seemingly reduces the cost of work performed and gets money to the business as fast as possible. Without a good understanding of operations science, organizations often corrupt their performance envelopes and work degrades into constant firefighting because everyone is trying to work as fast as is possible without proper attention to required sequences or available resources.

Operations science shows that working at high levels of efficiency, especially with high variability, drives task duration (cycle time) out of control. See Figure 19-3. The reader will note that the term *efficiency* here was called *utilization* in previous book chapters. This is an adaptation of operations science language to PSI's usage. Working at high levels of efficiency leads to expediting, chaos, project delays, and extra cost for the owner. In the short term, burn-and-earn does indeed lead to more money sooner in the pocket of the contractor. However, it is deadly to schedule attainment, ultimately increases cost, and leads eventually to owners getting smart and discontinuing cost-plus contracts.

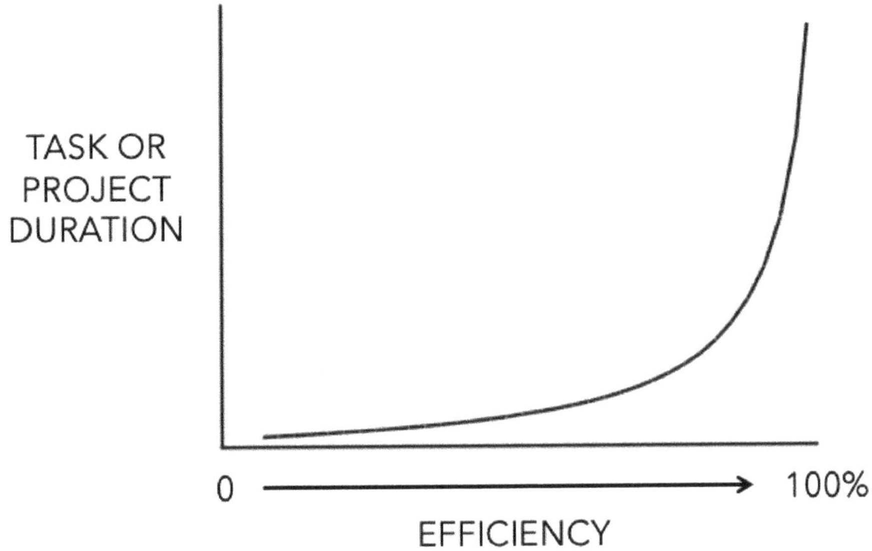

Figure 19-3: Operations science of duration versus efficiency

PSI analyzed demand drivers, collected market information, and used statistical process control (SPC) charts to refine their forecasting models and proactively target business opportunities. Being able to express operational elements visually provided insight into the level of variability in each of these areas.

Developing an operations strategy presented many challenges and took over a year to finalize. Other than the strategies PSI had in place to control cost and schedule on a project-by-project basis, design, implementation, and control at an operational level was difficult to envision in the beginning. PSI's operations strategy evolved over time by simply monitoring and then analyzing workload demand, cost, and schedule.

Operations Strategy to Implementation

Operations science will provide new insights into solutions for long-standing challenges. Operations science will inform and strengthen the intuition of every person who works for a company, enabling them to make better decisions faster. Operations science won't make your strategy for you. It won't implement changes for you. It won't control day-to-day work for you. What operations science did for PSI, and will do for you, is prevent wasted effort on either unrealistic strategies or implementation mistakes that consume resources.

Once PSI had settled on its operations strategy, implementing the strategy quickly was vital to achieve the growth it was chasing. With a sound knowledge of operations science, implementation was successfully achieved in two years. There was still plenty of trial and error, but the work was guided by an understanding of operations science and its insights into managing variability and variability's buffers of capacity and time.

Stock inventory is also a buffer for variability. For PSI, there was a strategic recognition that stock inventory was not a smart buffering option, so the company did not plan on stocking significant quantities of material inventory.

Weekly updates to workload demand curves (Figure 19-4) enabled PSI to forecast resource requirements more effectively and level-load the resource requirements across programs.

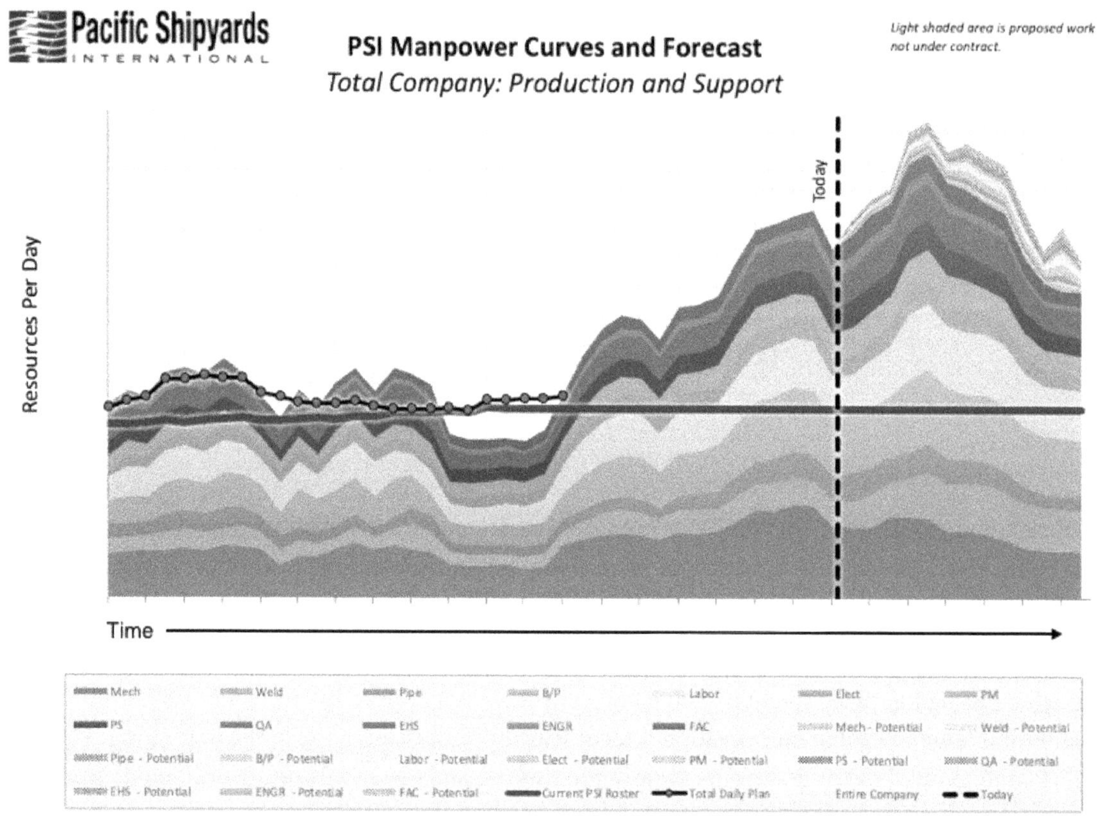

Figure 19-4: A sample plot of PSI workload demand curves

Designing a Ship Repair System

Successful execution of operations strategy through implementation is, simply put, a series of three tasks as described in Chapter 8—design, implement, and control.

1. Design of
   a. future state,
   b. control of the implementation
   c. control of the implemented processes
2. Implementation
3. Control of future state day-to-day operations

As part of the operations strategy, PSI formalized its ship repair system, a cradle-to-grave value stream that represents the set of actions that take place to add value to a customer from initial request through realization of value by the customer. This value stream was organized over the following five phases:

Phase 1.0 - Bid Notification to Contract Award

Phase 2.0 - Contract Award to Transition to Project Management Team (PMT)

Phase 3.0 - Project Management Team to Start of Availability (SOA)

Phase 4.0 - Start of Availability to Production Completion Date (PCD)

Phase 5.0 - Production Completion Date to Contract Closeout

The PSI ship repair system was formalized to

- provide a common framework across the organization and programs
- communicate the PSI way of estimating, planning, performing, coordinating, managing, controlling, and completing ship repair projects
- serve as an authoritative source of PSI's project-related processes
- train new employees on core processes and artifacts that enable processes
- enable future automation in constrained areas
- provide visibility of process constraints
- promote accuracy and completeness of deliverables
- communicate the robustness of our core processes to our customers

This effort produced 282 standard operating guides (two-page desk guides) of each process step in the value stream that describe the process, inputs, outputs, roles and responsibilities in an executable, repeatable manner. While this effort was accomplished over a two-year period, it serves as an authoritative source on process, baseline for continuous process improvement, and standardized business and production processes across programs.

Design, Implementation, and Control: Capacity

Operations science shows that capacity and capacity utilization are first-order concerns for any operations system. PSI leadership understood the effects of variability on capacity usage and planned a significant capacity buffer to accommodate both the increase in average demand and the variability that came with implementation of massive changes to its ship repair operations. The following capacity initiatives were implemented:

- Increased employee headcount from approximately 100 (2018) to approximately 300 (2024). This included increased production support and project management resources as well as direct labor
- Relocated shipyard to a larger facility
- Constructed new production shop
- Removed non-value-added time, spent more time on productive work
- Strengthened the Hawaii industrial base
- Leveraged subcontract/technical representatives
- Increased electrical capacity through solar-powered infrastructure to achieve energy independence

For day-to-day control of operations, PSI elevated the role of production. Its operations strategy reinforced the perspective that production was the revenue generator for the company. All other departments, including the senior management team, were categorized as production support. This was a paradigm shift across the workforce that continues to translate to high value-added time work for production.

Figure 19-4 shows a year's view of PSI's demand profile. PSI was able to flatten peaks and fill in valleys by overtly implementing strategies that leveled the demand around an operational sweet-spot level of resources and gradually raised that level over time. It was never perfect but served as a step towards proactively controlling demand at an operational level instead of being held hostage to the daily swings of project-level variability.

PSI shrewdly discarded a common ship repair control practice of trying to keep resources as busy as possible all the time. As mentioned previously, operations science shows that a dogmatic pursuit of high efficiency leads to high levels of stress, chaos, and waste. Operations science also provides effective control strategies to avoid these ill effects while working at best possible levels of efficiency. PSI has recently begun to exploit these strategies by measuring the number of tasks in process each week and tracking it as work in process. Since the task durations are unstandardized, WIP count is equalized based on daily WIP units, i.e., a three-day task has a calculated WIP of 3. PSI didn't begin with a goal of what the "right" level of task WIP was for the company. As the task WIP has decreased to the appropriate level needed for the available capacity, PSI has experienced a general *increase* of weekly task throughput. This seems counterintuitive to the reader now familiar with the flow performance graphs. However, the WIP overload effect described in Chapter 4 eventually leads to a decrease in throughput due to the increase in chaos and variability. For PSI, the data told the story over time.

Other activities to mitigate variability and make more productive use of capacity included

- 5S
- Project production management: Daily plans, one-week look-ahead, commitment reliability
- Visual factory
- ACFR and TIP flow-through implementation of Oxalis Ship Repair System, a tailored ship repair software solution that captures, processes, approves, and submits documentation directly to the customer[ix]

Design, Implementation, and Control: Inventory

During the early stages of PSI's transition, there wasn't much attention given to assessing the level of stock inventory, since by nature, physical inventory moves very quickly within the shipyard. Once the goods are received at the shipyard for ship repair, they are processed and transported to job site for installation within hours or a few days. While this is technically considered WIP, most of the material is direct charged to the project and consumed in the project, leaving very little physical materials left to stock, dispose, or transfer custody.

PSI doesn't maintain a lot of material stock, other than small batches of pipe and steel plates direct ordered for a given project. In shipbuilding, stocking a whole ship-set worth of steel plate and pipe results in a large inventory burden, in terms of size of inventory, process costs, and inventory holding costs. Conversely, a ship repair project is typically only replacing deteriorated or degraded areas of the ship, so the size of inventory is much smaller. Also, ship repair projects cannot be fully scoped until some areas of the ship can be accessed. For example, the work scope of tanks is not fully determined until they are drained and certified safe for workers by a shipyard competent person (SCP) or certified marine chemist. The short duration of repair projects inherently promotes just-in-time delivery; therefore, goods are often consumed shortly after being received by the shipyard. PSI assessed supply chain risks as a moderate risk and implemented long-term vendor strategies, increased their supply base, and added time buffers for early delivery where applicable to mitigate the risk of goods not being available on-time.

Design, Implementation, and Control: Time

The operations strategy emphasized the need to design, implement, and control procurement lead times, described as raw material queues, in PSI's operational model. For example, replacing unique engineered valves might require a two- to three-month lead time, which is likely to extend beyond the production need dates. To compensate for this, a time buffer may be added by adjusting the schedule or perhaps a capacity buffer by finding an additional supplier. Implementing supply chain and logistics strategies remains a critical focus of the materials group.

Another component of controlling time is the management of WIP as described previously in the capacity discussion. Operations science shows that WIP, whether physical material or virtual tasks, is visible cycle time. Through design, implementation, and control of the operations strategy, PSI decisively acted on this central operations science concept. It is described mathematically by Little's law, which can be stated as

$$Cycle\ Time = \frac{Task\ WIP}{Throughput}$$

As discussed previously in Chapter 4, flow performance curves show that decreasing WIP usually leads to decreased cycle time and to decreased throughput. In PSI's case, as described previously in the section on capacity, reducing WIP actually increased throughput. This is very common in shipbuilding and ship repair and in engineering and construction in general because project managers often flood processes with tremendous amounts of WIP, whether physical WIP or virtual task WIP, to keep resources busy. This in turn increases variability due to ever-changing priorities, looking for material or specs or rework due to design changes. The net effect of this WIP tsunami is to increase variability so much as to reduce throughput.

However, once WIP levels are set appropriately using operations science, WIP becomes a control lever for predictively managing task duration (cycle time). Task duration is different than the actual work time in a task. For example, if there are 20 tasks scheduled for a resource, and each task takes 5 minutes, the last task will be completed after 100 minutes if there is no variability caused by competing priorities, shared resource conflicts, and the like. In ship repair, there is tremendous variability at the deck-plate level, so wait time occurs, and task duration is often much longer than the sum of allocated task times. Since PSI has sized its WIP level appropriately for its capacity, WIP control using Little's law is a powerful strategy for determining response time and throughput. This approach takes the guesswork out of determining task completion dates.

Senior Management Reorganization

Prior to PSI's transformation, the senior management team consisted of the CEO/GM and experienced departmental managers of production, contracts, quality, environmental, safety, and health and human resources. This small management team wore multiple hats. PSI's CEO, armed with support from the board, invested heavily in creating a management structure that would enable operational performance and shift the organization away from managing through independent departmental silos. New positions were created for VP of operations, VP of programs, and VP of production. The VP of operations was later elevated to the position of chief operating officer. The talent and experience brought in with these positions at just the right time propelled the transition forward. It provided additional management capacity to champion the change efforts and focus the organization on accomplishing their normal ship repair activities while building the future state operations. This team understood the challenges of "changing the tire while the car is moving" and provided close coordination and accountability of activities on a daily basis. To reinforce how serious the shipyard was about improvement, they stood up a small process improvement department responsible for implementing and sustaining improved processes. This group is led by a director of process who provides a voice for process improvement among the senior management team when the pressures of ship repair tempt capacity and business decisions in the other direction.

Standing in the Light

PSI's transformation required senior management to make decisions that prioritized long-term decisions over short-term gains. For example, December 4, 2019, started like any other day in the shipyard. PSI employees and their subcontractor teams were performing ship repair activities on several ships at Pearl Harbor Naval Shipyard. At midmorning our teams received a shelter-in-place alert from the Navy responding to an active shooter aboard one of the ships. The sailor who opened fire killed two Department of Defense workers and wounded a third before taking his own life. While no PSI employees were harmed, the shelter-in-place lasted several hours. All contractors were allowed to submit a delays-and-disruptions claim to recover the lost time. However, PSI's customer engagement vision inspires the organization to enable the success of each customer, breakdown barriers, and promote goodwill at every transactional level. Keeping to this vision, PSI's CEO directed the project teams to implement recovery plans for each of their schedules and simply relayed to the Navy that they had lost enough on that day, so PSI would not be submitting a claim to recover costs. A few weeks later, Navy leadership expressed appreciation for PSI's response to this event.

A key success driver of PSI's transformation was the willingness to stay true to purpose when faced with adversity. The senior management team describes walking the walk as their biggest challenge. On paper, it was easy to envision how a future state would be a more productive, less chaotic business model. Right out of the gate, it was almost impossible to model the desired behaviors consistently, as the inertia of current projects and practices was powerful. However, PSI continued to invest in the time and energy of performing improvement actions, monitoring the data, and analyzing the impacts. Operations science didn't make or implement decisions, but management's understanding of operations science allowed them to make better decisions faster. It didn't take long for senior management to identify a few opportunities to model desired behaviors in their decision-making and priority-setting. This pick-your-battles approach allowed time for the organization to absorb change at a pace they could metabolize. The success continues.

Operations Science Application Summary Table: Pacific Shipyards

Table 19-1 provides a summary of the application of operations science to maritime operations.

ELEMENT	DESCRIPTION
BUSINESS STRATEGY	Grow business through successful management of fixed-price contracts for shipbuilding and ship repair (SBSR)
OPERATIONS STRATEGY	
DEMAND	Contracts for SBSR programs and their projects
TRANSFORMATION	SBSR labor and equipment
VARIABILITY	Demand: Ship repair requirements may be much different than thought before the ship arrives for repair. Transformation: Different skill levels in labor, especially when training to ramp up workforce. Weather.
BUFFERING OPTIONS: CAPACITY	Hire more people. Buy more equipment. Outsource services. Level loading. Process improvements.
BUFFERING OPTIONS: STOCK INVENTORY	Not much opportunity here. Very little stock inventory maintained.
BUFFERING OPTIONS: WIP INVENTORY	Established daily WIP units for task. Physical WIP required for completion of repair.
BUFFERING OPTIONS: TIME	To be avoided but can be used judiciously as reschedules.
TOOLS AND APPLICATIONS	CONWIP control for both physical and task WIP. Commitment reliability. Daily plans, one-week look-aheads, visual factory

Table 19-1: Pacific Shipyards' operations science application summary

---

[i] https://www.pacificshipyards.com/our-story/

[ii] "Economic Security," American Maritime Partnership, accessed June 11, 2024, https://www.americanmaritimepartnership.com/u-s-maritime-industry/economic-security/

[iii] "About," Ship Repair Association of Hawai'i, accessed September 20, 2025, https://srahawaii.org/about/

[iv] In ship repair projects, growth work refers to additional repairs, modifications, or upgrades that arise during the execution of the original project scope. These extra tasks are typically identified after the initial inspection or during the repair process

[v] See "Background of the Operations Science Institute" at https://opscience.org/about-us/ for a history of the term *operations science*

[vi] *OPTEMPO* is short for Operations Tempo and is the rate at which units are involved in all military activities, including contingency operations, exercises, and training deployments

[vii] In Naval procurement parlance, a project is called an availability

[viii] In shipbuilding and ship repair and in most construction projects it is common to use the term *efficiency* to mean "utilization." In operations science, utilization is simply the amount of time consumed by a person or a machine divided by the resource's available work time. Utilization may not be efficient as an optimum in that a resource can be busy doing tasks that don't need to be done or that are done incorrectly requiring rework

[ix] https://oxalis.io/oxalis-ship-repair-system-osrs/

20

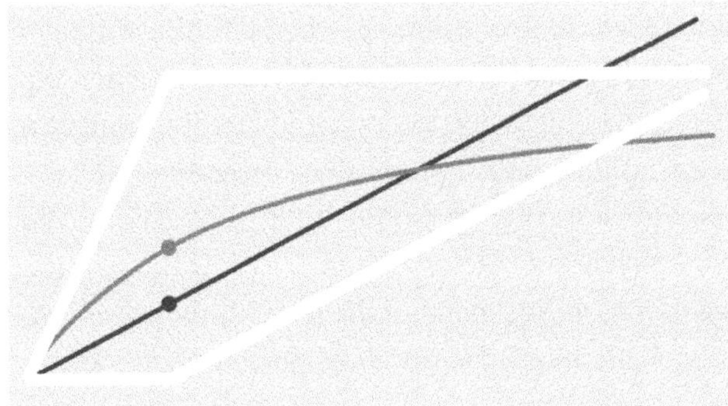

Operations Science and Healthcare Service Delivery

Jeff Bell

> As I reflect on my journey in applying continuous improvement principles to healthcare, I recall a graduate student project in the mid-1990s where I assessed RUSH University Medical Center's use of Total Quality Management tools like fishbone and Pareto charts. While I observed isolated successes, I noticed a lack of strategic connection. I wondered how much improvement in the overall service and cost of the patient experience resulted from the efforts.
>
> Over the years, I've come to realize that healthcare faces challenges similar to those in other industries when adopting initiatives without considering contextual differences. However, since it starts from first principles, operations science can be tailored to healthcare situations, from policy to practice, as this chapter demonstrates.

In this chapter, we introduce a high-level examination of buffer strategy in policy situations, culminating in a case study from *Hospital Operations*[i] by Wally Hopp and William Lovejoy, re-examined through the operations science framework.

Strategic Buffers in Healthcare Policy from the COVID Experience to Rapidly Evolving Business Models

In the wake of the supply chain shortages experienced during COVID, I was asked by one of my Arc Precision customers to participate in an improvement brainstorming session sponsored by the US Food and Drug Administration Office of Supply Chain Resiliency. As I listened to well-intended participants offering ideas to avoid shortages in the future without adding cost to the healthcare system, I had a flashback to my RUSH University Medical Center experience almost 30 years ago. It was clear that these well-intended ideas needed to be put into the context of the extreme variability situation of COVID, which would require larger buffers, meaning more cost. I recalled Hopp and Spearman, my factory physics professors from Northwestern's Master of Management in Manufacturing program, saying, "It is factory physics not factory magic." I knew that without a focus on a framework of buffers for a planned range of variability, many of the ideas to reduce shortages without increased cost were an exercise in supply chain magic, not grounded in operations science.

One chapter in one book cannot solve all healthcare service delivery policy problems or even recommend specific policy solutions, but it can put a spotlight on the fact the solutions to these issues must be viewed in the context of operations science to ensure they are feasible and productive. Policy involves many considerations, but if the operations mechanics of the solutions don't work, patient care will be degraded. Strategic policy choices grounded in operations science allow leaders to set SMART goals for investment, strategy, tactics, and controls that enable execution of chosen strategy. People may disagree on how to approach healthcare policy and which choices to make, but operations science helps make those choices—and their implications—clear.

The recent example of COVID showed the impact of increased variability when the time buffer—shortages and long lead times in many industries—appeared when there was not enough capacity and inventory in the supply chain to handle the variability. In fact, COVID revealed what happens when demand and demand variability increase greatly at the same time capacity, and any inventory it could provide, are in severe decline. This chapter does not need to recount all the shortages—masks, ventilators, hospital capacity, vaccines to distribute, as well as a wide range of industrial shortages ranging from toilet paper to cars. It certainly felt like a cascade of first-time events, with people doing their best to respond.

In my classes at Dominican University, we researched operations issues in COVID so the students could apply the operations science they were learning to the real-time experiences they were having during COVID. The results were a little surprising in that the students found many of the issues happening during COVID had been raised in the past.

For example, the following headline from a Jan. 12, 2006 *Wall Street Journal* article summarized the buffer trade-offs and ended up being predictive about what happened in COVID: "***Just-in-Time Inventories Make U.S. Vulnerable in a Pandemic; Low Stockpiles at Hospitals Boost Efficiency but Leave No Extras for Flu Outbreak.***"[ii]

Some quotes from the article were particularly descriptive and prophetic:

"Most if not all of the medical products or protective-device companies in this country are operating almost at full capacity," says Michael Osterholm, director of the Center for Infectious Disease Research and Policy at the University of Minnesota. "That's the reality of today's economy: just-in-time delivery with no surge capacity."

"You can't plan for a surge capacity in an emergency room of 500 or 1,000 patients from the 20 you see in a day," says Michael Bishop, a Bloomington, Ind., emergency physician who used to be on the board of directors for the national trade association for emergency physicians. "Nobody could afford to do that. You can't have 10 doctors and 100 nurses sitting around waiting for something to happen."

The article stated, ”In the event of a pandemic flu outbreak, that chain is almost certain to break. Thousands of drug-company workers in the US and elsewhere could be sickened, prompting factories to close. Truck routes could be blocked and borders may be closed, particularly perilous at a time when 80% of raw materials for US drugs come from abroad. The likely result: shortages of important medicines—such as insulin, blood products or the anesthetics used in surgery—quite apart from any shortages of medicine to treat the flu itself.”iii

The point is the pandemic experience demonstrated the interaction of variability and the buffers in real time. As leaders think about healthcare policy for the future, operations science can help inform and execute those strategic choices.

The authors share a similar perspective on an approach to evaluate the rapid evolution of business models in healthcare. Another well-known saying for students who took early factory physics classes is “It is better to experiment with the model then to experiment with the factory.” The point is that operations science can be used to avoid trial and error approaches to healthcare business models. This enables leaders to be proactive and timely with business model choices and avoid the delays and damages that often occur with trial and error. As Hopp and Lovejoy state in their book Hospital Operations,

Pundits gloss over the healthcare delivery process because they assume that if incentives (prices, rewards, costs) are set correctly, the rest will follow as people rationally respond by consuming more of this, less of that, and so on, reaching the desired allocation of resources and consumption. This faith is unfounded. ‘Correct’ incentives are necessary but insufficient for efficient operations.iv

Unfortunately, people usually find that out through trial and error, when operations science could help chart a more predictive path to desired outcomes. For example, consider the operations science–based differences between an emergency room and the clinic in the drugstore that offers repetitive services like strep throat tests and vaccines very efficiently. The emergency room is like a job shop in that it has expensive general capacity to handle a wide range of variable demand. As discussed in previous chapters, this scenario requires a capacity buffer to handle the variability unless one wants to deal with extremely long response time. Further, the cost of this expensive capacity addressing relatively straightforward issues like strep throat testing is very high compared to alternatives. The drugstore clinic is more like a repetitive assembly line that requires less capacity and can more easily use some of the concepts like CONWIP and scheduling based on virtual queues to reduce cost and response time. In thinking about another example, one can ask where urgent-care clinics or businesses fit in this mix. In the big picture of healthcare business models and policy, the authors acknowledge that the answers are not easy. For example, what type of financial stress will hospitals face when that expensive fixed-cost capacity only handles the highest variability and most difficult to treat cases? Healthcare systems face difficult choices about investment in different types of capacity and the gatekeepers who control access to the capacity. For this book, we will just point out the value of operations science in understanding and making these choices. We highly recommend the book *Hospital Operations* by Hopp and Lovejoy for those who want a more in-depth analysis.

The Operations Science Behind Success Stories in Practice

Providing examples of success stories in practice is a little more focused than the broad world of healthcare policy. For example, Purdue's *University News* referenced the factory physics–based work that the Krannert School of Management did with Grassy Creek Outpatient Medical Clinic to help the clinic serve 37% more patients on a given day.[v] Professors Suresh Chand and Herbert Moskowitz worked with students and Dr. Deanna Willis at the Indianapolis-based outpatient clinic serving low-income patients. They took the "strategies, tactics, and controls" approach. Chand said, "We decided to apply factory physics at a high level. Our goals were to increase the efficiency of the clinic in terms of patient time in the office, utilization of clinic resources, quality of care, and ultimately, revenue for the physician."[vi]

In terms of the previous policy conversation, the group took on the concept of revenue for the physician noting that many organizational healthcare arrangements focus on limiting physician revenue. Moskowitz stated, "If we organize a clinic so a patient spends less time there, then we are maximizing the utilization of the doctors. That means physicians are being more productive and giving better service while they also are generating more revenue."[vii] In terms of tactics, the group focused on finding and addressing processes with long queue times. Though they did not use it, the VU analysis mentioned in Chapter 12 would have been a powerful, practical analysis of queue times. The group's work included paperwork processing, single pooled queues for physicians, reserved capacity in a virtual queue, and reduced batching. Finally, from the people perspective, Moskowitz stated, "One of the keys to our success was establishing rapport with the medical personnel so we could work with them to solve problems. The physicians and staff were wonderful."[viii]

The above was a quick example, but let's take a look at applying the operations science framework introduced in Chapter 8 to staffing in a hospital. As Hopp and Lovejoy stated, "All the value in any conceivable system is only realized in the actual delivery when hands touch patients. Everything else is prelude."[ix]

St. Mary's Hospital

The following example is a synopsis of the case study "Nurse Scheduling at St. Mary's Hospital," excerpted from *Hospital Operations*, pp. 321–37.

Business Strategy

St. Mary's Hospital, a large tertiary care facility on the east coast, wanted to optimize labor costs without negatively impacting patient care. St. Mary's controllable labor costs were largely for nurse and staff salaries. Employee satisfaction in this group was important to avoid costly turnover. Patient care was the paramount concern.

Operations Strategy

The first step in applying operations science concepts to a process is describing the process and its associated resources and flows. A simple process illustration for St. Mary's patient placement process is provided in Figure 20-1.

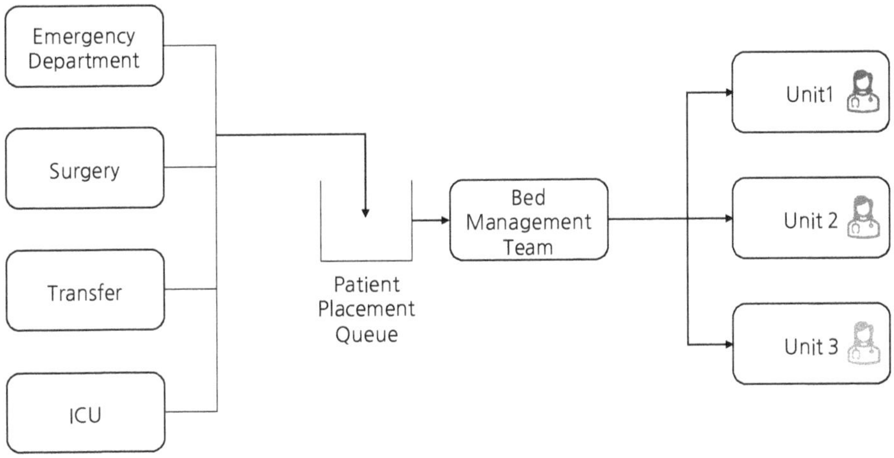

Figure 20-1: St. Mary's patient placement process

Demand

Demand for the nursing units came from sources such as referring physicians, the emergency department, transfers within the hospital, e.g., "step down" from an intensive care unit bed, or between hospitals. St. Mary's had a bed management team (BMT) of four people who took requests for beds and matched these demands against the bed availability in each unit. Bed availability was tracked on a bed board, which was a visual display of all the beds in each unit and the status of each (occupied, awaiting discharge, empty awaiting cleanout, or available). The BMT then matched patients to available beds, taking into consideration the referring physician's recommendation.

In some cases, a particular patient designation was mandatory due to the nature of the patient's disease. For example, patients with arrhythmias should go to a cardiology unit, and patients for chemotherapy should go to a unit equipped with infusion chairs and with a trained staff for that activity. In other cases, one or more units may be preferred but not mandatory. For example, there are preferred places for patients with bronchitis, pneumonia, and pulmonary medical problems. In still other cases, no particular unit was preferred over all others.

Demand and demand variability for nurses were generated by variability in both numbers and types of patients admitted to the hospital.

Transformation

Nurses on different units were trained for, and more experienced in, different types of diseases. The common intuition was that patients are best served by nurses most experienced with their particular ailment and in some cases data, such as length of stay, complications, or adherence to checklists of standard best practices, confirm this. However, patients did not always end up on St. Mary's recommended unit. When a patient was put in a non-ideal unit, it was known as boarding the patient on that unit. There were several reasons for boarding patients. First, each nursing unit had finite capacity, and if a unit was already full, it was not possible to place another patient there until someone was discharged, regardless of that patient's recommended designation. Figure 20-2 shows the initial practice followed by the BMT. The colors represent specific groups of nursing experience, for instance, cardiology, oncology, and labor and delivery. Note that Unit 2 has a boarded patient.

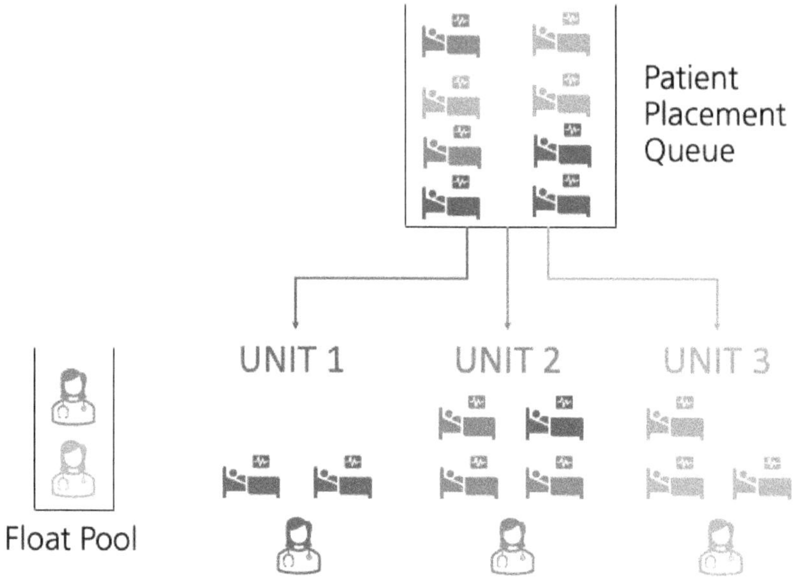

Figure 20-2: Float pool and boarded patient

Variability in transformation capacity (nurses) came from absenteeism and other issues that prevented a scheduled nurse from being available to handle patient demand.

Buffering Options

Since this is a service operation, stock inventory is not a buffering choice. St. Mary's performance improvement focus was on managing capacity to reduce cost and time to provide care. Frequently, due to the mismatch between demand and capacity, nurses would be scheduled but then not needed. In each nursing unit, when the mismatch of demand and capacity created a buffer, excess capacity in the unit was either shared with other units to shorten their time to provide care, with the excess nurses going into a float pool, or reduced by sending nurses home. Those nurses would then be on call for their particular unit at reduced pay. Other attempts were made to level load capacity with tools to allocate demand to capacity.

Tools and Applications

The team wanted to reduce variability in demand on individual nursing units to enable more efficient use of nursing capacity—operate with smaller capacity buffers and reduced cost. The question was, "How to level load the capacity?"

Level Loading Capacity

The first attempt at level loading was created by a consulting firm and employed a patient placement system that would begin each day by assigning patients to their "proper" unit (as determined by the referring physician), but after a certain target census in the unit, newly admitted patients (without a mandatory destination) would be assigned to the other nursing units in a fixed sequence, regardless of the patient's original designation. For example, the next five arriving patients might go to units 1, 2, 3, 4, and 5 in that order regardless of where they were "supposed" to go. The logic behind this was that the responsible physicians would still be from the "proper" unit even if the patients were housed in a different place in the hospital, and by honoring the mandatory assignments, this system would not compromise clinical outcomes. The belief was that nurses were skilled enough to follow the instructions of doctors from any discipline, and the rotated patients were not those for whom a "must" designation applied. This approach resulted in a mix of patients, illustrated in Figure 20-3.

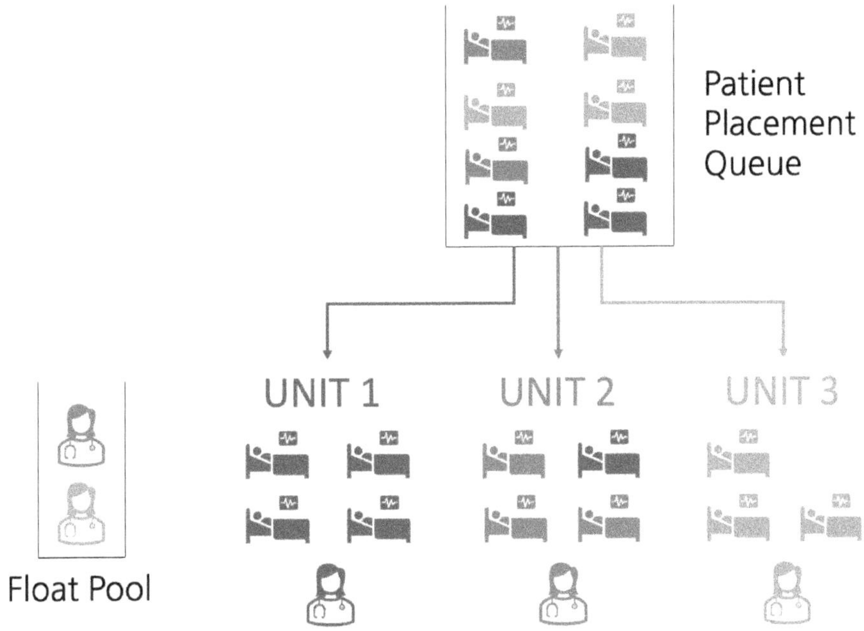

Figure 20-3: Patient mix resulting from demand leveling

Note that patient morbidities, as represented by color, are not well matched to nurse experience. The consulting firm did a simulation of their recommended solution and forecasted that its solution would result in significantly lower nursing labor costs. This is fairly typical of consulting analysis; a detailed simulation shows lots of effort and usually provides good eye-candy for the client. If the client is not familiar with operations science, they must take the consultant's recommendations at face value.

In this case, the team did not anticipate the consequences of the new patient placement system. After six months with the new system, St. Mary's had not yet realized the forecast cost savings. There were also an increasing number of complaints from physicians and nurses. Physicians did not like the increase in the number of boarded patients that resulted from the new system. Not only did doctors have to walk further on rounds, but they also had to deal with multiple nursing units, each with different staff and practices.

The physicians complained that this was inefficient and possibly even dangerous because medical outcomes might be compromised due to patients being placed with staff unfamiliar with their particular disease. Surprisingly, the nurses also complained. The hospital had anticipated that more reliable staffing, with fewer nurses being called in or sent home, would be well received, and indeed that aspect of the new system was not part of the nurses' concerns. However, the nurses felt that the clinical and administrative parts of their job were suffering.

Specifically, they reported that at times they were unsure what a particular patient needed and had trouble identifying and finding the responsible physician, who could be anywhere in the hospital. Also, the overhead activities for taking care of many different types of patients took time—time that came out of direct patient care. Upon further investigation, it also appeared that the new system was not being followed as prescribed. That is, even after six months of data gathering, it was impossible to tell how much the new system could potentially save because it was never fully implemented. That's because the BMT was not immune to emphatic input from doctors and nurses. Hence, although the BMT tried to implement something close to the desired cycling, in the case of any one patient there could be passionate voices arguing for specific nursing units, so the BMT frequently departed from the prescribed recipe.

The team went back to the drawing board for the best way to level load capacity using knowledge of operations science. Rather than focusing on artificially reducing demand variability by assigning to units regardless of the patient's original designation, they started to adjust the nursing capacity buffer to absorb demand variability. In looking at flexible capacity options, the team determined that a limited amount of cross-training, option c in Figure 20-4, would provide the best combination of patient care, nurse satisfaction and cost.

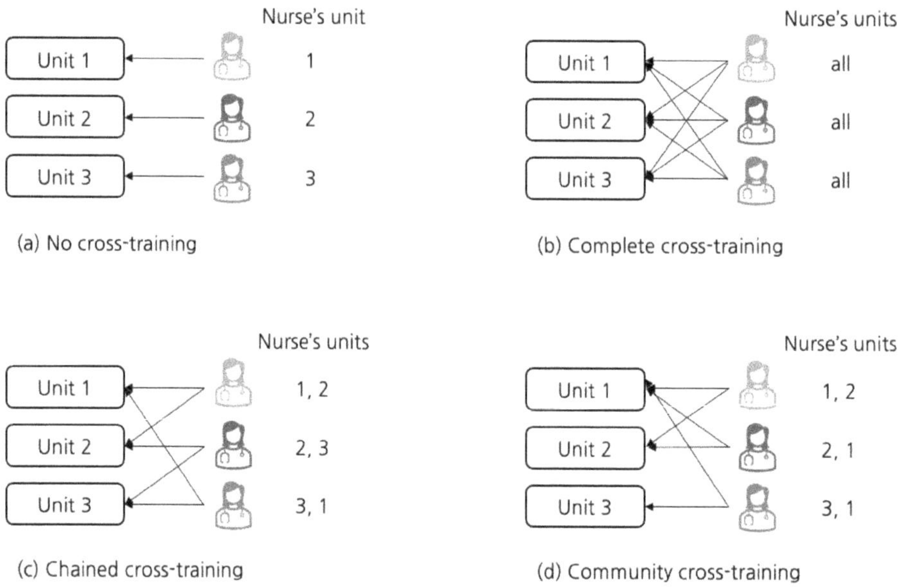

Figure 20-4: Cross-training options for capacity buffering of variability

With an understanding of operations science, the team developed a solution that worked best for the hospital, patients, and employees.

Operations Science Application Summary Table: Nurse Scheduling for Optimal Labor Cost

ELEMENT	DESCRIPTION
BUSINESS STRATEGY	Optimize labor costs while maintaining focus on patient care and employee satisfaction.
OPERATIONS STRATEGY	
DEMAND	Demand was generated by numbers and types of patients admitted to the hospital.
TRANSFORMATION	Nurses on different units were trained for, and more experienced in, different types of diseases
VARIABILITY	Demand variability in both numbers and types of patients admitted to the hospital. Transformation variability in the form of capacity to perform the transformation came from absenteeism and other issues that prevented a scheduled nurse from being available to handle patient demand.
BUFFERING OPTIONS: CAPACITY	Limited but focused cross-training to create flexible capacity provided the best combination of patient care, nurse satisfaction, and cost.
BUFFERING OPTIONS: STOCK INVENTORY	Not applicable
BUFFERING OPTIONS: WIP INVENTORY	Patients, not much control as the hospital was not in the practice of turning away patients.
BUFFERING OPTIONS: TIME	To be avoided in this healthcare setting; manifested itself in patients "waiting" for nursing capacity that could best serve needs.
TOOLS AND APPLICATIONS	Level loading with cross-training for staff flexibility.

Highland Hospital

Highland Hospital is a large, 500-bed hospital in the Midwest with 20 operating rooms (ORs). The OR suite is Highland's economic engine: By some estimates, 80% of revenues can be traced to surgical procedures (when one counts tests and pre and post procedure stays). At the time of this case, the OR suite was operating Monday through Friday from 8:00 AM to 4:00 PM, performing an average of 60 procedures each day (3 procedures per day per room). This, however, was not enough to keep up with demand, and the current wait for a patient to get on the schedule averaged 25 days, or 5 weeks. This concerned the administration greatly because some competitors were able to get patients into the OR for some very profitable procedures within 2 weeks.

The most straightforward way to increase capacity would be to add ORs. But this would be expensive, because the cost of adding each new OR is about $3.5 million. Furthermore, it would take several years to bring it online due to regulatory hurdles and construction delays. As a result, new ORs would not meet the hospital's requirement of a three-year payback for capital projects.

An alternative to construction would be to increase operating hours of the existing ORs. For a variety of reasons, opening on weekends was not considered feasible. Extending the day during the week was feasible, although physicians and surgical teams opposed it. The main reason for this opposition was that extending the day would mean that some surgeons would get late-afternoon starts, which were unreliable (due to the compounding variability of all the preceding cases). Nevertheless, because extending the day would help surgeons clear their queue of waiting patients and would enhance the hospital's bottom line, the administration was determined to consider it.

Under current policy, procedures were scheduled into the OR as follows. Surgeons were assigned block time, which constituted their allocated time in the OR. For example, Dr. Smith might be guaranteed an OR all day every Wednesday. On other days of the week, Dr. Smith would consult with patients in his clinic. If a patient required an operation, Dr. Smith would schedule that patient into the next open Wednesday. He would do this by estimating the length of the procedure and communicating the requirement to a central scheduling system, which would compute and schedule the procedure start times for each day in each room. For example, if Dr. Smith scheduled 3 procedures on a given Wednesday, with estimated durations of 2 hours, 1.75 hours, and 2.25 hours, the scheduling system would add 0.5 hour of cleaning time after each procedure and schedule start times of 8:00 AM, 10:30 AM, and 12:45 PM.

With the current case mix at Highland Hospital, the average procedure took 2.4 hours (including clean time), and there were on average 3 procedures scheduled in each room each day. Procedures still underway at 4:00 PM would not stop, of course, so the surgical team would stay until everything was complete. Physicians were paid by the procedure via their professional fees, which were billed separately from the facility fees for the hospital. The hospital provided the staff, rooms, and infrastructure and billed for these at a standard rate. The standard rate did not include overtime expenses for the staff, so the hospital had to absorb those costs when cases ran long. Surgeons were not permitted to schedule so many procedures into a single day that significant overtime would result.

To plan the length of a regular work day, the administration assumed 2.4 hours for each procedure it wanted to perform and rounded that time up to the nearest hour. Under the current standard of 3 cases per day, this meant a day length of $3 \times 2.4 = 7.2$ hours, which rounded up to an 8-hour day. If, instead, the hospital wanted to schedule an average of 4 cases per day per room, the day would be extended to 4×2.4

= 9.6 hours, or 10-hour days. To process 5 cases per day per room, they would plan for 12-hour days, and 6 cases per day per room would imply 15-hour days.

The main appeal of extending the OR day was that patient delays to get on the surgical schedule would decline because more procedures could be performed each day. For example, at 4 procedures per day per room, the average delay to get on schedule was estimated to be 21 days. At 5 procedures per day per room, it would be 14 days. At 6 procedures per day per room, it would only be 7 days.

Scheduling more cases into a day would have some significant disadvantages, however. The procedures in each room were run back-to-back, so if the first procedure was estimated to take 2 hours, and the scheduler added 0.5 hour of clean time, the second procedure would be scheduled to begin 2.5 hours later. Because procedures were highly variable, 50% of the time the first patient would be done early and the second patient would go into the OR on time, but 50% of the time the first procedure would run longer than expected and delay the start of the second patient. Such a delay would likely lead to a delay in the completion of the second patient and hence delay the start of the third patient. The third procedure of the day, for example, could finish up to 2 hours late, and this would only get worse if more procedures were scheduled into a room on any given day. Because delays tend to stack up, a fourth procedure added to the end of the day would not only have a good chance of being delayed, but that delay could be substantial, sometimes to the point of requiring cancellation. The bottom line was that scheduling more back-to-back procedures would lead to less reliable start times and higher risk of cancellation.

As we've mentioned, in addition to wasting physicians' time, delayed or cancelled cases are emotionally draining for patients and their families, some of whom take vacation days or drive long distances to come to the hospital.

To keep things simple, we highlight the negotiations between the hospital administration and the physicians. The other members of the surgical teams had preferences similar to the surgeons.

Physicians

The block-scheduling system was efficient for the physician, who stayed in the OR for continuous blocks of time. However surgeons believed it was unsafe to work longer than they were currently working on any single day, due to personal fatigue. So, although they would gladly have accepted more block time, they did not want it on the same day as their current block time. This meant that any extension of the day would require that at least two surgeons work in the same room over the course of the day.

This presented a serious problem for surgeons. Each would gladly accept a block of time with an 8:00 AM start, but none wanted a block starting in the afternoon. They knew afternoon start times would be unreliable, and cancellation of late-afternoon cases was frequent. Hence, although surgeons would clearly value the increased professional fees they could earn by performing more cases during the added time and would also value reduced wait times to get their patients on schedule, the combination of safety concerns and resistance to unreliable afternoon schedules outweighed these advantages in the surgeons' minds. Consequently, they opposed extending the operating day and instead strongly recommended the more expensive option of building new ORs.

Administration

Using the average cost per hour to keep an OR open and staff on hand and the average costs and revenues per procedure, the administration computed the expected profit per day for any regular time day length and any average number of procedures per day. The result is shown in Table 20-1.

Day Length (Hours)	Day Ends	Number of Procedures Scheduled per Room per Day 3	4	5	6
8	4:00 PM	$ 2,020			
9	5:00 PM	$ 1,926			
10	6:00 PM	$ 1,801	$ 2,786		
11	7:00 PM	$ 1,653	$ 2,711		
12	8:00 PM	$ 1,491	$ 2,605	$ 3,549	
13	9:00 PM	$ 1,322	$ 2,474	$ 3,490	
14	10:00 PM	$ 1,149	$ 2,325	$ 3,401	
15	11:00 PM	$ 975	$ 2,164	$ 3,286	$ 4,264

Table 20-1: Profit($) per room per day for alternative OR times

Negotiations

Effective negotiations always begin with each party understanding the perspective of the other. A good rule of thumb is to continue to ask "Why?" when diagnosing somebody's position. Here, the basic tension was that the hospital wanted to more highly utilize their existing ORs by extending the operating day. Why? Because it would be profitable and would make the hospital more competitive in terms of time to get patients on the operating schedule. Physicians, however, did not want to start in the afternoon or to do more cases in a single day. Why? Because it would compromise safety and because afternoon starts would be unreliable.

Negotiations could easily have stalled at this point of impasse. The key to forward progress was to recognize that it was not the afternoon start, per se, that physicians disliked; it was the unreliable start and finish times that they associated with afternoon starts. If the administration could offer an afternoon block time with a reliable start, would physicians be happy? Probably, because the other consequences for additional block time in the afternoon—shorter delays to get their patients on schedule and more professional fees—were positive.

To manage the unreliable start times that physicians disliked so much, the administration applied the operations science of variability buffering. Note that this was not a situation of establishing an optimal CONWIP level. The level of WIP (which was patients) was determined by the number of ORs. In this situation, increasing WIP would have been very expensive. Adding ORs was not a viable option. As a service, there was no inventory of completed operations so buffering with inventory was not an option. This meant that the only buffering options for dealing with variability were capacity and time. One of the primary objectives of the initiative was to reduce the time buffer—so time was not a viable buffer either. This left capacity as the only viable buffer for a solution.

The data suggested that delays longer than 2 hours were highly unlikely for the third procedure. This implied that the hospital could operate at 15-hour days broken into 2 segments separated by 2 hours to absorb all the variability. That is, the hospital could schedule 3 procedures in the morning segment, have a 2-hour fallow period, and then schedule 2 more procedures, all in a 15-hour day. Because 5 procedures (each at 2.4 hours) require 12 hours, adding a 2-hour fallow period results in a total time of 14 hours, which fits well within a 15-hour day.

Under this policy, the hospital would be operating 15-hour days, performing 5 procedures per day, and making $3,286 per room per day (compared to the current level of $2,020 per room per day). This added up to an increase in annual profits of about $317,000 per room per year. Patients would wait on average 2 weeks to get on schedule, as compared to the current 5 weeks. The net result was that the hospital would be both more competitive relative to the competition and more profitable. At the same time, surgeons would be able to do more cases in a timely fashion.

Note that the hospital was giving up something: The ORs would not be used for two hours in the middle of the day to operate on patients. This was the capacity buffer because the OR capacity would be planned at an overall lower utilization rate. It was not increasing the time buffer because the time patients would wait to get on schedule would actually decrease.

Highland Hospital initiated an afternoon surgical shift with a two-hour break between shifts. At first, not all procedure types were allowed into the afternoon shift because not all the support labs remained open into the later hours. But a sufficient number of labs were activated, and a sufficient number of surgeons voluntarily accepted afternoon starts in addition to their regular block time to dramatically reduce the average delay to get patients on schedule. An afternoon surgical shift is now standard at the hospital.

Highland Hospital and St. Mary's Hospital both provide great examples of innovative insight provided by operations science to provide highly productive solutions for service operations in the healthcare industry.

Operations Science Application Summary Table: Operating Room Capacity Optimization

ELEMENT	DESCRIPTION
BUSINESS STRATEGY	Maximize productivity of operating rooms while providing more rapid response for patients and reduced wait time for patients.
OPERATIONS STRATEGY	
DEMAND	Demand was generated by numbers and types of patients admitted to the hospital.
TRANSFORMATION	Operating rooms, doctors, nurses and administrative staff
VARIABILITY	Demand variability in types of operations performed and in number of patients. Transformation variability: length of surgeries depending on patient and surgical complications
BUFFERING OPTIONS: CAPACITY	Schedule less time available for operating rooms. Build more hospital rooms.
BUFFERING OPTIONS: STOCK INVENTORY	Not applicable
BUFFERING OPTIONS: WIP INVENTORY	Hard-coupled to the number of operating rooms. Patients are the active WIP but can't have more patients than ORs.
BUFFERING OPTIONS: TIME	Making patients wait longer for procedures to be scheduled. Not a desired outcome.
TOOLS AND APPLICATIONS	Capacity utilization calculations. Planned time buffer.

---

[i] Wallace J. Hopp and William S. Lovejoy, *Hospital Operations*, FT Press Operations Management (Pearson FT Press, 2012), 321–37,. Kindle.

[ii] Bernard Wysocki Jr. and Sarah Lueck, "Just-in-Time Inventories Make U.S. Vulnerable in a Pandemic. Low Stockpiles at Hospitals Boost Efficiency but Leave No Extras for Flu Outbreak," *The Wall Street Journal*, January 12, 2006, https://www.wsj.com/articles/SB113703203939544469?reflink=desktopwebshare_permalink.

[iii] Ibid.

[iv] Hopp and Lovejoy, *Hospital Operations*, 8.

[v] "Profs employ industrial methods to re-engineer health-care clinic," *Purdue University News*, August 22, 2006. https://www.purdue.edu/uns/html4ever/2006/060822.Moskowitz.clinic.html.

[vi] Ibid.

[vii] Ibid.

[viii] Ibid.

[ix] Hopp and Lovejoy, *Hospital Operations*, 7.

PART FIVE

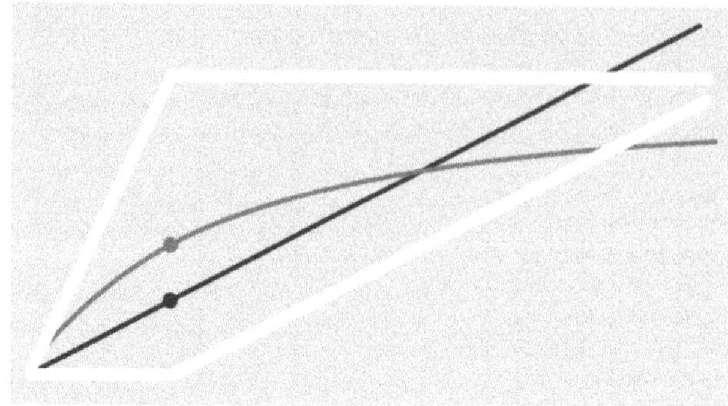

Profiles in Leadership

To conclude our exploration of operations science and its power to drive performance and career growth, we share a series of vignettes from operations leaders and practitioners who have applied these principles to transform the way they see work, sharpen their performance, and advance their careers.

Bill Van Sant

Emile Tohme

Joan Tafoya

Mike Severson

Larry Doerr

Lisa Elles

Bill Fierle

21

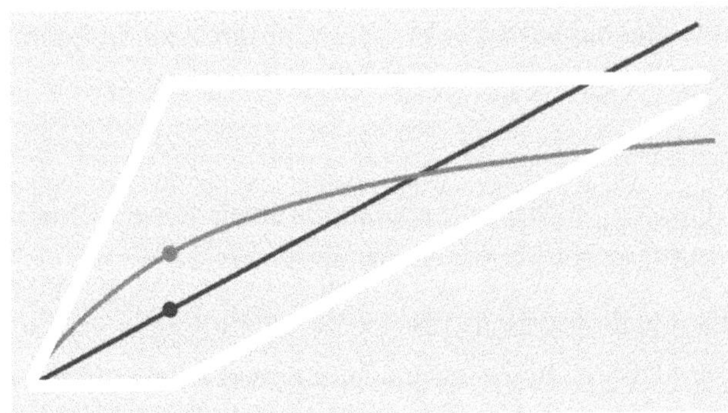

Operations Science Profiles in Leadership

Bill Van Sant

General Manager, Director of Operations, Plant Manager, Network Operations Manager, Program Manager

DeZURIK, Inc.; Quality Ingredients Corp.; Bolt Industries; Qwest Communications International, Inc.; Honeywell International, Inc.

Bill first learned about factory physics and operations science when he met Jeff Bell and Mike Severson in his work as an engineer at Bolt Industries. He later worked with the co-authors of this book in several different business situations. He leveraged the knowledge of operations science to achieve business results that helped drive his career progression in operations, ultimately landing him roles in general management.

Bill stated,

> Factory physics and operations science is a differentiator that you can bring to any organization to not only improve cost and customer service but also achieve career recognition.

Bill applied operations science successfully in a variety of different situations in multiple industries. These included high-mix machining and food-contract manufacturing environments as well as product-based businesses. His first application of operations science involved a machining and metal-fabrication environment similar to the one described in Chapter 16. He recalled,

> We had a focus on due-date quoting and virtual queues. We also leveraged an opportunity to collaborate across the supply chain for mass customization with a customer that had an assemble-to-order product. It was a great way to start to learn the factory physics–based cadence of strategies, tactics, and controls that lead to day-to-day execution.

As he progressed in his career, Bill appreciated that operations science helped quantify strategic choices. He said,

> I've had experiences in my career where the stated high-level strategy did not match tactical direction, day-to-day activity, and incentives. Specifically, at one company, the strategic mantra was all about customer service and responsiveness. Unfortunately, the incentives were all about capacity utilization and earned hours. Not surprisingly, we got what we measured with high utilization of 97%+ and poor responsiveness. The great thing about operations science was it helped quantify the benefits of investing in a capacity buffer with utilization of about 92%. The data showed this investment would result in lead-times being cut in half based on the utilization versus cycle time curve. The value of the ability to quantify strategic choices was demonstrated when our CEO indicated the company was not going to invest in the cost of a capacity buffer unless marketing could show the benefits of improved responsiveness in terms of increased sales and/or higher pricing.

As he progressed into general management, Bill focused not only on making sure strategy aligned with tactics and controls, but also on developing organizational knowledge of operations science so goals were understood and feedback loops were clear. He cited inventory-position-based monitoring as one example of clear, real-time feedback loops that drive needed action before the weekly management or monthly operations report shows a gap in performance. He recalled a situation where all buyers had access to an inventory position monitor that showed whether or not their parts were following policy because their inventory position was between reorder point + 1 and reorder point + Q. Bill said,

> I was able to ask objective, data-based questions that focused on results and team development. When a buyer had an opportunity to reduce cost with a blanket order, we proactively discussed implications in terms of dollars saved in cost and the impact of cash tied up in inventory. One of the benefits of a team focused on checking compliance with policy included visibility to well-intended changes from policy that could be discussed to avoid unintended consequences that can take a long time to correct.

Bill credited the direct tie of operations science to the financial statements for helping his career advance to general management levels. Once again, he pointed to the integration across strategies, tactics, and controls. At a strategic level, he specifically cited the benefit of tying capacity and inventory plans directly to the costs and assets on the financial statements. At a tactical level, he recalled making great use of the concept of contribution margin at the bottleneck to influence product management decisions like product mix as well as supply chain decisions for outsourcing.

Finally, in terms of using operations science as a key enabler for organizational leadership, Bill said,

> Once you build a core group that both understands and relies upon the operations science–based framework, successful outcomes begin to flow. A team that understands operations science can set up strategies, tactics, and measures for their unique business situation. I relied on my network of factory physics education and support resources to help me set things up in several of my leadership roles so I could then work with my teams to use operations science to get results. As a leader, you need to be engaged and visible when running your department, manufacturing facility, or company using these methods. It was really helpful to know and track the operations science–based leading indicators like schedule adherence. With lagging indicators, the only natural tendency is to lose momentum before the results become obvious. This is one of the biggest challenges in implementing any systems or changes. I know Mike Severson is mentioned in this book, and he helped us with the analytics so I could focus on leading the group based on data. This led to my teams achieving results and building organizational capability.

Emile Tohme

General Manager, UL Solutions

Previous experience in small manufacturing companies: President, Chief Operations Officer, Vice President of Operations, Director of Operations, Plant Manager, Product Development Engineer

Emile took Wally Hopp's Factory Physics course in the executive MBA program at Northwestern University's Kellogg School of Management in the 1990s. It was a three-day course designed for learning and networking with executives from all over the world. The curriculum showed how operations science could work for executives from different backgrounds and industries. Emile found that the key insights of the Factory Physics course, and operations science in general, validated work he and others in the class had done or been exposed to in the past. It also offered a great framework for the future. Emile said,

> Taking this training back to real life was exciting, and the challenge was to understand what to implement and where and how to ultimately achieve and measure success.

The participants in the course achieved success in their careers by leveraging the knowledge of inventory and process flow (response time and capacity). Emile said,

> A key tenet of the course was to avoid initiative by imitation and instead use the science to develop and execute strategy specific to your own operation. The group discussed that we had a portfolio of inventory, capacity, and time options, as opposed to just trying to react with capacity. Specifically, for high-run products, it made sense to build inventory to level load capacity to be responsive and cost effective. For other higher-mix products, a mass customization approach worked, with a focus on where to put the inventory/order interface.

The course showed how strategic component inventory could level load capacity while building common components and then customize to order when the components took on personality, which was very helpful for those in an engineer-to-order business. Emile continued:

> I love that the science tests and validates intuition. This came up time and time again in the course, with examples ranging from the value of older, slower, cheaper machines that could help reduce load on the bottleneck to the trade-offs between scrap caused by dumping the truck in a concrete business versus the cost of labor to unload blocks one by one to avoid scrap. The science enables us to test, and

> often validate, the intuitive ideas and actions of employees. It creates ownership and buy-in for the strategies, tactics, and measures that will drive results.

Emile recognized the value of applying the principles of the science of operations in service functions as well as manufacturing. He said,

> The basics apply in service industries. In fact, it can be more important to clarify things in scientific terms. Take for example the Taiichi Ohno statement that went something like "All we are doing is looking at the timeline from the moment the customer gives us an order to the point we collect the cash, and we are reducing that timeline by removing the non-value-added wastes." As Ed Pound points out in his training about the value-added fantasy, it is difficult and subjective to define and quantify what is value-added. This is especially true in services. The science clarifies in quantified terms the things that drive response time and cost. For example, utilization is Load/Capacity and Little's law says Response Time = WIP/TH. Instead of arguing about subjective opinions on what is value-added and non-value-added, the key is to focus on finding bottlenecks and controlling WIP to improve performance. This is very helpful in setting direction for organizations and project teams.

Emile said,

> It is important to think of strategy first. The portfolio of buffers approach reinforces strategy and avoids unintentional suboptimization that sometimes occurs when people leap to tactics. I knew this intuitively and even coauthored a paper, "A Multiple Period Capacitated Inventory Model for Airline Fuel Management: A Case Study" published in the *Journal of the Operational Research Society*, about the dangers of suboptimization.[i] The operations science framework codifies a strategic first approach.

A key part of Emile's leadership approach was education. He said,

> I believe and have supported teams going through basic operations science training. With this knowledge, the team is empowered to look at its unique business strategy and situation to determine the right portfolio of response time, capacity, inventory, and product mix to achieve business goals. The team can define strategies, tactics, and controls to drive operational execution. With operations science, one can inspire and energize employees to solve problems in the context of their business. In complex and dynamic organizations, it can be dangerous to do initiative by imitation. One has to be careful about copying strategies and key performance indicators that might not be relevant to the current situation. The key to leadership is not blindly applying someone else's historical strategy to your current situation but instead helping teams understand why, based on operations science, those strategies worked in the past and what needs to be tailored to develop and execute strategy in the current environment. Strategies and geographies change over time, but the operations science principles provide a consistent framework so that operations strategy can evolve with business strategy. My advice would be to start with the business strategy and the science—then do the math!

The views and opinions expressed in this work are those of the author and do not represent the official position of UL Solutions.

Joan Tafoya

Director Global Data Center Operations, Director, Senior Manager, Senior Director UAV Segment, Senior Principal Engineer, Director Supply Planning Operations, Manufacturing Manager, Industrial Engineering Manager, Semiconductor Fab Engineer

Meta, Sandia National Laboratories, Intel Corporation

The Society of Women Engineers recognized Joan with its Global Leader Award in 2015, Intel recognized Joan with its Highest Achievement award for Game-Changing Technology and Leadership in Drone Fleet Management in 2017, and she is an appointed fellow of the Institute of Industrial and Systems Engineers.

Joan was introduced to the operations science framework (at the time, the term *Factory Physics* was used) nearly 25 years ago when she met Dr. Mark Spearman and Ed Pound. It changed the paradigm and approach she took to leading the production of one of Intel's premier semiconductor fabs. Joan remembered,

> We were having challenges delivering product with the speed, volume, and timing needed to meet customer requirements. Informed by the science of operations, our views on inventory, variation, and, more broadly, queuing theory and Little's law, allowed us to change our approach. These changes led to a new way of planning and running the fabs and extended to the packaging and testing of devices. It transformed the way we worked in operations at Intel.

Operations science provided a foundation for Joan's work and career advancement. She said,

> Inventing the first of something is special. It breaks paradigms and often solves an important problem for someone. I fully appreciate that coming up with a new idea and building a few working samples can be hard. I would suggest, however, that building that same new thing, repeatably, consistently, with high quality, is often the more challenging task. There are thousands of problems to solve when making operations work. This is one of many reasons I love operations, I am challenged to solve problems in new and different ways, every day. Operations science provides a

> foundation for this work because it adds clarity and precision to areas I naturally gravitate to. Moreover, it presents as a baseline from which to further improve operations.

As Joan advanced into executive leadership positions in companies that are on the leading edge of technology, she deployed operations science in physical systems, knowledge systems, and leadership systems. She pointed out people were key to success, stating,

> All operations systems involve people. The way they are organized, rewarded, and developed, all in service of what the organization needs, both short term and long term. This is where I personally combine the rigor of operations science with the key people principles of Lean tenets championed by Taiichi Ohno.

Joan believes educating people on the basics of operations science helps inform their intuition so they can achieve results. She cited the example of WIP control, stating,

> In their courses, Ed Pound and Dr. Mark Spearman often ask participants about the basic relationship between WIP and Response Time. It quickly becomes clear that the audience does not have a shared understanding of what happens to throughput time if you increase the inventory in your system. This was the situation for Intel's chip manufacturing. The prevailing thought before our engagement with Factory Physics Inc. and operations science principles was that more inventory is good—in our case, more wafer starts. This idea of controlling WIP has been around for some time, introduced by Taiichi Ohno in the 1940s, when he used simple kanban boards. However, it seems that we forgot that basic principle. The resurgence of Agile/Scrum practices in the early 2000s brought that important concept back to life. I often tell people that the math and science behind Agile/Scrum is the *Factory Physics* textbook. It seems to me that the key idea to getting more done requires doing less, and doing it with focus, is now widely accepted; the challenge is to implement this in practice.

Joan has succeeded with operations science in physical systems many times and believes its use in these systems is fairly straightforward. She said,

> Operations—whether we are talking about manufacturing, data centers, call centers, or retail—require a series of steps, both automated and with human intervention, to work together to deliver the final product. Maintaining the flow of that work determines the success of the operation. Nothing works as predicted for extensive periods; that is variation, and it is all around us, whether it is the planned and known variation sources or the unplanned/unknown variation sources. Operations science makes it possible to intentionally measure, reduce, and address variation, ensuring that the flow of the work continues. My approach to planning work changed completely when I better appreciated the variation of the system that operations science taught me.

Applying operations science principles in leadership is a current focus for Joan, In a recent blog post for the Operations Science Institute titled "Operations Science is for Leaders," she wrote,

> I want to share with you that executive leaders can also apply operations science thinking at their level to dramatically increase their organization's focus and output. I call this a leadership operating system; this too is a system of inputs, outputs, feedback loops, and buffers, and is most successful when leveraging operations science methodologies.[ii]

She explained further, saying,

Applying operations science in a framework of strategies, tactics, and controls quantifies choices so leaders can develop strategy and execute it with predictive control. This empowers my basic leadership advice:

1. State your goal or where you are going. Make sure your goal is in complete alignment with the company's direction and needs.
2. Take actionable steps to get there.
3. Continually monitor and adjust along the way.

Recognize that goals can change based on where your business is going and how your team is performing. Consequently, being agile and aware is a critical behavior of leaders, and this system ensures that the feedback loops are in place to continually make progress and add value.

This advice worked when I started my career in manufacturing systems, and it is more critical today in a fast-paced world of knowledge systems powered by artificial intelligence.

Mike Severson

Operations Scientist, Data Analyst, Consultant, Manufacturing Engineering

Severson Consulting, Arc Precision, Factory Physics Inc., Emerson Electric Co.

Mike learned and leveraged the science of operations to shape a career with full flexibility to work remotely. He enjoys his work as an operations scientist, stating,

> I like getting the chance to really understand a specific business and its operations. The framework of operations science enables me to run the numbers to find the real issues and opportunities in a business. It is not about office politics and opinions. It is about strategy, science, and data-based decision-making. I help companies to focus on the signal and not the noise so they can make step change improvements. Operations science applies across all fields, not just production. I have helped companies in a wide range of businesses. That keeps things interesting to me, and I love the fact I can do this work while living where I want to live with minimal travel.

After graduating with a mechanical engineering degree from North Dakota State University, Mike started his career as a design and manufacturing engineer in the Rosemount Flow Division of Emerson Electric in Minneapolis, Minnesota. His responsibilities included the flow labs, machine shop, product engineering, and Lean manufacturing. He first came into contact with operations science because his colleagues at Emerson Electric, Jeff Bell and Dale Nugent, had recently graduated from the aforementioned MMM program at Northwestern University. They were students of Spearman and Hopp during the time when the *Factory Physics* textbook was being written. Nugent and Bell were bringing the concepts to Emerson's Rosemount Flow Division as leaders of the magnetic and vortex flowmeter business units. Mike also attended Wally Hopp's Factory Physics class in executive education at Northwestern. As Mike became known for his ability to apply the factory physics framework to achieve business results, he had opportunities to become a contributor to the Harvard Business School case titled *The Rosemount Vortex Flowmeter Plant* (see Chapter 15 in this book) and a guest lecturer in Wally Hopp's MMM class. He became a member of the Emerson Lean Leadership Council.

From Mike's perspective, one downside to the Emerson career path was global travel and the potential need to relocate to advance. He charted a path to consulting, using his operations science expertise to support businesses remotely, gaining control over his work location and travel schedule. The mix of intellectual challenge, ability to help achieve business results, and the lifestyle he wants is a perfect fit. Further, he is able to advance the science of operations across a wide range of businesses, where he trains and supports leaders to achieve business goals. He is a thought leader and key influencer at the companies he serves.

Mike credited the science of operations with enabling him to lead through influence, stating,

> It creates a framework for accountability at the organization level. There is data-based accountability and ownership at all levels that are consistent with strategy and operations science. From the executive leadership teams to middle management to supervisors to the shop floor, there is alignment between strategies, tactics, and controls. The executive leadership team needs to decide its portfolio of buffers in terms of planned customer service supported by investments in capacity and inventory. Middle managers and supervisors establish tactics and controls so the shop floor can execute the

> strategy. For example, in the Rosemount case, the executive leadership team determined and communicated planned responsiveness for customer service to the sales channel. Management put in the tactic of a WIP cap, and the production team was accountable to achieve planned TH. We had alignment and predictive control based on the organization's understanding of Little's law. We knew if we capped WIP and got our TH, we would have predictive control of customer service versus lead-times. This built up everyone's confidence, resulting in an engaged sales channel driving profitable growth.

Mike, in his role as a consultant, offers insights into the importance of leadership in achieving results. Mike explains,

> One of Dr. Spearman's famous statements is that "It is called factory physics, not factory magic." I have seen leaders who did not engage in understanding operations science or who chose to defer to individuals and teams who did not want to get on board with the direction made clear by science and data. When leaders deferred to opinion and office politics instead of building awareness of operations science so teams could develop and execute strategy based on science, I would become frustrated because I knew they would fail. This always reminds me of a statement used in Ed Pound's Operations Science Institute classes: "You can choose to ignore the law of gravity if you want, but when you walk off a skyscraper it does not end well." I want my consulting engagements to achieve results so I am very careful to choose assignments where I think the leadership team will engage, learn, and lead based on operations science that can support their business strategy. Effective leaders translate operations science–based strategies, tactics, and controls into terms the organization will understand. There is no need to fight about the name of an initiative. Operations science explains why stuff works—kanbans, Lean, Six Sigma, theory of constraints—so let the organization use familiar terms, allowing them to focus on execution of tailored strategies, tactics, and controls for their business. Ultimately, the leaders of an organization are responsible for developing a culture that understands and deploys operations science. They must cultivate and train so people have some idea about what to do, how to do it, and why it will work.

In terms of the most common and successful tactics at clients, Mike offered,

> It all starts with understanding demand and variability. Many people jump straight to operations initiatives, but since operations is really about servicing demand with the lowest cost possible, it is really important to understand demand and its variability. Only then can you assess the best operations strategies, tactics, and controls to meet the demand. I find a Pareto chart of demand pays great dividends. Business strategy should target a certain type of demand with a certain type of responsiveness. If marketing and sales strategies drive demand that is different from what business strategy desires, the operations will struggle to support it profitably. The operations strategy set up to support a business strategy works best when demand reflects that business strategy. If the demand is different, then a different operations strategy will be needed. I always recommend starting with a significant effort to understand demand and its variability.

In terms of operations-focused tactics to meet demand, Mike offered,

> It starts with leadership decisions on the portfolio of buffers in terms of planned customer service supported by investments in capacity and inventory. There is no one "right answer" but instead evaluating what is best for the business based on scenario analysis of investments in capacity and inventory. Once those investments are made, I see organizations find great power in Little's law and

> applying a WIP cap in a way that makes sense for their operations—might be Kanban, CONWIP, or a cap on hours of work released. This was the essence of what we did at Rosemount (Chapter 15) and what we currently do at Arc Precision (Chapter 16). Capping the WIP and getting planned throughput enables due-date quoting with time buffers and capacity triggers so you can tell customers when they will get their stuff. Capping the WIP and getting planned throughput are key to customer service and responsiveness.

Mike also observed that profit through the bottleneck was a critical tactic. He said,

> Understanding the bottleneck and how to manage it is critical. All operations max out at throughput of the bottleneck, so it is important to manage contribution margin through the bottleneck. This applies both internally and in deciding what products to send out when a company needs to access capacity in the supply chain. Ultimately, focusing on the bottleneck helps drive results and gets the biggest return on continuous improvement initiatives.

To pull it all together, Mike emphasized the importance of metrics to understand current status and a path forward to meet goals. He said,

> The science of operations makes it easier to set and align goals. Instead of just throwing goals out there, the science enables leaders to understand what is possible and set goals at all levels that align. From long-term goals for customer service and profitability for the organization to daily goals for the individual on the shop floor, the science helps set goals and track performance with both leading and lagging indicators. The organization and the individual have predictive control of planning and execution.

Predictive control is the key element of Mike's advice on how to use operations science to achieve career goals. He said,

> To get things done, it is vital that leadership have confidence in you and your analysis. This is true whether it is about getting initiatives funded or achieving your career goals. Operations science and the predictive control it enables inspire confidence to get executive buy-in and approval. Never underestimate the value of being able to say what you are going to do and then doing it.

Larry Doerr

Chief Executive Officer, Chief Revenue Officer, Chief Operating Officer, President of Research and Development, Vice President of Operations, Supply Chain Leader, Manufacturing Manager, Engineering

Infinite Materials Solutions LLC, Rev Materials, Stratasys, Nilfisk Advance, Detector Electronics, Emerson Electric Co., Northrup Grumman Corporation

As chief operating officer of the 3D printing company Stratasys, Larry was quoted on the back cover of *Factory Physics for Managers: How Leaders Improve Performance in a Post-Lean Six Sigma World*, stating,

> *Factory Physics for Managers* is a proven path to flawless execution and results. Leading versus following in our industry is predicated on the relentless pursuit of putting order to chaos. Factory Physics science has given our organization the ability to plan, predict, model, and execute based on explosive growth and rapid-fire, dynamic changes to our business model. In our case, history is not a good predictor of the future, so we need to deploy our resources wisely, and the Factory Physics approach has helped us do just that.

As a CEO and advisor to rapidly growing entrepreneurial technology businesses, Larry stated,

> Sometimes the nucleus of the business is chaos, and if you put too much process into the business, it won't be a business. Initiative by imitation won't work, and you have to leverage the science of operations to chart the path that makes sense for your business. There is no one answer on how to be successful or run a successful operation. I have made the speech many times about the buffers (inventory, time and capacity) and making the choice to manage them before they manage you... and know there is solid science and tools available to help."

Larry's career path from engineer to chief executive officer in high technology companies required the ability to understand and manage variability essential to innovation. Larry started his career predominantly in research and development and product design roles at Northrop Grumman and Honeywell. He first came into contact with operations science because Jeff Bell and Dale Nugent, his colleagues at Emerson Electric, had recently graduated from the aforementioned Master of Management in Manufacturing (MMM) program at Northwestern University. They were students of Spearman and Hopp as the *Factory Physics* textbook was being written. Nugent and Bell were bringing the concepts to Emerson's Rosemount Flow Division as leaders of the magnetic and vortex flowmeter business units. Larry was in supply chain at the time and filed away the conversations on operations science for future reference and application, stating,

> I made note of the fact that I had not encountered ANY attempt to scientifically describe and manage operations/manufacturing prior to learning about the *Factory Physics* textbook.

Larry had the opportunity to apply the key principles at Stratasys, where he was COO and ultimately Executive VP of North America for the world's largest 3D printing company at that time:

> We were growing 25%–50% per year and trying to scale operations while managing lead times, capacity, and working capital (inventory). We engaged with Factory Physics, Inc. to bring the science to our filament manufacturing, which had high mix and high volumes with product availability paramount to business success. The concepts were embraced and put to work, with the support of executive leaders and operations leaders to the ultimate benefit of the business.

Business success led to personal success and opportunity for Larry. He noted,

> Factory physics concepts helped me make sense of complex operational challenges, not to mention being a resume builder and nice icebreaker during interviews. When you boil things down to their simplest concepts, operations science helped me understand and manage variability inherent in all operations. It was a differentiator skill that helped me as a leader who excelled in technology-based environments with explosive growth. I frequently find myself at the white board talking about inventory, time, and capacity as the basis for deeper conversations. It has been a very useful conversation to drive robust sales and operations planning meetings, giving leadership in the company the chance to have "what if" conversations when deciding how to manage these easily defined buffers. It helps organizations make specific choices on where to take the risk as they plan and execute. Finally, the science provides predictive control that results in organizational credibility. When you achieve the results you planned on and committed to, based on the tactics and controls chosen to support a strategy, it frees the leadership of the organization to focus on strategy and how it needs to evolve as things change in dynamic business environments. Variability happens every day, but if that variability can be managed so the revenue provided by market-facing change outweighs the cost of the buffers to deal with the variability, you make money.

Larry currently coaches and consults with entrepreneurial executives. With regard to this role, he said,

> I am convinced that my career progression was favorably influenced by my exposure to factory physics and operations science. I share this with my clients and the students I teach in a capstone master's level course at St. Thomas University. I focus on factory physics and operations science for one week each semester to plant the seeds for these future business masters. The basis for my class is to instill curiosity about all things business, so introducing factory physics and operations science seemed logical to me. My final advice is that I have found that every organization is unique in some way, but if you are manufacturing "stuff," it is important to understand the things that are the same (operations science) whether you are making plastic pellets, experimental marketing props, or aluminum bed covers for pickup trucks."

Lisa Elles

Chief Executive Officer, President, Distribution Manager, Facilities Manager, Industrial Engineering Manager

Hepinstall Consulting Group, Inc., Atlantic Marine Inc., Florida Power and Light

In her 20 plus years improving large complex government and commercial programs, Lisa has had phenomenal success, including 10-fold increases in throughput, 50% reductions in project cycle time, and cost savings of more than 50%. She observes,

> Large, complex projects are typically burdened with high levels of variability throughout a program's life cycle—in demand, design, procurement, production, testing, commissioning, operations, maintenance, modernization, and disposal. Wastes are hidden in this variability, making it difficult for organizations to effectively and methodically improve what they can't see or measure. When viewed through the operations science lens, variability and its effects are easy to see and can be addressed to achieve predictable and profitable project performance.

Lisa's career path in construction and project management has journeyed through roles in engineering and management to being chief executive of a consulting company. It has required an ability to understand facility and process constraints, construction planning, and organizational transformation. She started with a strong foundation in Total Quality Management by leading Florida Power and Light teams in Deming audits. Florida Power and Light was the first non-Japanese organization to win the Deming Prize for quality improvement. Lisa built on this foundation when she moved to shipbuilding and ship repair, where she innovated by developing a Lean Production System based on Lean principles. The system was documented in the form of a policy and checklist that sequenced all activities from the beginning of precontract phase to ship delivery.

She co-developed and implemented the Lean Shipbuilding Model and Lean Ship Repair Model for the U.S. shipbuilding industry through the National Shipbuilding Research Program (NSRP). This standard served as a baseline for continuous improvement for the shipbuilding industry.

> Our Lean Shipbuilding Model included a Rapid Engineering Model (REM) Line for Ship Design that governed the product definition and development process for ship construction programs. Implementing the REM on a four-vessel commercial shipbuilding program resulted in 100% on-time delivery of all engineering products.

She catalyzed these best practice efforts when she learned about operations science, recalling,

> In the early 2000s, I joined the Project Production Institute (PPI). At one of PPI's annual symposiums in San Francisco, where thought leaders from around the world were gathered to discuss the complexities of large-scale project delivery and share best practices, Dr. Mark Spearman presented factory physics/operations science theory, concepts, and case studies. Given my industrial engineering background and experience improving large, complex government and commercial programs, Dr. Spearman's description of the foundational concepts of a general science of operations resonated perfectly with my experience and gave me a powerful new lens for viewing my work. I've been a student and practitioner of operations science ever since.

In her current role as President and CEO of Hepinstall Consulting, Lisa leads maritime organizations through transformational change to achieve significant bottom-line improvements. She recognizes the value of operations science in helping organizations achieve those results, stating,

> Developing an operations strategy using operations science provides a common framework to the program and project teams. These concepts are used to drive effective decision-making and controls during execution. Operations science principles can be applied in all environments. For example, performance envelopes can be defined for the products and services a business will offer. Demand forecasts can be approximated using targeted average and variance of demand.

Lisa credits operations science with helping catalyze Lean, Theory of Constraints, Project Production Institute Methodologies, and Continuous Improvement initiatives. She said,

> Being able to propose a course of actions, based on science, promotes early buy-in and adoption. As a consultant specializing in performance improvement of complex systems, I apply operations science tools and methodologies routinely to help my clients achieve increased throughput and bottom-line cost reductions."

Finally, in terms of advice she has for others interested in using operations science to get results and enhance career opportunities, Lisa says,

> You don't have to be an operations science expert to begin practicing operations science principles and best practices. The more I understood operations science, the easier it became to apply the practices tailored to each opportunity. One thing I recommend is taking the Operations Science Certification offered by the Operations Science Institute. It is a great way to accelerate your operations science proficiency.

Bill Fierle

President, Chief Operating Officer, General Manager, Sales Manager, Product Manager, Total Quality Management Leader, Production Control Supervisor, Production Supervisor

Conax Technologies, Emerson Electric Co., American Precision Industries

Bill Fierle was a Lean practitioner and a self-described "JIT head" at a mid-sized industrial manufacturer, where he drove what he later called Lean 101 practices to eliminate waste in production lines. Bill was introduced to the science of operations during his time in the Master of Management in Manufacturing (MMM) program at Northwestern University. The MMM program was a joint degree from the Kellogg School of Graduate Management and the McCormick School of Engineering and Applied Sciences. The curriculum was created for those desiring an MBA for use in companies where products were produced. The MMM program taught the science of Lean manufacturing. He internalized the principles of operations science and applied them in different environments to drive results and develop his career.

A key insight deployed by Bill in the executive suite is that business strategy drives operations strategy. This is true in marketing, sales, and operations. It sounds straightforward, but Bill had to deploy it to manage variability, capacity, and response times for customer support teams while educating his organization on why customer support teams got more resources. The leader of the manufacturing team was complaining about the sales and marketing team. She was focused on productivity in manufacturing and the idea that everyone needed to be busy, while her counterpart in sales and marketing ran his teams like a country club, meaning everyone was not always busy. Bill said,

> What she didn't see was the strategy of the business, the corrupting influence of variability, and the need for a capacity buffer for responsive customer service.

In this case, the business made significant investments for dedicated production lines, thus reducing the c_e^2 in each production line. In addition, Bill's manufacturing team buffered with time, which enabled a smaller capacity buffer in production. Customer service had visibility to see the next available slot in the ERP system and date the order accordingly for the customer. The result was 50% of the orders shipped within five days after receipt of order (ARO) with 95% on-time delivery (OTD). By reducing variability and buffering with time, they could run production lines at high utilizations. The average time to build a product on the production lines was typically about 90 minutes.

At the heart of Bill's decision for a capacity buffer in sales and marketing was the strategy to serve the market for his electrical products. The buying process included four decisions makers: the architect, the professional engineer, the electrical contractor, and the building owner. The products could be configured in tens of thousands of different SKUs. The go-to market strategy was to be the best (i.e., fastest) at coordinating the application engineering between all the parties. With quick responses, Bill's company gained market share.

Customer service was paramount, so customer support teams targeted a four-hour cycle time, on average, for the response time. The average time a customer service inquiry took was 10 minutes, but each inquiry was highly variable in time. That is, some inquiries took very little time, while others were quite involved. Also, there was little to no control as to when a new inquiry would come in, adding to the variability.

For manufacturing to achieve a five-day lead time, they could run at high utilization rates, as shown in Table 21-1:

CT	c_a^2	c_e^2	u	t_e
5 days	0.50	0.50	99%	90 minutes

Table 21-1: Manufacturing Analytics

Customer support teams, with higher variability, needed to run at lower utilization rates (Table 21-2):

CT	c_a^2	c_e^2	u	t_e
4 hours	2.50	2.50	80%	10 minutes

Table 21-2: Customer Support Analytics

Bill drove home the strategy of a capacity buffer in marketing and sales, stating,

> Why would an organization pay to have a department to sit on their hands for 20% of the day? The answer is very simple: To maximize profits.

Bill continued,

> If you want to delight customers with the same lead times for high-variability processes as they get for lower-variability processes, you need to invest in a capacity buffer to handle that variability. He cited another example from a production environment where the key insight was planning the capacity buffers needed to handle variability inherent in the process. Bill recognized that different processes had differing levels of variability, which required a different approach with the capacity buffer. Some processes have very high process capability indexes (Cpk), where parts are produced consistently and with minimal scrap. Other processes challenge the limits of materials science, so the process will be known to have less consistent outcomes and will require inspection, sorting, rework, and sometimes even scrap. Specifically, he knew he had to run plasma welding stations at lower utilizations and with a bigger capacity buffer compared to CNC lathes to maintain the same lead times for products from both processes.

Finally, Bill excelled in managing the interactions between variability and the buffers to drive a virtuous cycle of continually improving results instead of vicious cycles where declining results lead to expedites and more variability. Specifically, he recalled an inventory optimization effort, stating,

> When we initially embarked upon applying inventory optimization in our operations, we knew we would get three benefits:
>
> 1) shorter lead times to customers, 2) improved OTD, and 3) less inventory.
>
> But what actually happened was a virtuous cycle of continuous improvement that resulted in five benefits: 1) shorter lead times to customers; 2) improved OTD; 3) less inventory and, unexpectedly, 4) cost reductions; and 5) less expediting of the supply base.

Bill observed,

> In what I call Lean 101 initiatives; they typically focus on work in process (WIP) pull systems within a production line and flow within a factory. The basic thinking here is that finished goods inventory (FGI) is not desirable and WIP turns should be very high, and thus there is an effort to push inventory investments to raw material (RM). Our initial pilot program was for a specific product line. The situation prior to inventory optimization was very similar to what is described above, with little to no FGI. We were good at converting RM to FGI with approximately 50% of orders fulfilled within five days ARO from the customer and a 95% OTD performance.

This product line had hundreds of RM SKUs and several assembly lines that could be configured into hundreds of FGI SKUs. Bill's organization applied the 80/20 rule and created an FGI policy, using inventory optimization for order quantities (Q) and reorder points (r) and to build to stock. This was done as a strategic move in that if they could ship even faster than five days ARO, they would gain market share. They also used inventory optimization for the raw materials. The production lines were already "Leaned" out, so they made no changes there. The result of this initial pilot project is shown in Table 21-3

	Before Inventory Optimization	After Inventory Optimization
Lead Time	50% within 5 days ARO	95% within 0–1 days ARO
On-Time Delivery	95%	99%
Inventory $ (RM+WIP+FGI)	—	35% reduction

Table 21-3: Pilot Project Results

After the initial pilot project, Bill applied inventory optimization to the entire business, which had more than 2,000 active RM SKUs and more than 2,000 active FGI SKUs each year. The team ran the inventory optimization monthly in order to account for the following:

Seasonality

Changes in annual demand from

- market conditions,
- market share gains,
- special promotions,
- new product introductions, and

- products migrating through their product life cycles

Changes in lead times for RM

They found the business needed fewer personnel in production control and purchasing to execute the new inventory planning model. These resources were redeployed to help suppliers improve lead times and OTD, resulting in the need to run a new inventory optimization monthly; the efficient frontier curve incrementally improved in an evergreen manner. The number of times they needed to call the supply base to expedite raw materials dropped by a staggering 90%. It was a virtuous cycle.

Bill did offer a cautionary tale in understanding operations science, using predictive control to communicate expectations. He said,

> The key lesson we had to learn through experience is that before you get a reduction in inventory, an inventory bloom occurs. With the new Q,r for 2,000 SKUs, the newly optimized policy required us to buy the raw materials that we did not have enough of, while we could not easily or practically return the oversupply of raw materials to suppliers. The bloom results in a one-time inventory increase that can be significant but temporary and that has a negative impact on cash and trade working capital. It takes several months to burn the excess inventory down before settling into the efficient frontier curve. It was important to communicate that reality and set expectations, so everyone stuck with the plan long enough to achieve our 30%+ inventory reduction.

Bill reflected on the success he had with the above examples of applying operations science and its impact on his career, saying,

> Deep experiential knowledge and the application of the science of operations has allowed me to gain second-order intuition. By this, I mean the ability to easily see things in operations that others would not observe. This has made me a better leader of people at the several businesses I led—and lead—on a global basis. Strategy is about products, markets, and geographic regions, while operational excellence adds to a competitive advantage beyond strategy.

When asked what advice he had for others interested in applying operations science in their businesses, Bill quoted Malcom Forbes, the founder of *Forbes* business magazine, "With all thy getting, get understanding."

Bill then went on to clarify,

> Too many Lean practitioners are stuck in pull-one thinking that disregards the interactions of inventory, utilization, variability, and time. I embrace pull systems in factories but limiting one's thinking to only pull-one systems often results in less throughput. Any rising star in manufacturing will need deep knowledge about the science that goes into world-class operations.

[i] P. P. Zouein, W. R. Abillama, and E. Tohme. “A Multiple Period Capacitated Inventory Model for Airline Fuel Management: A Case Study.” *Journal of the Operational Research Society*, 53, no. 4 (2002): 379–86, https://doi.org/10.1057/palgrave.jors.2601315.

[ii] Joan Tafoya, “Operations Science Is for Leaders,” *Operations Science Institute*, accessed June 26, 2025, https://opscience.org/operations-science-is-for-leaders/.

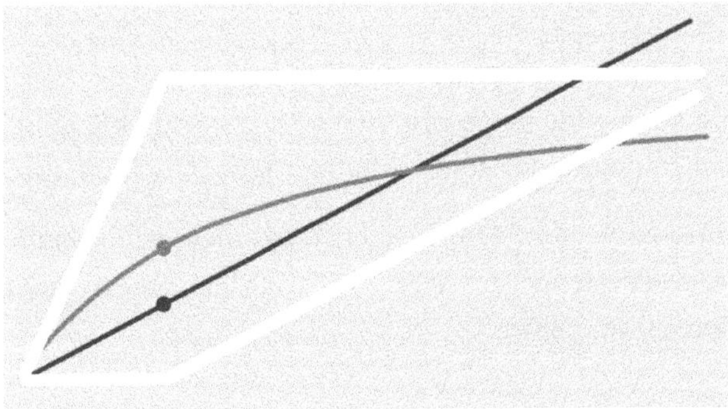

APPENDIX

Support Resources

For Academia

Professors using the book as a required text in their courses can have free access to support material for teaching the material in the book. That material includes

- Study questions
- Case Studies
- Problems and solutions
- Spreadsheets
- PowerPoint lecture files for chapters

To gain access to these teaching materials,

1. Create a free account at opscience.org
2. Send an email to info@opscience.org
   - Subject: Academic access
   - Include a copy of the course syllabus or a link to the course syllabus where the book is listed as a required resource for the course
3. Once academic access is approved, the opscience.org account will be updated to allow access of the support materials from the website
   - Or-email the publisher at dick@hercherpublishing.com

As provision for those professors who are evaluating the book for their course, a detailed listing of support materials will be provided.

For Industry

There are a wide range of free resources available at opscience.org for those who want to harness the power of operations science in their jobs. Click on opscience.org/resources to explore further.

Additionally, at the OSI Insider portal there is a progressive scale of opportunities to learn more about operations science and tune it for your business to quickly accelerate operations success and rapidly advance your career.

Go to insider.opscience.org to find out more

www.ingramcontent.com/pod-product-compliance
Ingram Content Group UK Ltd.
Pitfield, Milton Keynes, MK11 3LW, UK
UKHW051118290926
15471UKWH00010B/1747

9 781939 297310